Cooperation and Conflict

The Interaction of Opposites in Shaping Social Behavior

Understanding the interaction between cooperation and conflict in establishing effective social behavior is a fundamental challenge facing societies. Reflecting the breadth of current research in this area, this volume brings together experts from biology to political science to examine the cooperation–conflict interface at multiple levels, from genes to human societies. Exploring both the exciting new directions and the biggest challenges in their fields, the authors focus on identifying commonalities across species and disciplines to help understand what features are shared broadly and what are limited to specific contexts. Each chapter is written to be accessible to students and researchers from interdisciplinary backgrounds, with text boxes explaining terminology and concepts that may not be familiar across disciplinary boundaries, while being a valuable resource to experts in their fields.

Walter Wilczynski was Professor of Psychology and Director of the Neuroscience Institute at Georgia State University (GSU), USA. With more than forty years of teaching and research experience, his work focused on the study of the neural origins of social behavior in animals, with particular interest in communication, behavioral endocrinology, comparative vertebrate neuroanatomy, and sensory processing. In June 2020, Professor Wilczynski passed away after a second battle with cancer.

Sarah F. Brosnan is Distinguished University Professor of Psychology, Philosophy, and Neuroscience and Co-Director of the Language Research Center, Georgia State University (GSU), USA. She studies decision-making in humans and other primates, particularly decisions relating to cooperation and inequity, and how these decision processes evolved. The editors collaborated closely for several years on the Center for Behavioral Neuroscience's (GSU) research into the neurobiology of cooperative behavior.

Cooperation and Conflict

The Interaction of Opposites in Shaping Social Behavior

Edited by

WALTER WILCZYNSKI
Georgia State University

SARAH F. BROSNAN
Georgia State University

CAMBRIDGE
UNIVERSITY PRESS

CAMBRIDGE
UNIVERSITY PRESS

University Printing House, Cambridge CB2 8BS, United Kingdom

One Liberty Plaza, 20th Floor, New York, NY 10006, USA

477 Williamstown Road, Port Melbourne, VIC 3207, Australia

314–321, 3rd Floor, Plot 3, Splendor Forum, Jasola District Centre, New Delhi – 110025, India

79 Anson Road, #06–04/06, Singapore 079906

Cambridge University Press is part of the University of Cambridge.

It furthers the University's mission by disseminating knowledge in the pursuit of education, learning, and research at the highest international levels of excellence.

www.cambridge.org
Information on this title: www.cambridge.org/9781108475693
DOI: 10.1017/9781108671187

First published 2021

Printed in the United Kingdom by TJ Books Limited, Padstow Cornwall

A catalogue record for this publication is available from the British Library.

ISBN 978-1-108-47569-3 Hardback

In memory of Walter Wilczynski (1952–2020) –
much loved scientist, husband, colleague, mentor,
and friend who is missed greatly by us all.

To the students, postdocs, and lab techs who have
been part of my lab over my long career. They formed
a true family that I valued all my life.

Walter Wilczynski

To Alex

Sarah F. Brosnan

Contents

List of Contributors   *page* ix
Acknowledgments   xi

Introduction: Understanding for Relationship between Cooperation and Conflict   1
Walter Wilczynski and Sarah F. Brosnan

Part I Broad Insights from Political Science to Molecular Behavior   5
1   **Cooperation and Conflict in International Relations**   7
William J. Long

2   **Internalizing Cooperative Norms in Group-Structured Populations**   26
Erol Akçay and Jeremy Van Cleve

3   **Reputation: A Fundamental Route to Human Cooperation**   45
Junhui Wu, Daniel Balliet, and Paul A. M. Van Lange

4   **Finding the Right Balance: Cooperation and Conflict in Nature**   66
Elizabeth A. Ostrowski

Part II Neural Mechanisms   87
5   **Social Living and Rethinking the Concept of "Prosociality"**   89
Heather K. Caldwell and H. Elliott Albers

6   **The Role of the Temporal Lobe in Human Social Cognition**   104
Katherine L. Bryant, Christina N. Rogers Flattery, and Matthias Schurz

7   **Role of Oxytocin and Vasopressin V1a Receptor Variation on Personality, Social Behavior, Social Cognition, and the Brain in Nonhuman Primates, with a Specific Emphasis on Chimpanzees**   134
William D. Hopkins and Robert D. Latzman

Interim Summary   161

Part III Species Comparisons   165
8   **Understanding the Trade-off between Cooperation and Conflict in
           Avian Societies**   167
           Amanda R. Ridley and Martha J. Nelson-Flower

9   **Cooperation and Conflict in Mutualisms with a Special Emphasis on Marine
           Cleaning Interactions**   185
           Redouan Bshary

10   **The Fundamental Role of Aggression and Conflict in the Evolution and
           Organization of Social Groups**   212
           Clare C. Rittschof and Christina M. Grozinger

           Index   234

Color plates can be found between pages 180 and 181.

Contributors

Erol Akçay
Department of Biology, University of Pennsylvania, USA

H. Elliott Albers
Center for Behavioral Neuroscience, Neuroscience Institute, Georgia State University, USA

Daniel Balliet
Department of Experimental and Applied Psychology, Vrije Universiteit Amsterdam, The Netherlands

Sarah F. Brosnan
Departments of Psychology and Philosophy, Center for Behavioral Neuroscience, Neuroscience Institute, Georgia State University, USA

Katherine L. Bryant
Donders Institute for Brain, Cognition, and Behaviour, Radboud University Nijmegen, The Netherlands

Redouan Bshary
Department of Zoology, University of Neuchâtel, Switzerland

Heather K. Caldwell
Department of Biological Sciences, Laboratory of Neuroendocrinology and Behavior, School of Biomedical Sciences, Kent State University, USA

Christina M. Grozinger
Department of Entomology, Center for Pollinator Research, Huck Institutes of the Life Sciences, Pennsylvania State University, USA

William D. Hopkins
Department of Comparative Medicine, The University of Texas Anderson Cancer Center, USA

Robert D. Latzman
Department of Psychology, Georgia State University, USA

William J. Long
Department of Political Science, Georgia State University, USA

Martha J. Nelson-Flower
Department of Biology, Langara College, Vancouver, Canada

Elizabeth A. Ostrowski
School of Natural and Computational Sciences, Massey University, Auckland, New Zealand

Amanda R. Ridley
Centre for Evolutionary Biology, School of Biological Sciences, University of Western Australia
and
Percy Fitzpatrick Institute of African Ornithology, University of Cape Town, South Africa

Clare C. Rittschof
Department of Entomology, College of Agriculture, Food, and the Environment, University of Kentucky, USA

Christina N. Rogers Flattery
Department of Human Evolutionary Biology, Harvard University, Massachusetts, USA

Matthias Schurz
Department of Experimental Psychology, Wellcome Centre for Integrative Neuroimaging, University of Oxford, UK

Jeremy Van Cleve
Department of Biology, University of Kentucky, USA

Paul A. M. Van Lange
Department of Experimental and Applied Psychology, Vrije Universiteit Amsterdam, The Netherlands

Walter Wilczynski
Department of Psychology, Center for Behavioral Neuroscience, Neuroscience Institute, Georgia State University, USA

Junhui Wu
Institute of Psychology, Chinese Academy of Sciences, China

Acknowledgments

The contributing authors each acknowledge grant support or other assistance where appropriate. The editors would like to thank the John Templeton Foundation for a grant to the Center for Behavioral Neuroscience (CBN) that established a consortium of Georgia State University and Emory University researchers and trainees to study the mechanisms of pro-social behavior. This group met regularly under the guidance of Dr. Elliott Albers to discuss this research project and larger issues related to cooperation and aggression in social groups. Several members of that consortium contributed chapters, along with others that were solicited to fill out the topics of this book. We are also indebted to the CBN staff who assisted in organizing the group meetings and facilitating the production of the manuscripts. Finally, we are very grateful to the staff at Cambridge University Press for their advice and their patience with the editors as they overcame unexpected setbacks in the later stages of finishing the book.

Introduction

Understanding for Relationship between Cooperation and Conflict

Walter Wilczynski and Sarah F. Brosnan

Finding a balance between cooperative or prosocial behavior – such as social bonding and empathy – and conflict – or competitive-aggressive, self-interested behavior – is the fundamental challenge to the operation of societies and to the behavior of individuals in a social setting. But how do these apparent opposites relate to one other? As would many social or behavioral scientists, we initially approached this with the idea that they are two separate functions that need to be balanced against each other to varying degrees to construct a functioning social entity, and to some extent this holds true. But independently of one another, the contributing authors to this volume advanced a more sophisticated view of the relationship; that the poles of social interaction are in fact interconnected to the extent that what we view as antisocial or aggressive behavior are fundamental to establishing and maintaining positive or prosocial behavior within groups or individuals.

Societal structures, whether those of modern humans or the eusocial insects, depend on some level of cooperation among participating individuals. This cooperation must occur despite an individual's immediate self-interest, which might put it in conflict with others, and might be at odds with the longer-term goals of the social group, or even that individual's own long-term benefit. On the other hand, while self-interest leads to conflict between individuals within a society, self-interest at the societal level (combined with the cooperative bonds between individuals within a social group) can lead to conflict between social groups. The balance between cooperation and conflict therefore is an issue that applies to the individual, but transcends the individual to apply to groups and societies as well. It is also a mechanistic issue for psychology, which must understand the cognition and emotion underlying these transitions, as well as neuroscience and molecular biology, which must consider how the molecular, cellular and systems operations of the nervous system can and do mediate switches between conflicting states.

But what is the true relationship between these apparent poles of social relationship? Why and how the balance between cooperation and conflict is achieved, maintained, and modulated is a concern across disciplines. The humanities, social sciences, and biology all contribute to understanding how groups of individuals or, on a larger scale, societies, emerge and how individuals operate within them. In every discipline, studies of the cooperation–conflict balance span many levels and consider both the proximate mechanisms by which cooperative behavior and aggression are regulated as well as the ultimate mechanisms leading to the evolution of the capacity to express and control cooperation and conflict. The biological sciences clearly define these two types of behavior in their attempts to understand the mechanism and evolution of cooperative (including altruistic) and competitive (including aggressive) behavior, and how both are regulated and expressed in an appropriate way. Proximate mechanisms are the focus of a great deal of research in neuroscience and behavioral biology in studies that investigate the underlying genomic, neural, hormonal, or behavioral processes by which an individual adjusts the balance between its cooperative and its aggressive behavior. Proximate mechanisms also concern much of the psychological research investigating how cooperative and competitive behavior are controlled at the cognitive and emotional levels. Disciplines such as political science that focus on the operation of human societies place an understanding of the mechanisms triggering or resolving intra- and inter-societal cooperation and conflict as fundamental to the field.

Biological research is pursued within the framework of evolution and its own focus on reproductive fitness in order to gain insight into the ultimate causation of a particular trait. At the most fundamental level, fitness considerations are based on the inherent conflicts between individuals. Evolutionary biology has therefore considered the paradox of how cooperation can evolve, given the focus of selection on individual fitness. Furthermore, there are clearly differences across species in the frequency of and capacity to express cooperative behavior. A comparative analysis of these differences represents the intersection of mechanism and evolution: under what selective processes have these differences emerged, and what are the genetic and neural changes that accompany the enhancement of cooperation? This question is an important one for human evolution. Humans have an enormous capacity for cooperative behavior, and with it a vastly complex social structure. Can we find the biological features that differ among primates, and more broadly across species, that support the diversity of social behavior, and thereby gain new insights into human evolution?

The tension between cooperation and conflict is also an area of interest in disciplines outside biology. Cognitive and social psychology are concerned with understanding different aspects of the underpinnings of complex social decisions. There is a strong focus in many areas of contemporary psychology on the cognitive and emotional processes behind prosocial traits such as cooperation, altruism, and empathy. Political science and related disciplines consider human behavior at a higher level of organization by examining the emergence of societies and how people operate within them. Although historically lacking an evolutionary perspective, like biology, these fields nonetheless approach the problem within the framework of ultimate and proximate. Questions regarding how the cooperative interactions that define social structure are reconciled with the competition among individuals and the conflict that competition can generate are important at both the individual and the societal levels. At the proximate level, psychology research asks how an individual's social judgement and resultant behavior emerges from cognitive and emotional responses interacting with the environment, whereas questions in fields examining the operation of societies include what are the mechanisms that enforce a particular level of cooperation, or at least tolerance and other prosocial behaviors, to counteract potential conflict among individuals or groups, and equally important, how can entrenched conflict be overcome to yield a functioning society? Even the comparative approach important to evolutionary biologists finds parallels in the humanities and social sciences in examinations of how societies differ in their approach to the problem of balancing cooperation and conflict, and whether there are common features across these mechanisms that tell us something fundamental about social behavior.

The goal of this book was to provide a thoughtful consideration across disciplines of the cooperation–conflict relationship to generate insights into this fundamental issue. To this end, we enlisted researchers whose works range from an analysis of human society, to the psychological processes behind cooperation, to the biological mechanisms at the behavioral, neural, and genomic levels to provide their perspectives on this question. The resulting chapters converged to a remarkable degree on the same key themes, particularly mechanistic ideas and implications: For example, how does the brain regulate and balance the expression of the prosocial and aggressive behaviors? How do behavioral interactions lead to decisions balancing the expression of these behaviors in individuals and in societies? How do natural or sexual selection regimes lead to behaviors emphasizing one or the other in social situations? What do comparative studies of behavior and the brain tell us about the evolution of cooperation and the capacity to regulate conflict? The most important theme, however, consistent across all chapters, is that cooperation and conflict should not be seen as opposing forces that must be balanced, but inextricably linked from the societal to the molecular genomic level. One cannot exist without the other, and focusing on one without the other gives us an incomplete picture.

About This Volume

Because the interplay between cooperation and conflict is an area of inquiry that spans many disciplines, researchers may not be familiar with advances and ideas in other areas. To address this, our volume represents a multidisciplinary collaboration among biologists, psychologists, neuroscientists, and political scientists. Each author in this volume prepared scholarly, synthetic contributions highlighting key issues and recent contributions in their field. To make the chapters accessible to the broadest audience, we have included text boxes to clarify terms or provide information necessary for nonspecialists to understand issues outside their fields. We also liberally referred to other chapters within each contribution. We hope that this approach will result in a more approachable and stimulating book that allows readers to consider the important problem of the cooperation–conflict balance from multiple perspectives, as well as how these intersect.

This volume is divided into three parts. In Part I, we focus on more general issues of anti- and prosocial interactions. Chapter 1 considers how cooperation and conflict interact at the global level when considering domestic and international conflicts, while Chapter 3 considers the role of reputation as a mechanism to stabilize human cooperation and Chapter 2 models the role of internalizing such social norms in maintaining human cooperation. These chapters introduce themes that repeat through later chapters focusing on nonhuman species, particularly the role of aggression or punishment in maintaining prosocial behavior. Chapter 4 provides a big-picture view from the biological sciences, considering the evolutionary dynamics of social conflict and coming to the conclusion that even at the genomic level, there has been concurrent pressure for both cooperation and mechanisms to protect cooperation from self-interested outcomes. Indeed, she argues that focusing exclusively on cooperation will limit our understanding of these dynamics and the role of conflict in shaping society.

Part II focuses on the neural substrates underpinning cooperation and conflict. Chapter 5 takes a broad view on prosocial behavior, using as a model the interplay between two hormones linked to cooperation and conflict, oxytocin and vasopressin. This chapter echoes Chapter 4, pointing out that when considering prosocial behaviors, we only get half the picture if we focus exclusively on the positively valenced behaviors and ignore conflict and agonistic behavior. Chapter 6 considers the role of the temporal lobe, which is the neural basis for many of the cognitive abilities essential for both cooperation and conflict, while Chapter 7 provides a neural perspective on the roles of oxytocin and vasopressin, focusing on receptor distribution in the brains of nonhuman primates, primarily focusing on chimpanzees.

In Part III, we move to non-primate organisms. There is a strong tendency to focus on nonhuman primates when considering the evolution of human behavior, but considering other species provides a novel perspective that may highlight previously unrecognized issues and allow for an exploration of specific ecological or social factors that may be influencing behavior. Chapter 8 discusses long-term work on pied babblers, a cooperatively breeding bird, and emphasizes that we must understand both inter and intra-group conflict or we cannot understand the dynamics of cooperation. Chapter 10 echoes this, using social insects to make the case that the focus on cooperation has obscured the fundamental role of conflict in shaping their social structures. Chapter 9 considers mutualisms, or cooperative interactions between individuals of different species, including the well-studied cleaner fish–client mutualism. Mutualisms differ from intraspecific cooperation in both the level of conflicts (i.e., no reproductive conflict) and the potential for complimentary resources, and are an important addition to the conversation because the diversity of mutualisms makes them a good model of the different types of cooperation seen in humans. Finally, one key conclusion that can be drawn from all three of these chapters, again echoing Chapter 4, is that many of the cooperative and competitive behaviors that we see in humans and primates do not require complex cognition. Although humans are certainly capable of complex cognition, it would be useful to look for circumstances in which humans may be using simpler mechanisms.

There is nearly fifty years of research on the evolution of cooperation, and at least as much on conflict. What becomes clear in this volume is the astonishing consensus across each of these disciplinary boundaries and species of the fundamental role of the interplay between cooperation and conflict in shaping social systems across a variety of species.

Part I

Broad Insights from Political Science to Molecular Behavior

1 Cooperation and Conflict in International Relations

William J. Long

Introduction

The most fundamental questions in international relations are: "Why do states go to war?" "How can interstate conflict be prevented or ameliorated?," and "What are the pathways to greater international cooperation?" In considering these questions, the dominant paradigm in international relations, *political realism*, emphasizes the enduring propensity for conflict among self-interested states seeking their security in an "anarchic" international environment, that is, one where there is no central authority to protect states from each other or to guarantee their security. Hence, international cooperation is thought to be rare, fleeting, and tenuous – limited by enforcement problems and each state's preference for larger relative gains in any potential bargain because of its systemic vulnerability (Morgenthau, 1949; Waltz, 1979). At the extreme, states find themselves in a security condition of mutual distrust that resembles a prisoner's dilemma game. (See Box 3.1 in Chapter 3 for a description of various games.) Maintaining an equilibrium in the international system through a balance of power and limited cooperation are all that can be hoped for; a situation where war, large-scale violent conflict, is natural and merely "diplomacy by other means" (von Clausewitz, 1989). This is not to argue that international relations are in a constant state of war, rather that they exist within the shadow of war as a final arbiter.

The major alternative paradigm, *political liberalism*, focuses on identifying ways to mitigate the conflictual tendencies of international relations. Liberals argue that shared economic interests such as international trade that produces divisible, absolute gains from cooperation (win-win situations where relative gains are thought to be less essential) are a restraint on bellicosity. Liberals also note that common or cosmopolitan norms (such as democratic values and respect for human rights) can restrain the resort to violence and provide a means for settling disputes peacefully, particularly in disagreements between democracies (Box 1.1). Liberals also believe that international institutions (multilateral organizations, regimes, and laws) can help address the problems of creating and enforcing agreements in anarchy and help solve cooperation dilemmas. Finally, liberals urge the development of collective security to keep the peace through the coordinated actions of the entire international community (such as the United Nations [UN]) or a federation of like-minded states rather than relying on the war-prone balance-of-power system (Kant, 1983; Doyle, 1997).

Box 1.1 Democracies and International Conflict

Although the idea that democracies are more pacific in their international relations than states with more authoritarian forms of government can be traced to the writings of Immanuel Kant, it was not until the 1980s that scholars developed systematic evidence that mature democracies almost never go to war with other democracies. This finding led to the theory of the "democratic peace," which variously argues that (1) peaceful norms of dispute resolution that prevail within and between democracies, (2) citizen enfranchisement that constrains and slows the martial ambitions of leaders to take their citizens into wars, and (3) transparency in communication aided by a free press that reduces miscalculation in international relations all contribute to the democratic peace. There are several important caveats to the democratic peace. Most importantly, democracies are not more peaceful than other states generally; they are as likely to get involved in wars as authoritarian states, either against authoritarian states or via imperial wars. Further, democratizing states, those transitioning to democracy, are often unstable and war prone, especially multiethnic states that struggle to meet popular demands while protecting minority rights.

More recently, *political constructivist* approaches to international relations, drawing from sociological and linguistic theory, have emphasized that non-material, ideational factors, not just material interests and national and international institutions, are critical to understanding the formation of preferences and the possibility of cooperation. As the name implies, for constructivists, the interests and identities of states are highly malleable and context-specific and the anarchical structure of the international system does not dictate that conflict is the norm and cooperation the exception. Rather, the process of interaction between and among agents shapes how political actors define themselves and their interests: "self-help and power politics do not flow logically or causally from anarchy …. Anarchy is what states make of it" (Wendt, 1992, pp. 394–395). Because identities and interests are not dictated by structure, a state's purely egoistic interests can be transformed under anarchy to create collective identities and interests by intentional efforts and positive interaction; think, for example, of the transformation in French–German relations from enmity to amity with the creation of the European Union after a century of warfare between the two states.

What's Science Got to Do with It?

What has been the relationship between these theories and the scientific understanding of the physical and biological world? Since the Enlightenment, classical international relations scholars have developed their theories of the social world consistent within the dominant vision of physical reality. They have implicitly relied on the ontology, epistemology, and conception of human nature that predominated in the natural sciences for their social theories. In the field of politics and international relations, for example, the prevailing materialist/positivist/realist paradigm assumes the following:

- Ontologically, a Cartesian duality exists between observer and observed, between physical events and subjective human consciousness: each exists from its own side independent of the other and an unbridgeable gulf separates matter from mind.
- Epistemologically, objective truth is knowable through third-person methods of science focused exclusively on the material world, what some call "foundationalism" – the rational, self-directed search for permanent and authoritative principles of human knowledge (Toulmin, 1992). In physical and social science, the goal is to uncover deterministic patterns of behavior. As such, scientific facts are necessarily separate from values.
- Behaviorally, human nature, governed principally by reason, is essentially self-interested, and human action reflects the rational pursuit of one's preferences.

The predominant political conclusions flowing from these tenets are that insecurity and conflict naturally arise in groups of independent, materially real, self-interested actors. Thus, the benefits of cooperation are unlikely without a fear-based, hierarchical social contract domestically and, by extension, a balance of power among self-interested states acting in the anarchic international environment (Hobbes, 1651/1979; Morgenthau, 1949).

As noted above, liberal social thinkers working within this physical paradigm see greater possibilities for international cooperation if (1) institutional arrangements or international society can mitigate systemic anarchy, and (2) rational actors, over time, seek reciprocal cooperation (Jervis, 1978; Axelrod, 2006). Fundamentally, liberal schools of thought do not challenge the existing ontological or epistemological assumptions.

Although Enlightenment thinkers drew their conclusions about human nature from empirical observations of action within the social realm, since the time of Darwin, some materialist social thinkers have attempted to connect with the natural world through the life sciences to explain behavior – a different sort of naturalism. These approaches explicitly incorporate humans' biological legacy and psychological makeup into their understanding of human nature. They recognize that human beings and human brains can be understood as shaped by an evolutionary process and that events occurring within individuals may be as important to understanding their social behavior as events occurring between them. For example, the evolutionary paradigm suggests that the human brain has developed reliable, specialized functions that help it understand and navigate social interactions – such as interpreting threats or weighing value in exchange – and that this biological inheritance must be accounted for, along with environmental factors – such as culture and institutions – in explaining behavior (Barkow et al., 1992). Psychological approaches likewise focus on human agents and ground their theories on the innate nature of humans and the human mind and the limits and possibilities of behavior. Overwhelmingly, these approaches support the notion of self-interest as the norm in human behavior and altruism the exception, and they endorse the materialism and objectivism of physical scientists. Human behavior is considered the emergent product of "indifferent physiochemical processes" where "selfishness and a ready penchant toward violence are the principal elements of our nature" (Davidson and Harrington, 2002, p. 20). Consequently, social theorists should feel comfortable asserting that human behavior can be assumed to be largely rational and self-interested, and they should be free to focus their attention on social variables, like ideas and institutions, in explaining state and interstate interactions.

The third paradigm for understanding international relations, political constructivism, maintains that social agents create their own contexts and that these contexts shape social agents (Onuf, 1989). Thus, it rejects the subject–object dualism of objectivist, materialist science and views human nature as socially, linguistically, or normatively created (constructed); infinitely malleable; and largely detached from the physical world. Human behavior can be understood, or, more accurately, interpreted through the subjectively determined categories, institutions, and ideas of humans' own making. So, for example, although the human organism is not exempt from the evolutionary process that shaped it, because reality is overwhelmingly socially determined, humans can be effectively divorced from their biological inheritance. The resulting explanations for social behavior therefore focus on the importance of culture, language, institutions, and other human artifacts alone. Social phenomena can be understood from the outside in. Conclusions about the "nature" of social behavior derived from this approach are necessarily contingent and provisional and not generalizable claims. By emphasizing the role of shared beliefs and social forces that shape reality, constructivism mostly disregards agents' internal motivation and innate biological and psychological constraints and capabilities.

Beyond this tacit, arm's length relationship with the natural sciences, mainstream political science and international relations theories have not ventured and the reasons for the separation are historical and practical more than philosophical. Historically, since the mid-nineteenth century through the mid-twentieth century, for example, biological theories and concepts have been used in nefarious ways to promote or justify social pathologies such as colonialism, racism, and fascism. It is understandable, therefore, that social scientists have avoided considering explicitly the biological sciences in part because of the pseudo-certainty these approaches historically bestowed on several social evils. A practical reason to avoid new developments in the physical and biological sciences are the issues of complexity and uncertainty. Social scientists lack the specialized knowledge to connect with the natural sciences generally, and revolutionary changes in scientific understandings of the physical and biological worlds instigated by quantum physics and neuroscientific findings, regarding the plasticity of the human brain and behavior specifically, are highly esoteric and not yet fully agreed upon or understood. Thus, social thinkers have little ability or appetite to delve into these fields to consider their political implications until after the dust settles. In disengaging from the natural sciences, however, social theory may be missing an opportunity to find insights into important questions, a topic addressed later in the discussion of national reconciliation.

The Changing Nature of Global Conflict

In contemporary international relations, the realities of conflict and cooperation have changed over the last half century and there is good news, and bad, to report. First the good news: interstate conflicts, the historical focus of most international relations theory, have declined substantially and all but disappeared. In 2011, for example, the lone interstate conflict was a low-intensity war between Thailand and Cambodia. Putting aside the possibility that certain interstate conflicts could result in thermo-nuclear exchange and end planetary life as we know it, some say we should rejoice in our good fortune and the better angels of our nature (Lebow, 2010; Pinker, 2011). Such a celebration may be premature, however, as the recent conflict between Russia and Ukraine reminds us or as growing tensions in the South China Sea portend. Further, as discussed below, contemporary civil wars often embroil other states and create regional and global problems. Nonetheless, Figure 1.1 illustrates the waning of interstate wars over the past several decades.

There are many competing and complementary explanations for the decline of interstate war. Some of the more notable assertions reference the following factors as contributing to the diminishing of interstate war:

- States are more economically interdependent.
- The increase in democratic states reduces the number of potential belligerent dyads.
- Europe, the historical location of many interstate wars, is at peace.
- The possibility of nuclear escalation constrains interstate violence.
- Increases in communication technologies reduce the likelihood of war caused by misperception or misunderstanding.
- The benefits of war are lower than in the past and the costs are higher.
- War fatigue from World War I and World War II exists.
- International norms have changed and violence is no longer a legitimate means for settling disputes.

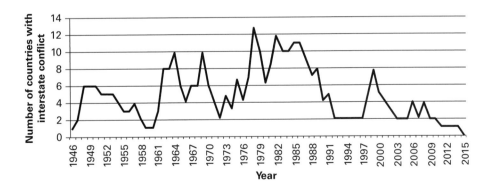

Figure 1.1 Declining interstate conflict. *Source*: Monty G. Marshall, Major Episodes of Political Violence (MEPV) and Conflict Regions 1946–2012, Center for Systemic Peace, 2013.

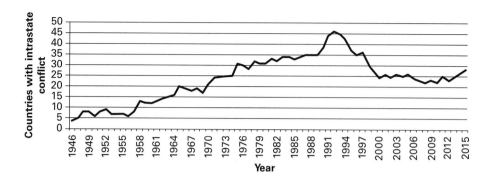

Figure 1.2 The rise of intrastate conflicts. *Source*: Monty G. Marshall, Major Episodes of Political Violence (MEPV) and Conflict Regions 1946–2012, Center for Systemic Peace, 2013.

- Greater institutionalization of international politics creates venues for negotiation and enhanced cooperation.
- More effective conflict management techniques – such as conflict prevention, peacebuilding, and peacebuilding – defuse or shorten wars.

Failure may be an orphan, but success in reducing interstate wars has many claimants.

While interstate wars have declined, Figure 1.2 illustrates that intrastate (civil) wars have become more numerous, peaking at the end of the Cold War and remaining the dominant feature in global conflict. Today, a much greater percentage of conflict in the international system takes place within states rather than between states.

Civil wars now constitute the major threat to international peace and security because they frequently embroil other states and non-state actors in their conflicts, and they destabilize wider regions by creating weak, fragile, or failing states that foster a host of transnational problems such as refugee flows, genocide, terrorism, environmental degradation, interstate crime, and infectious disease contagion, among others. This is the bad news regarding conflict trends. The current "civil" war in Syria is an object lesson in how civil conflicts are internationalized, as this war also serves as conflict between Saudi Arabia and Iran over dominance in the Islamic world and among Russia, America, and Turkey for influence in the Middle East. Syria also quickly became a haven for the Islamic State of Iraq and Syria (ISIS) and related terrorist groups, and the forced migration of millions from Syria has profoundly impacted neighboring Turkey, the European Union, and North America.

Civil wars also enmesh the international community when the failure of a state to protect the human rights of its own citizens results in gross violations of international norms, such as the prohibition of genocide and prompt "humanitarian intervention" by other states or international governmental organizations such as the UN or the North Atlantic Treaty Organization (NATO). Recent conflicts in the horn of Africa and in the Balkans are examples of internationalized civil conflict caused by a state's failure to protect its own people.

Making matters worse, many of these intrastate conflicts possess features that make them more intractable and protracted than interstate conflicts. Civil conflicts have been estimated to last more than ten times longer than interstate wars, and are more likely to return to violence after settlement efforts than interstate wars (Bennett and Stam, 1996; Collier et al., 2001; Fearon, 2004). Various estimates of recidivism in civil conflicts range from 40 to 90 percent depending on the methodology employed (Walter, 2009; World Bank, 2011). Lasting peaceful settlement of civil conflicts is rare – a subject I return to below – absent military conquest by one side, a "victor's peace," such as the defeat of the Liberation Tigers of Tamil Eelam in Sri Lanka in 2009, or massive third-party intervention, such as the interminable UN peacekeeping operation in Cyprus. In short, most civil conflicts are never formally resolved and often reoccur.

Peaceful settlement of civil conflicts is problematic because of acute "security dilemmas." The notion of a security dilemma comes from the realist understanding of international relations wherein actions taken by a state to increase its own security (expanding its military expenditures, for example), cause reactions from other states (such as an arms race), which in turn lead to a decrease rather than an increase in the original state's security and instability in the system. This dilemma flows naturally from the posited nature of the international system in which self-interested states pursue their security in an anarchic environment. As applied to civil conflicts, the security dilemma facing warring factions can be more acute because a war settlement requires one side to lay down its arms (state sovereignty means a monopoly in the use of force by the government within its territory) and then live together with its former adversary. Conflict between states at least allows for the parties to withdraw behind national borders and for both sides to provide for their own defense against future aggression in most war settlements (Walter, 2009).

Further, civil conflicts are difficult to resolve peacefully because they often concern "existential" values that go to the heart of the identities of the warring factions and are, by nature, difficult to compromise relative to a dispute over material issues such as a border disagreement between nations. Civil conflicts are often defined by disputes between parties of different ethnicities, religions, and races. These intrastate differences reflect, in part, the legacy of colonialism, which left behind multiethnic states where a tradition of playing one group against another and where the ruling group often monopolizing political authority and being unresponsive, if not exploitative, toward the needs of other groups, had been part of a divide-and-rule strategy of the colonial power. After formal decolonization, new national elites often continue to manipulate these cleavages for their personal and communal advantage by perpetuating negative images of the other and stoking communal antagonisms. The connection between colonialism and current civil strife can be seen in the geographic distribution of recent civil conflicts, which are predominantly in former colonial territories in the Middle East, Africa, and Asia. Because postcolonial conflicts often cleave along ethnic, tribal, religious, and/or racial lines, these existential values become part of the conflict and, when such values are threatened, they do not lend themselves to obvious compromises and settlements in situations characterized by animosity, distrust, and acute security dilemmas.

In some ways, these new forms of conflict harken back to the pre-Westphalian, medieval period, before the formation of distinctive nation-states and the firm establishment of the principal of state sovereignty, in that the boundaries between state and society, internal and external, and war and criminality are blurred. In describing these new wars, Mary Kaldor has noted that they are characterized by consolidation of power along ethnic lines, based on tribalist or communal identities rather than international ideologies (communism or fascism for example); external support by diasporas more than superpowers or ex-colonial powers; irregular and privatized warfighting strategies often involving terrorism and the use of atrocity rather than standing armies and traditional military strategies; and financial sustenance from criminal trade in natural resources or contraband such as drugs rather than through state taxation and mobilization (Kaldor, 2006).

In explaining the causes of civil conflicts, scholars fall primarily into one of two camps. First, there are those who focus on the economic opportunity for insurgents to grab political power and economic resources through inter-elite power struggle, criminality, and warlordism in situations where the state control and viability are weak and insurgencies have a good chance of success – the so-called greed explanation (Fearon and Laitin, 2003; Collier and Hoeffler, 2004). Alternatively, there are those who focus on "grievances" as the underlying cause of civil strife, particularly the failure of the state to provide access to political participation, economic opportunity, and expression of identity for certain social groups or communities, again often based on colonial-era ethnic, tribal, or religious distinctions (Azar, 1991; Cederman et al., 2013). The failure to provide for the basic political, material, and identity needs of excluded groups is offered as the underlying source of protracted social conflict. Neither explanation is mutually exclusive, and on inspection, several civil conflicts involve a mix of both types of motivation while others begin as one type of conflict, say over political exclusion, and devolve into a conflict over who owns and exploits national resources or secures the benefits from a "war economy" (Keen, 1998).

At a broader level, the growth in civil conflict can be traced to larger social processes such as decolonization, which, as noted, created many new and unstable states with weak central governments and a history of ethnic division. Also, the end of the Cold War, which coincided with the peak number of civil conflicts, is thought to have removed the constraint of superpower influence on intrastate violence. During the Cold War era, one or the other superpower often propped up an authoritarian regime of its liking that suppressed underlying national grievances; Yugoslavia was perhaps the most dramatic example. Finally, the process of globalization itself – the rapid growth of trade, transport, finance, and communication – may have weakened state authority and subjected emerging states to forces largely beyond their control. These nascent states often failed to meet the challenges of globalization, thus creating opportunities for insurgents to challenge the state for control of power and wealth.

Returning to the Prospects for Peaceful, Internal Resolution of Civil Conflicts: The Process of Reconciliation

Despite the general intractability of many civil conflicts and the likelihood of recidivism in civil violence, research by my colleague Peter Brecke and me suggests that the process of reconciliation provides an intrinsic way for states to overcome the intractability of today's dominant form of conflict (Long and Brecke, 2003). We were first drawn to this possibility by many formal and informal observations suggesting that *reconciliation events* – public displays of mutually conciliatory accommodation between antagonists – was somehow integral to mitigating future violence and maintaining social order after violent conflict. Consider four descriptions of "reconciliation events" in very different societies.

- In primate society, Frans de Waal described a fight in the chimpanzee colony of the Arnhem Zoo:

  It was the winter of 1975 and the colony was kept indoors. In the course of a charging display, the dominant male attacked a female, which caused screaming chaos as other chimpanzees came to her defense. When the group finally calmed down, an unusual silence followed, with nobody moving, as if the apes were waiting for something. Suddenly the entire colony burst out hooting, while one male worked the large metal drum in the corner of the hall. In the midst of the pandemonium, I saw two chimpanzees kiss and embrace ... the embracing individuals were the same male and female of the initial fight. (de Waal, 1989)

- In subnational tribal relations, the letters of Samuel Sewell captured the following ceremony of Native Americans in the northeast colonies in 1630:

  Meeting with the Sachem they came to an agreement and buried two axes in the ground ... which ceremony is to them more significant and binding than all the Articles of Peace, the hatchet being the primary weapon. (Hendrikson, 1989)

- In the national society of South Africa, the *Telegraph* reported on a public handshake and raised arms of President F. W. de Klerk and Nelson Mandela in Cape Town, on May 4, 1990, after announcement of an agreement on steps that would lead to talks ending white-minority rule (Alleyne, 2012).
- In the realm of international politics, contemporary historian Hendrick Smith described the signing of a peace treaty and pubic joining of hands among President Anwar Sadat of Egypt, Prime Minister Menachem Begin of Israel, and President Jimmy Carter of the United States:

  The elusive, unprecedented peace treaty that Egypt and Israel signed today has enormous symbolic importance and the potential for fundamentally transforming the map and history of the entire region.... the best diplomatic estimate here is that the treaty reduced the risk of major war in the Middle East. (Smith, 1979)

Each anecdote contains the same implicit or explicit hypothesis: future violence is less likely to occur, and social order more likely to be restored, if principals to a conflict engage in a formal, public reconciliation event indicating a desire for improved relations.

We wondered, "Is there systematic empirical support for this alleged link between a 'reconciliation event' and the resolution of large-scale conflict such as intrastate wars?" We define a reconciliation event as one with the following elements: direct physical contact or proximity between opponents, usually senior representatives of the respective factions; a public ceremony accompanied by substantial publicity or media attention that relays the event to the wider national society; and ritualistic or symbolic behavior that indicates that the parties consider the dispute resolved or resolvable and that more amicable relations are expected to follow. We questioned, first, whether these events correlate with successful civil war settlement, and, second, if so, how, that is, "What is the process by which reconciliation events contribute to restoring order and affinity in relations?" Ultimately, we discovered that reconciliation events often correlate with successful civil conflict resolution and that, in certain instances, they were part of a forgiveness-like process that could break the intractability of even the most severe civil conflicts.

To explore these questions, we examined within-country (civil) violent conflicts in the twentieth century (430 conflicts in 109 countries) to see if any had experienced a reconciliation event. We unearthed 11 cases (in 10 countries) with reconciliation events meeting the definition given above, all occurring in the second half of the century. Importantly, we found that 7 of the 11 (64 percent) of them did not experience a return to violent conflict in the five years following the reconciliation event – a result at odds with the recidivist rates expected of civil conflicts. In contrast, among the countries that experienced civil conflict but did not experience a reconciliation event, only 9 percent avoided recurrence of violence during the time-period demarcated by our reconciled cases (1957–2000). Table 1.1 summarizes the results of our investigation. (For an explanation of the definitions, databases, methodology and

Table 1.1 Reconciliation after civil conflict

Country	Date	Subsequent outcome
Colombia	1957	War
North Yemen	1970	War
Chad	1971	War
	1992–1993	War
Argentina	1984	Peace[a]
Uruguay	1985	Peace
Chile	1991	Peace
El Salvador	1992	Peace
Mozambique	1992	Peace
South Africa	1992–1993	Peace
Honduras	1993	Peace

[a] "Peace" refers only to absence of large-scale civil violence; what some call "negative peace" (this threshold was set at 32 political fatalities per year) rather than "positive peace," i.e., the establishment of life-affirming and life-enhancing values and structures (Galtung, 1969).

classification of which countries did and did not experience a reconciliation, see Long and Brecke, 2003, chapter one and appendix A.)

With this preliminary indication that reconciliation events correlated with an improved chance of enduring peaceful settlement of civil conflicts, we turned to the more interesting question: "Why might reconciliation events matter?" To explore that question, we conducted a comparative case study of the eleven cases guided by two possible explanatory models and hypotheses derived from them, each connected to a different underlying set of assumptions about human nature and human rationality, their respective "microfoundations." The first we call a "signaling model," and the second a "forgiveness model." We derived hypotheses about the behavior to be expected from each model, and we grounded each model in distinctive paradigmatic assumptions about human decision-making and problem-solving – rational choice and evolutionary psychology.

With respect to civil conflicts, our study found that reconciliation events contribute to lasting social order when they were part of a forgiveness-like process characterized by public truth-telling, redefinition of the identities of the former belligerents, partial justice, and a call for a new relationship (as symbolized by the reconciliation event). We also found that this finding is more consistent with an understanding of human rationality and problem-solving that argues we possess evolutionarily evolved, func- tionally specific, emotionally assisted problem-solving mechanisms rather than a mind characterized by universal rationality. We take a closer look at the models, hypoth- eses, and empirical results in the next section.

Two Models, Two Hypotheses, Two Notions of Rationality

The first model attempting to account for the role of reconciliation events in civil conflict resolution we call the "signaling model." It is derived from the dominant approach in international relations for understanding conflict resolution, rational choice/ game theory, that seeks to specify possible outcomes from the interaction of rational actors seeking to "win," that is, achieve desired strategies and satisfy their preferences. The model describes a mechanism or process consistent with a general signaling hypothesis, to wit, the best strategy for breaking a pattern of hostile interactions is by sending signals that provide a measure of *commitment* to the pursuit of improved relations. According to this theoretical perspective, reconciliation events or gestures are particularly effective forms of this type of signal because they are almost always politically costly to leaders of the opposing sides, and costly signals are more reliable determinants of a leader's true intention for improved relations than low- cost or cost-free signals, so-called cheap talk.

Reconciliation offers or attempts are costly because they are undertaken before domestic and international audiences and thus are potentially exploitable by other party or a rival domestic constituency. A reconciliation event (and the possible reconciliation it symbolizes) is a costly (or potentially costly) signal that the other party is likely to interpret as a genuine offer to improve relations and thus may break a deadlocked conflictual situation. By sending costly (and therefore trustworthy) signals indicating a less-hostile intent, reconciliation events reduce the perception of threat between actors (other things being equal) and permit improvement in relations.

This hypothesis maintains that, because of the attendant costs, reconciliation events potentially reduce threat perception between adversaries and permit improvement in future relations, because continuing the conflict in such an environment and with this new information may no longer maximize the utility of either party. Reconciliation events may break through a conflictual relationship with its conditions of high ambiguity, high mistrust, and low credibility. Based on the signaling literature, we should expect a successful reconciliation to emerge from cases with signals that possess the following characteristics:

(1) *Costliness*. The offer of a reconciliation must impose a cost on the initiator and its reciprocation a cost on the other party; the higher the cost the better.
(2) *Vulnerability*. Reconciliation initiatives should involve risk and be vulnerable to exploitation.
(3) *Novelty*. Reconciliation attempts are most likely to break established conflict patterns when they are dramatic, positive (not merely refraining from negative action), unexpected, and thought-provoking or newsworthy.
(4) *Voluntariness*. Reconciliation signals are best when made unilaterally, rather than as the result of pressure or coercion. The offer of conciliation from the stronger party is prima facie evidence of voluntariness.
(5) *Irrevocability or noncontingency*. Making noncontingent and irrevocable offers that are likely to be understood as conciliatory, rather than quid pro quo, contributes to the success of a reconciliation attempt.

As noted, each model rests on a distinctive set of assumptions about human rationality. The signaling model is fully consistent with the paradigmatic assumptions of the dominant social science approach of rational choice; that is, humans apply universal, general reasoning rules to all problems in making choices in their current environment, including interpreting and acting on signals in reaching a negotiated settlement of conflicts. The mind is assumed to be content-independent, taking its cues from the environment, and domain-general, its rational processes operating similarly in all aspects of human activity.

A second model for accounting for the power of reconciliation events we call the "forgiveness model." Societal reconciliation in this model is understood as analogous to a process of interpersonal forgiveness, transforming certain emotions (moving from anger to affinity) and transcending certain beliefs about oneself and the other that opens the possibility of new, beneficial relations.

This model maintains that reconciliation is part of an evolved, specialized, problem-solving mechanism to deal with the recurring problem of how to restore social order and the benefits of affiliation despite inevitable societal conflicts. National reconciliations are, hypothetically, modern manifestations of functionally specialized, emotionally assisted, and human problem-solving mechanisms that we possess to explicate ourselves from this recurrent dilemma (Box 1.2). Without such a mechanism, Hannah Arendt supposed, "Our capacity to act would, as it were, be confined to one single deed [conflict] from which we could never recover; we would remain the victims of its consequences forever, not unlike the sorcerer's apprentice who lacked the magic formula to break the spell" (Arendt, 1989, p. 237). Unlike general rationality approaches, the mind is considered an organ that has evolved to include a large collection of functionally specialized, domain-specific mechanisms that are adaptations constructed by natural selection and other evolutionary processes over time to cope with regularly occurring, reproduction-threatening problems. This is a very different understanding of the function of mental processes than that posed by rational choice theorists, who posit a general problem-solving mind whose function is to maximize an individual's goals or well-being in response to its *existing* environment. Also, unlike rational choice, which treats emotions as exogenous to, or impediments of, reason, an evolutionary approach incorporates emotion as well as reasoning in explaining human behavior because emotions are products of an evolutionary process – the result of functional adaptation. Specifically, emotions identify priorities and help solve regulatory problems in a mind filled with many functionally specialized mechanisms as well as general reasoning ability. In ways only partially understood, emotions animate and help coordinate among problem-solving techniques and their appropriate application to situations. According to this approach, the brain's systems were not designed for cool rationality but for hot cognition, to respond to critical events related to survival and reproduction. As such, affect and cognition generally complement each other. Reconciliation, this model suggests, is a manifestation of patterned, emotively driven, problem-solving behavior writ large, not merely rational calculations.

Box 1.2 Considering Evolutionary Factors in Contemporary Social Science

The universality of a problem such as sociality, or evidence of a ubiquitous problem-solving mechanism such as reconciliation, is not proof of an evolved human capability, but it does allow for generating hypotheses about behavior and designing observations and tests that are plausibly consistent with psychology and biology and otherwise would not have been thought of. Procedurally, the method of deriving and examining social science hypotheses from an evolutionary perspective begins by noting the existence of complexly articulated and recurrent behavioral trait, in this case reconciliation events. Second, one can ask, deductively, whether the trait could reasonably be the expression of an adaptation; that is, a response to a species-typical problem encountered over several million years of human evolution. If so, we might be witnessing a contemporary manifestation of an evolutionarily engineered, emotionally influenced problem-solving capability rather than simply the exercise of general reasoning. Third, armed with a plausible hypothesis, the posited behavioral characteristic must be linked with and understood in its cultural, social, and political system.

Behaviorally, the process of forgiveness and reconciliation as described across many disciplines invariably includes four phases. First, parties to a conflict must recognize shame or anger from a perceived wrong, injustice, frustration, or injury; they must acknowledge the harm. In large social settings, this means public truth-telling via official investigations, testimonies, judicial proceedings, and mass-media reporting. Second, reconciliation requires a changed understanding of oneself and the other party to a conflict that transcends the narrow roles of victim and perpetrator. Often this process involves reframing the other by separating the wrongdoer from the wrong committed. Without absolving the wrongdoer of responsibility, redefinition nonetheless implies seeing that party in a more complete, well-rounded way. Third, the parties must forego the option of revenge, however natural, desirable, or justifiable. This forbearance does not require abandonment of all versions of punishment, redress for wrongs or injuries, or abandonment of justice, only a willingness to break the cycle of injury and counter-injury. Retribution must be less than total. Fourth, one or both parties make an offer that results in contact between them and a public expression of the likelihood of improved relations, what we have called the reconciliation event. The goal is to reestablish mutual affiliation, coexistence, and mutual toleration – in short, a "civil" relationship between those who share the same community.

In sum, forgiveness requires recognition of harm – truth-telling, development of a new understanding of oneself and the other, and willingness not to prolong hostility through acts of revenge. It can also include the offer of a renewed community in the future – a reconciliation event. Although cognitive strategies and bargaining are involved in this process, it fundamentally represents an emotionally animated practice entailing a specific problem-solving technique that exists to restore relations in one's social group. In contrast with universal rational choice, the forgiveness model is more consistent with the assumptions of evolutionary psychology and affective neuroscience. Those assumptions – that humans possess numerous, patterned, specific problem-solving capabilities because of interactions with past environments, and that those capabilities rely not just on reason alone but work in sync with our emotional repertoire – are different from those underlying rational choice.

Applying the Hypotheses to the Cases

On empirical investigation of the cases, we found that those countries that reconciled successfully, that is, restored lasting social order and did not return to large-scale violence for at least five years after the reconciliation event, did so through a protracted process of recognition of harm and public truth-telling, redefinition of identities and social roles of the antagonists, and partial justice short of revenge, not merely through signal-sending in a negotiated bargain. An untidy, seemingly idiosyncratic, but undeniably patterned process consistent with the forgiveness model was the foundation of successful national reconciliations. The three instances confined to a negotiated bargain – Colombia, Yemen, and Chad – did not lead to a restoration of peace.

Table 1.2 Presence of signaling factors and outcome

Country	Costliness vulnerability	Novelty	Voluntary	Irrevocable/ noncontingent	Outcome
Colombia	Yes	Yes	No	Yes	Conflict
North Yemen	Yes	Yes	Yes	No	Conflict
Chad	Yes	Yes	Partial	Yes	Conflict
Argentina	Yes	Yes	No	No	Peace
Uruguay	No	No	No	No	Peace
Chile	Yes	Yes	Yes	No	Peace
El Salvador	Yes	No	Partial	No	Peace
Mozambique	No	Yes	Yes	No	Peace
South Africa	Yes	Yes	Yes	Yes	Peace
Honduras	Yes	Yes	Partial	No	Peace

Although negotiation was part of all the cases, the mere existence of, or conditions surrounding, negotiations was not a good predictor of possibilities for long-run restoration of social order. In the three cases where the signaling factors were largely favorable – Colombia, Yemen, and Chad – conflict was not successfully resolved. In contrast, secret, low-cost, and highly contingent negotiations, as was the case in Uruguay, for example, can be part of a successful national reconciliation. In short, the presence of a negotiation, in itself, is not predictive of a successful reconciliation (see Table 1.2).

Instead, a pattern of recognition of harm through public truth telling, redefinition of self and other, limited justice, and a symbolic call for a new relationship were part of all successful national reconciliations. Extensive truth-telling was a part of each successful national reconciliation and absent from the three unsuccessful cases. Moreover, with only two exceptions, it was one of the first acts of a new or interim government, thus making the truth officially sanctioned. The methods of truth-telling included the publication of books, radio and television broadcasts, documentaries, sermons and newspaper exposés, and the formation of commissions to hold public hearings. Truth-telling was also directly linked to the pursuit of justice and redefinition of the identities and roles of parties to the conflict. It contributed to reshaping social and political roles of key actors. Armed with the authority of official truth, truth commissions and fact-finding bodies were often empowered to make detailed institutional recommendations and push fundamental reforms and a redefinition of societal relations. Truth-telling did not complete the process of reconciliation. Instead, it opened a public space for reconciliation by allowing a formerly taboo subject to become amenable to the action of political bodies and future policies.

Evidence of redefinition of parties to a conflict whereby the narrow identities of victim and perpetrator or repressor and insurgent are replaced with a new sense of self and other that makes a new relationship possible appears throughout the cases of successful reconciliation. The process begins with recognition and dialogue and truth commissions often provide the first sympathetic hearing for victims that helps them restore a sense of wholeness. Successful reconciliations also redefined the role and relationships of important social groups and institutions. Former belligerents were often brought into the political process, the military was often depoliticized and its former impunity removed. The cases vary in their forms of institutional redefinitions, but in every case, countries that successfully reconciled established a set of new identities for key social actors.

In every instance of successful reconciliation, save one (Mozambique), justice was meted out, but never in full measure. This fact may be lamentable, even tragic from certain legal or moral perspectives, yet it is consistent with the requisites of restoring social order postulated in the forgiveness hypothesis. In all cases of successful reconciliation, retributive justice could neither be ignored nor fully achieved. The reasons are practical – for example, the danger of provoking still-powerful security forces – as well as profound – it allows those who suffered, if they so choose, to give the gift of mercy. Whether for reasons of force majeure or other considerations, imperfect justice is often tolerated in the name of societal order. What justice is secured is usually achieved through extensive truth-telling, material reparations for some victims, and limited prosecutions of individuals, with punishment being loss of impunity, reputation, moral standing, office, and privileges, more often than incarceration.

Finally, each case of order-restoring reconciliation was accompanied by a national commitment to a new social relationship that transcended the antagonism of the war years. Legislatures passed solemn resolutions, peace accords were signed and embraces exchanged by heads of formerly rival groups, statues and monuments to the tragedy were erected, textbooks were rewritten, and a thousand actions, large and small, were undertaken to underscore the notion that the past was different and the future more hopeful.

To appreciate the respective "fit" between the forgiveness model in the eleven cases, see Table 1.3.

Table 1.3 Presence of forgiveness factors and outcome

Country	Public truth-telling	Partial justice	Redefinition of social identities	Call for a new relationship	Outcome
Colombia	No	No	No	Yes	Conflict
North Yemen	No	No	Partial	Yes	Conflict
Chad	Partial	No	No	Yes	Conflict
Argentina	Yes	Yes	Yes	Yes	Peace
Uruguay	Partial	Yes	Yes	Yes	Peace
Chile	Yes	Yes	Yes	Yes	Peace
El Salvador	Yes	Yes	Yes	Yes	Peace
Mozambique	Yes	No	Yes	Yes	Peace
South Africa	Yes	Yes	Yes	Yes	Peace
Honduras	Yes	Yes	Yes	Yes	Peace

Conclusions

These findings are both practically and theoretically significant. Conventional wisdom regarding civil conflicts argues that intrastate wars rarely end in a peaceful settlement. Our study offers a hopeful caveat to this otherwise pessimistic conclusion. It suggests that states wracked by civil conflict can reach a peaceful settlement through an arduous process of national reconciliation that resembles a process of forgiveness. In seven of the eleven cases involving a reconciliation event, the combatants reached a peaceful solution that produced lasting social order that did not devolve into further conflict. This result is a significant and very positive qualification to the general proposition that peaceful resolution of civil conflict is extraordinarily rare. Further, contrary to the notion that large-scale, third-party involvement is necessary for civil war settlements, these forgiveness and reconciliation processes were substantially "home grown" rather than imposed from outside or under the tutelage of a more powerful third-party country or organization.

For policy and practice, the civil war settlement cases presented here are both hopeful and cautious. Hopeful, in the sense that the process of national forgiveness holds great promise in resolving such conflicts, and although the steps involved are extraordinarily difficult, they are both knowable and possible. The findings are cautious in the sense that third-party intervention may be less important than previously believed. The belligerents themselves must do the heavy lifting to reach the truth, redefining themselves and the other party to the conflict, and stopping short of vengeance. Third parties can help in this process, but they primarily can help those that help themselves.

As evidence of the promise of reconciliation as a means of civil war settlement, there has been a proliferation of truth and reconciliation efforts during the first decade of the twenty-first century. Although scholars disagree on their definition of what constitutes a genuine truth commission case, Table 1.4 notes only those recent cases where there is substantial agreement of a genuine truth and reconciliation effort.

Table 1.4 Contemporary truth and reconciliation cases 2000–2010

Country	Year(s)
Nigeria	1999–2002
South Korea	2000–2004
Uruguay	2000–2003
Panama	2001–2002
Peru	2001–2003
East Timor	2002–2003
Ghana	2002–2003
Serbia and Montenegro	2002–2003
Sierra Leone	2002–2003
Paraguay	2003
Democratic Republic of the Congo	2004
Morocco	2004–2005
Liberia	2005–2006

Source: Brahm, 2009.

For purposes of this volume, which considers conflict and cooperation from the perspective of the natural and social sciences, these findings regarding reconciliation have an additional significance. The study illustrates that the crucial challenge in conflict resolution today, resolving civil conflict, cannot be understood using a signaling model based on a general rationality assumption used by most social scientists. Negotiation among self-interested players was a part, but only a part, of a much deeper, more protracted, more emotive pattern of behavior that can be recognized and understood only through a different model based on different assumptions about human nature – one informed by biology and neuroscience.

Drawing from natural science insights – a theory of mind and human behavior that looks at evolved, emotionally directed mental processes and patterns of behavior – allows us to generate a forgiveness hypothesis that offers greater explanatory power as to why and how national reconciliation serves as a mechanism for restoring social order after civil war. This approach integrates human rationality and human emotion in a hypothesis of social behavior that is more efficacious regarding understanding the most important problem in global conflict today.

An evolutionary biological foundation for generating political hypotheses has its own strengths and weaknesses, some theoretical, some methodological, and some historical, although some critiques are misguided. But it is equally or more scientific than rational choice approaches in the explicitness and accuracy of its assumptions and its amenability to the scientific method. Moreover, it is consistent with established facts and predominant theories in the natural sciences and allows for the application of our best understanding of the human brain and behavior instead of relying on the folk psychology of universal rationality that is divorced from our emotions and evolutionary inheritance. In sum, social science should move carefully, but remain open to the insights of the natural sciences if it is to advance its understanding of conflict and cooperation.

References

Alleyne R. (2012) Famous reconciliations sealed with a handshake. [online] Telegraph.co.uk. Available at: www.telegraph.co.uk/news/uknews/9359476/Famous-reconciliations-sealed-with-a-handshake.html [Accessed October 23, 2018].
Arendt H. (1989) *The Human Condition*. Chicago: University of Chicago Press.
Axelrod R. (2006) *The Evolution of Cooperation*. New York: Basic Books.
Azar E. (1991) Protracted social conflict: Theory and practice in the Middle East. *Journal of Palestinian Studies*, 8(1): 41–60.
Barkow J., Cosmides L., and Tooby J., eds. (1992) *The Adapted Mind: Evolutionary Psychology and the Generation of Culture*. New York: Oxford University Press.
Bennett D. S., and Stam A. C. (1996) The duration of interstate wars, 1816–1985. *American Political Science Review*, 90(2): 239–257.
Brahm E. (2009) What is a truth commission and why does it matter? *Peace and Conflict Review*, 3(2): 1–14.
Cederman L. E., Gleditsch K. S., and Buhaug H. (2013) *Inequality, Grievances, and Civil War*. Cambridge: Cambridge University Press.

von Clausewitz C. (1989) *On War*, indexed edition. Trans. Michael Eliot Howard and Peter Paret. Reprint edition. Princeton, NJ: Princeton University Press.

Collier P., and Hoeffler A. (2004) Greed and grievance in civil war. *Oxford Economic Papers*, 56(4): 563–595.

Collier P., Hoeffler A., and Soderbom M. (2001) On the Duration of Civil War. Policy Research Working Paper Series 2681. The World Bank.

Davidson R. J., and Harrington A., eds. (2002) *Visions of Compassion*. Oxford: Oxford University Press.

De Waal F. (1989) *Peacemaking among Primates*. Cambridge: Harvard University Press.

Doyle M. W. (1997) *Ways of War and Peace: Realism, Liberalism and Socialism*. New York: W. W. Norton & Co.

Fearon J. D. (2004) Why do some civil wars last so much longer than others? *Journal of Peace Research*, 41(3): 275–301.

Fearon J. D., and Laitin D. D. (2003) Ethnicity, insurgency, and civil war. *American Political Science Review*, 97(1): 75–90.

Galtung J. (1969). Violence, peace, and peace research. *Journal of Peace Research*, 3(6): 167–191.

Hendrikson R. (1989) *Encyclopedia of Word and Phrase Origins*. London: Macmillan Press.

Hobbes T. (1651/1979) *Leviathan*. New York: Penguin.

Jervis R. (1978) Cooperation under the security dilemma. *World Politics*, 30(2): 167–214.

Kaldor M. (2006) The "New War" in Iraq. *Theoria*, 53(109): 1–27.

Kant I. (1983) *Perpetual Peace and Other Essays on Politics, History, and Moral Practice*. Indianapolis, IN: Hackett Publishing.

Keen D. (1998) The economic functions of violence in civil wars. London: International Institute for Strategic Studies. Adelphi Paper, 320: 119–150.

Lebow R. N. (2010) *Why Nations Fight: Past and Future Motives for War*. Cambridge: Cambridge University Press.

Long W., and Brecke P. (2003) *War and Reconciliation: Reason and Emotion in Conflict Resolution*. Cambridge, MA: MIT Press.

Marshall M. G. 2013. Major Episodes of Political Violence (MEPV) and Conflict Regions, 1946–2012. Vienna, VA: Center for Systemic Peace.

Morgenthau H. (1949) *Politics among Nations: The Struggle for Power and Peace*. New York: Knopf.

Onuf N. G. (1989) *World of Our Making: Rules and Rule in Social Theory and International Relations*. Columbia: University of South Carolina Press.

Pinker S. (2011) *The Better Angels of Our Nature: The Decline of Violence in History and Its Causes*. New York: Penguin.

Smith H. (1979) Treaty impact still unknown. *New York Times*, March 17, p. A1.

Toulmin S. (1992) *Cosmopolis: The Hidden Agenda of Modernity*. Chicago: University of Chicago Press.

Walter B. F. (2009) Bargaining failures and civil war. *Annual Review of Political Science*, 12: 243–261.

Waltz K. (1979) *Theory of International Politics*. Reading, MA: Addison-Wesley.

Wendt A. (1992) Anarchy is what states make of it: The social construction of power politics. *International Organization*, 46(2): 391–425.

World Bank (2011) *World Development Report 2011: Conflict, Security and Development*. Washington, DC: The World Bank.

2 Internalizing Cooperative Norms in Group-Structured Populations

Erol Akçay and Jeremy Van Cleve

Introduction

The success of humans in spreading through all of Earth's ecosystems and transforming them at planetary scale is directly dependent on our capacity to cooperate in large groups and self-organize in complex social structures that sustain such cooperation. One of the main components of such large-scale cooperation is the human capacity and propensity for inventing and following social norms (Ostrom, 2000; Fehr and Schurtenberger, 2018). Social norms influence almost all aspects of human behavior, providing a "grammar of society" (Bicchieri, 2005, 2010) that constrains and enables different kinds of individual behaviors, coordinates collective behavior, and sustains cooperation in the face of conflicts of interests.

Although commonly discussed as a single phenomenon, social norms are best seen as a diverse set of emergent phenomena that result from the interaction of mechanisms at multiple scales, from individual-level cognition to population-level gene-culture coevolution (Gintis, 2003). Some social norms turn social dilemmas into coordination games (Bicchieri, 2005, 2010) by prescribing particular behaviors and inducing individuals to expect others to behave the same way while other norms are signals that coordinate individual behavior to implement outcomes (i.e., correlated equilibria) that improve on the outcomes possible without such signals (i.e., Nash equilibria) (Gintis, 2010; Morsky and Akçay, 2019).

Prescriptive norms will have little impact on actual behavior unless there are some mechanisms that enforce them. Some norms are self-enforcing in the sense that once they are established, it is in the self-interest of agents to follow them (Binmore, 1998). Other social norms, however, may require the threat of institutional or peer punishment to make individuals adhere to them (Gintis, 2003). The idea that social aggression and punishment are instrumental in maintaining prosocial norms is found in multiple chapters in this book, including Chapters 3, 5, 8, and 10. It is well established that by sufficiently punishing individuals who deviate from the norm, one can enforce a wide range of outcomes in social dilemmas (Boyd and Richerson,

We thank W. Wilczynski, S. Brosnan, H. Gintis, and B. Morsky for comments that improved the manuscript. EA acknowledges support from Army Research Office (W911NF-17-1-0017) and Defense Advanced Research Projects Agency Next Generation Social Science Program (Grant D17AC00005). JVC acknowledges support from National Science Foundation (Grant #1846260).

1992; Ostrom et al., 1992; Fehr and Gachter, 2000; Gintis, 2000; Henrich et al., 2010). However, since costly punishment is itself a public good whose provision requires overcoming another social dilemma (the "second-order free-rider problem" Heckathorn, 1989), explaining the evolution of social norms also requires explaining the evolution of the mechanisms that sustain them.

Finally, many social norms are followed by individuals because the norms are internalized; in other words, individuals have acquired intrinsic preferences to comply with norms even if such compliance is costly to their material interests. Internalization of norms is a long- and widely-recognized fact of human social life (Chudek and Henrich, 2011). Intuitively, we are all familiar with the concept: we follow countless norms daily, at varying inconvenience to ourselves, even when we run little risk of detection or punishment for not complying. Experimental evidence suggests that people's behavior in laboratory games are affected by their beliefs about others' expectations (Dufwenberg et al., 2011) (see also Chapter 3) and by variation in their sensitivity to norms (Kimbrough and Vostroknutov, 2016). Such intrinsic preferences for norm compliance may be modulated by particular neural circuitry in the brain (Spitzer et al., 2007; Ty et al., 2017). (See Chapters 5 and 6 for reviews of neural systems related to social decisions.) Theoretical accounts of intrinsically motivated compliance with social norms have modeled how parents might invest into socializing their offspring to internalize different preferences (Bisin and Verdier, 2001), how guilt from failing to live up to others' expectations can drive individual behavior (Battigalli and Dufwenberg, 2007), and how natural or cultural selection might favor intrinsic preferences for norm compliance in social interactions (Gintis, 2003; Gavrilets and Richerson, 2017). Our chapter contributes to this literature by modeling the coevolution of social norms and their internalization.

Social decision-making involves cognitive mechanisms that evaluate the direct benefits and costs of potential social behaviors. In the context of norms, these benefits and costs will include how different behaviors compare with the norm. A simple way to summarize how these cognitive mechanisms might work is to assume that social behaviors generate internal (i.e., neurophysiological) signals of reward or punishment, and that individuals behave in such a way as to increase their internal reward signals and decrease internal punishment signals. For example, in the presence of a contribution norm to a public good, an internalized norm (or beliefs of others' expectations) can induce a subjective reward for complying with the norm or subjective displeasure for falling short (Axelrod, 1986; Bicchieri, 2005; Dufwenberg et al., 2011; Kimbrough and Vostroknutov, 2016). These internal rewards or punishments (e.g., feelings of guilt) can create intrinsic motivations to follow norms and reduce or obviate the need for external punishment or reward.

The internal signals driving decision-making are sometimes called "preferences" and their evolution can be studied using mathematical models (Güth, 1995; Dekel et al., 2007; Akçay et al., 2009; Alger and Weibull, 2013). In these models, individuals have genetically or culturally transmitted traits that determine their preferences, represented by a utility or objective function. This function typically depends on individuals' material payoffs but need not be identical to payoffs. Individuals then choose behaviors to maximize their utilities or preferences given others' behaviors. These behavioral choices in turn lead to material payoffs, and the traits affecting the utility functions evolve according to these material payoffs. If individuals do not know whom they are interacting with (and therefore cannot distinguish between opponents with different preferences), evolutionarily stable preferences coincide with individual material payoffs (Ely, 2001) or a linear combination of one's own and others' payoffs when there is assortment (Alger and Weibull, 2013). In many cases, however, individuals can respond to others with different preferences because the behavioral game will be played repeatedly over time allowing individuals to indirectly learn each others' preferences. In such a case, prosocial preferences can evolve to stabilize cooperation by generating positive behavioral feedback between individuals (Akçay et al., 2009). Importantly, this behavioral feedback acts synergistically with genetic relatedness in sustaining cooperation (Akçay and Van Cleve, 2012; Van Cleve and Akçay, 2014).

Internalized social norms can thus be seen as an evolved component of individual preferences that biases, but does not necessarily dictate, individual behavior (Gintis and Helbing, 2015; Gavrilets and Richerson, 2017). Here, we use the theoretical framework for preference evolution to ask when and how much norm internalization will be selected for, and how the presence of external punishment affects the coevolution of norm internalization and the social norm itself. Specifically, we model internalization as a subjective disutility experienced by individuals who deviate from the norm. In this setting, whether internalization evolves or not turns out to depend on whether this disutility is an accelerating or decelerating function of the deviation from the social norm. We show that in the absence of external punishment, internal punishment functions that decelerate can evolve whereas internal punishment functions that accelerate cannot. When external punishment is present, however, accelerating internal punishment functions yield stronger norms and more cooperation than decelerating ones. These results highlight the important role that the proximate cognitive and psychological mechanisms play in shaping whether and how norm internalization evolves.

Modeling Framework

External Enforcement Only

We model a population composed of groups of n individuals who play a public goods game with the possibility of punishment. (See Box 3.1 in Chapter 3 for additional information about game approaches.) In the first model, each individual is endowed with two traits: a normative "opinion" about what individuals in the group should contribute to the public good, denoted by α_i, and an investment p_i into a punishment pool $p = 1/n\sum_{i=1}^{n}p_i$, which determines the amount of punishment that can be inflicted on individuals who deviate from the norm of the group. The norm of the group, denoted by α, is a function of the individual opinions α_i regarding the value of the norm. For instance, we can imagine that the group norm α is simply the average of individual opinions:

$$\alpha = \frac{1}{n}\sum_{i=1}^{n} \alpha_i.$$

We assume individuals first make a one-time contribution to the pool that will mete out punishment for deviations from the group norm and then play a public goods game with each other. Individuals in the public goods stage follow a behavioral dynamic where they adjust their behaviors given their preferences and the behaviors of their groupmates (Akçay et al., 2009; Akçay and Van Cleve, 2012). Specifically, in our first model we assume that a focal individual chooses its action, a_i, to maximize its own payoff, denoted by u_i and given by

$$u_i(a_i, a_j, \alpha, p) = b(a_i, a_j) - c(a_i) - \varepsilon_p(a_i, \alpha),$$

where the first term is the public good benefit from the contribution of the focal individual a_i and the contributions of other individuals in the group a_j, the second term is the private cost of contributing, and the last term ε_p represents the material punishment inflicted to the focal individual due to its deviation from the social norm and the mean level of punishment p where $\varepsilon_p(a_i, \alpha)$ is an increasing function of the absolute deviation $|a_i - \alpha|$. Note that this payoff function reflects only the gain from the public goods stage, and thus does not include the cost of contribution to the punishment pool, reflecting our assumption that individuals make this contribution before the public goods game starts; hence it is a sunk cost at that point. We assume that all individuals adjust their behaviors dynamically until they reach a behavioral equilibrium, which happens at a sufficiently fast timescale such that their total payoff from the public goods game is given by the equilibrium contribution levels, which we denote with an asterisk.

The payoff of a focal individual at the end of the public goods game minus the cost of the punishment contribution is the fitness of that individual. For a focal individual with normative opinion α_i and punishment investment p_i in a group with opinion α_j and punishment investment p_j, the fitness w_i is given by

$$w_i(\alpha_i, p_i, \alpha_j, p_j) = u_i(a_i^*, a_j^*, \alpha, p) - p_i,$$

where the fitness cost of a unit of punishment is assumed to be unity. Below, we proceed to analyze this model. We first characterize the behavioral equilibrium of the public goods game given a punishment pool, then derive the first-order conditions for the evolutionary stability of the normative opinion α and the punishment contribution p.

The Behavioral Equilibrium

The first-order condition for the (monomorphic) behavioral equilibrium is given by

$$\frac{\partial u_i}{\partial a_i} = \frac{\partial b}{\partial a_i} - c'(a^*) - \frac{\partial \varepsilon_p}{\partial a_i} = 0, \tag{2.1}$$

where all the partial derivatives are evaluated at $a_i^* = a_j^* = a^*$. From this condition, we can read a couple of things. First, assuming that the benefit function is decelerating $\left(\frac{\partial^2 b}{\partial a^2} < 0\right)$ and the cost function is accelerating ($c''(a) > 0$), increasing the punishment pool p will have the effect of increasing the equilibrium contributions. Second, the equilibrium contribution level a^* will only exactly match the normative expectation α when the latter is equal to the individually optimal or "selfish" contribution level, which occurs when the marginal benefit equals the marginal cost in the absence of any punishment. This can be seen from the fact that when $a^* = \alpha$, the third term vanishes, and the behavioral equilibrium conditions reduce to the equilibrium condition without punishment $\frac{\partial b}{\partial a_i} = c'(a^*)$. Third, any equilibrium contribution level greater than the purely selfish one can only occur when the contribution norm is at an even higher level or $\alpha > a^*$. The first two terms in Equation (2.1) will be negative since a^* is above the selfish level and benefits decelerate and costs accelerate. When $\alpha > a^*$, the third term in Equation (2.1), which measures the effect of a change in punishment on payoff, will be positive and can cancel the first two terms since increasing a^* closer to α decreases punishment. If the opposite is true and $\alpha < a^*$, the third term will be negative since increasing a^* further from α increases punishment. In other words, if the social norm exceeds that of individually optimal behavior, players will shade their contributions to the public good to be somewhere between the individually optimal level and the normative prescription.

The behavioral equilibrium a^* given by solving Equation (2.1) is a stable rest point of the behavioral dynamics whenever (Akçay and Van Cleve, 2012, appendix A3)

$$\frac{\partial^2 u_i}{\partial a_i^2} < \frac{\partial^2 u_i}{\partial a_i \partial a_j} < -\frac{1}{n-1}\frac{\partial^2 u_i}{\partial a_i^2},$$

which translates to

$$\Omega_p < 0 \quad \text{and} \tag{2.2}$$

$$\Omega_p - n\frac{\partial^2 b}{\partial a_i \partial a_j} < 0, \tag{2.3}$$

where $\Omega_p = \dfrac{\partial^2 b}{\partial a_i^2} + (n-1)\dfrac{\partial^2 b}{\partial a_i \partial a_j} - c''(a^*) - \dfrac{\partial^2 \varepsilon_p}{\partial a_i^2}.$

First-Order ESS Conditions

We can write the first-order evolutionarily stable strategy (ESS) conditions with population structure as in Akçay and Van Cleve (2012). The first-order conditions for p and α are as follows:

$$\frac{\partial w_i}{\partial \alpha_i} + (n-1)r\frac{\partial w_i}{\partial \alpha_j} = 0 \tag{2.4}$$

$$\frac{\partial w_i}{\partial p_i} + (n-1)r\frac{\partial w_i}{\partial p_j} = 0. \tag{2.5}$$

Working out the partial derivatives in Equation (2.4) first, we have

$$\frac{\partial w_i}{\partial \alpha_i} = \left[\frac{\partial u_i}{\partial a_i}\frac{\partial a_i}{\partial \alpha} + (n-1)\frac{\partial u_i}{\partial a_j}\frac{\partial a_j}{\partial \alpha} + \frac{\partial u_i}{\partial \alpha}\right]\frac{\partial \alpha}{\partial \alpha_i} \tag{2.6}$$

$$\frac{\partial w_i}{\partial \alpha_j} = \left[\frac{\partial u_i}{\partial a_i}\frac{\partial a_i}{\partial \alpha} + (n-1)\frac{\partial u_i}{\partial a_j}\frac{\partial a_j}{\partial \alpha} + \frac{\partial u_i}{\partial \alpha}\right]\frac{\partial \alpha}{\partial \alpha_j}. \tag{2.7}$$

The only difference between the two equations are the partial derivatives $\frac{\partial \alpha}{\partial a_i}$ and $\frac{\partial \alpha}{\partial a_i}$ at the end. Since we are considering homogenous groups, we can assume each individual's normative opinions have the same effect on the group norm; in other words, $\frac{\partial \alpha}{\partial a_i} = \frac{\partial \alpha}{\partial a_j}$, and therefore $\frac{\partial w_i}{\partial \alpha_i} = \frac{\partial w_i}{\partial \alpha_j}$. This means that Equation (2.4) can be written as

$$(1 + (n-1)r)\frac{\partial w_i}{\partial \alpha_i} = 0.$$

This condition implies that the evolutionarily stable contribution norm will maximize a focal individual's fitness, taking into account the behavioral responses of the whole group to the contribution norm. Expanding the partial derivatives in Equation (2.6) using the definition of u_i and using the fact that $\frac{\partial \alpha}{\partial \alpha_i} > 0$, we can write the first-order ESS condition for the group norm as follows:

$$\left[n\frac{\partial b}{\partial a} - c'(a^*) - \frac{\partial \varepsilon_p}{\partial a_i}\right]\frac{\partial a^*}{\partial \alpha} - \frac{\partial \varepsilon_p}{\partial \alpha} = 0. \tag{2.8}$$

This equation yields some immediate insights: the first term on the left-hand side (LHS) is positive (because it is equal to the LHS of Equation (2.1) plus $(n-1)\frac{\partial b}{\partial a}$, which is positive, and contributions increase with increasing α). This means that at ESS, the second term has to be negative, i.e., $a_i^* < \alpha$, meaning that individuals underinvest relative to the normative expectation of the group. That implies that individuals are experiencing some punishment at ESS.

Likewise, for the punishment investment p, we can write

$$\frac{\partial w_i}{\partial p_i} = \frac{\partial u_i}{\partial a_i}\frac{\partial a_i^*}{\partial p_i} + (n-1)\frac{\partial u_i}{\partial a_j}\frac{\partial a_j^*}{\partial p_i} + \frac{\partial u_i}{\partial p_i} - k \tag{2.9}$$

$$\frac{\partial w_i}{\partial p_j} = \frac{\partial u_i}{\partial a_i}\frac{\partial a_i^*}{\partial p_j} + (n-1)\frac{\partial u_i}{\partial a_j}\frac{\partial a_j^*}{\partial p_j} + \frac{\partial u_i}{\partial p_j}. \tag{2.10}$$

The derivatives with respect to p_i and p_j in Equations (2.9) and (2.10) evaluate to $-\frac{1}{n}\left(a_i^* - \alpha\right)^2$. Thus, we can write the first-order ESS conditions for p as

$$\left[\left(\frac{\partial u_i}{\partial a_i} + (n-1)\frac{\partial u_i}{\partial a_j}\right)\frac{\partial a^*}{\partial p} - \frac{\partial \varepsilon_p}{\partial p}\right]\frac{dp}{dp_i}(1 + r(n-1)) - k = 0, \tag{2.11}$$

where for Equation (2.11), we used the fact that both the focal and non-focal individuals' punishment contributions go to a common pool that affects each individual in the same way, and hence each individual's equilibrium contribution reacts to a change in any individual's punishment contribution in the same way (i.e., $\frac{\partial a_i^*}{\partial p} = \frac{\partial a_j^*}{\partial p} = \frac{\partial a^*}{\partial p}$).

To calculate the partial derivatives of a_i^* and a_j^* with respect to α, and p, we differentiate the behavioral equilibrium condition Equation (2.1) with respect to the evolving variables, and solve for the relevant partial derivatives to obtain

$$\frac{\partial a^*}{\partial \alpha} = \frac{\dfrac{\partial^2 \varepsilon_p}{\partial a_i \partial \alpha}}{\Omega_p}$$

$$\frac{\partial a^*}{\partial p} = \frac{\dfrac{\partial^2 \varepsilon_p}{\partial a_i \partial p}}{\Omega_p}.$$

With Internalized Punishment

In this section, we model the evolution of a psychological mechanism for internalizing norms in the presence of (fixed) external punishment. We operationalize internalization as an inherent motivation to follow the prescribed behavior. Specifically, we assume agents have an evolving trait, τ, that determines how much internalized "discomfort" they feel due to deviations from the prescribed norm. Mathematically, we assume now that our agents maximize the following objective function during the behavioral dynamics:

$$x_i(a_i, a_j, \tau_i, \alpha, p) = u_i(a_i, a_j, \alpha, p) - \varepsilon_{\tau_i}(a_i, \alpha),$$

where $\varepsilon_{\tau_i}(a_i, \alpha)$ is an increasing function of the deviation from the norm, $a_i - \alpha$. The $\varepsilon_{\tau_i}(a_i, \alpha)$ term is analogous to the external punishment term included in the payoff u_i, but is only felt subjectively, with no direct effect on the material payoff of the individuals. However, the presence of such subjective discomfort can alter the behavior of the focal individual and can therefore have an indirect effect on its payoff.

The behavioral equilibrium condition for a focal individual is now given by

$$\frac{\partial x_i}{\partial a_i} = \frac{\partial u_i}{\partial a_i} - \frac{\partial \varepsilon_{\tau_i}}{\partial a_i} = 0. \tag{2.12}$$

This condition implies that the effect of internalized discomfort (positive τ) is similar to external punishment: as long as the group norm α is higher than the individually optimal contribution maximizing u_i, the effect of τ is to increase contributions. The stability of the behavioral equilibrium is again determined by the same conditions as Equations (2.2) and (2.3), except Ω_p is replaced by $\Omega_{p\tau} = \frac{\partial^2 b}{\partial a_i^2} + (n-1)\frac{\partial^2 b}{\partial a_i \partial a_j} - c''(a^*) - \frac{\partial^2 \varepsilon_p}{\partial a_i^2} - \frac{\partial^2 \varepsilon_{\tau_i}}{\partial a_i^2}$.

The partial derivatives of the equilibrium contribution a^* with respect to α and τ are given by

$$\frac{\partial a^*}{\partial \alpha} = \frac{\dfrac{\partial^2 \varepsilon_p}{\partial a_i \partial \alpha} + \dfrac{\partial^2 \varepsilon_{\tau_i}}{\partial a_i \partial \alpha}}{\Omega_{p\tau}} \tag{2.13}$$

$$
\left[
\begin{array}{l}
\dfrac{\partial a^*}{\partial \tau} = \dfrac{\dfrac{\partial^2 \varepsilon_{\tau_i}}{\partial a_i \partial \tau}}{\Omega_{p\tau}} \left(\dfrac{\Omega_{p\tau} - \dfrac{\partial^2 b}{\partial a_i \partial a_j}}{\Omega_{p\tau} - n\dfrac{\partial^2 b}{\partial a_i \partial a_j}} \right) \\[30pt]
= \dfrac{\dfrac{\partial^2 \varepsilon_{\tau_i}}{\partial a_i \partial \tau}}{\Omega_{p\tau}} \left(1 + \dfrac{(n-1)\dfrac{\partial^2 b}{\partial a_i \partial a_j}}{\Omega_{p\tau} - n\dfrac{\partial^2 b}{\partial a_i \partial a_j}} \right)
\end{array}
\right]. \tag{2.14}
$$

The ESS condition for τ is given by

$$
\left[
\begin{array}{l}
0 = \dfrac{\partial w_i}{\partial \tau_i} + (n-1)r\dfrac{\partial w_i}{\partial \tau_j} \\[12pt]
= \dfrac{\partial u_i}{\partial a_i}\dfrac{\partial a_i}{\partial \tau_i} + (n-1)\dfrac{\partial u_i}{\partial a_j}\dfrac{\partial a_j}{\partial \tau_i} + (n-1)r\left[\dfrac{\partial u_i}{\partial a_i}\dfrac{\partial a_i}{\partial \tau_j} + \dfrac{\partial u_i}{\partial a_j}\dfrac{\partial a_j}{\partial \tau_j} + (n-2)\dfrac{\partial u_i}{\partial a_k}\dfrac{\partial a_k}{\partial \tau_j} \right] \\[12pt]
= \dfrac{\partial a_i}{\partial \tau_i}\left(\dfrac{\partial u_i}{\partial a_i}(1 + (n-1)r\rho) + \dfrac{\partial u_i}{\partial a_j}(n-1)(r + \rho + (n-2)r\rho) \right),
\end{array}
\right] \tag{2.15}
$$

where $\rho \equiv \frac{\partial a_j}{\partial \tau_i} / \frac{\partial a_i}{\partial \tau_i} = \frac{\partial a_i}{\partial \tau_j} / \frac{\partial a_j}{\partial \tau_j} = \frac{\partial a_j}{\partial a_i}$ denotes the responsiveness of an individual's contribution to the change of another individual's contribution (Akçay et al., 2009; Akçay and Van Cleve, 2012). At the monomorphic equilibrium, the responsiveness of any individual to any other individual is the same, which allows us to write the last line in Equation (2.15). So long as $a_i \neq \alpha$, then $\frac{\partial a_i}{\partial \tau_i} \neq 0$, and satisfying the first-order ESS condition requires that the term in the parentheses in Equation (2.15) has to vanish. By setting this term to zero and rearranging, we obtain

$$
\frac{\mathcal{B}}{\mathcal{C}} = \frac{1 + (n-1)r\rho}{(n-1)(r + \rho + r\rho(n-2))}, \tag{2.16}
$$

where the marginal benefit to the focal individual from the investments of others is $\mathcal{B} = \frac{\partial b}{\partial a_j}$ and the net marginal cost from its own investment is $-\mathcal{C} = \frac{\partial b}{\partial a_i} - c'(a_i) - \frac{\partial \varepsilon_p}{\partial a_i}$. Because τ does not affect the payoffs of individuals directly, the response coefficient ρ can be written as in Akçay and Van Cleve (2012, eq. 13):

$$
\left[
\begin{array}{l}
\rho = -\dfrac{\dfrac{\partial^2 x_i}{\partial a_i \partial a_j}}{\dfrac{\partial^2 x_i}{\partial a_i^2} + (n-2)\dfrac{\partial^2 x_i}{\partial a_i \partial a_j}} \\[28pt]
= -\dfrac{\dfrac{\partial^2 b}{\partial a_i \partial a_j}}{\dfrac{\partial^2 b}{\partial a_i^2} + (n-2)\dfrac{\partial^2 b}{\partial a_i \partial a_j} - c''(a_i^*) - \dfrac{\partial^2 \varepsilon_p}{\partial a_i^2} - \dfrac{\partial^2 \varepsilon_{\tau_i}}{\partial a_i^2}} \\[28pt]
= -\dfrac{\dfrac{\partial^2 b}{\partial a_i \partial a_j}}{\Omega_{p\tau} - \dfrac{\partial^2 b}{\partial a_i \partial a_j}}
\end{array}
\right]. \tag{2.17}
$$

Further, from the behavioral equilibrium conditions (Equation A27) in Akçay and Van Cleve (2012), the response coefficient ρ must satisfy the following inequality: $-1/(n-1) < \rho < 1$.

Since τ does not affect the payoffs directly, the ESS condition for α stays the same as Equation (2.8) except that all derivatives are evaluated at the behavioral equilibrium solving Equation (2.12).

Representative Functions

In order to numerically analyze the model, we need to specify a few particular functions for the cost, benefit, and punishment functions. These functions are meant to capture our intuitions about public goods payoffs, and different notions of how external and internalized enforcement of norms might work.

First, we introduce a public good benefit function b:

$$b(a_1, a_2, \ldots, a_n) = (1-v)\sum_{i=1}^{n}\sqrt{a_i} + v\sum_{i\neq j}\sqrt{a_i a_j}. \tag{2.18}$$

This function captures two important intuitions about public goods benefits. First, the fact that all contributions are in square roots ensures that there will be (eventually) diminishing results from contributions to the public good so that a finite contribution is socially optimal. Second, it allows individual contributions to be synergistic or anti-synergistic (or complements or substitutes in economic terminology), as modulated by the parameter v, which we assume is in the range $[-1, 1]$. Specifically, the second sum of Equation (2.18) corresponds to individuals' contributions interacting pairwise. Positive v can be interpreted as representing collaborative interactions that contribute positively to the public good. Negative v, on the other hand, represents agonistic or competitive interactions that diminish total public goods provision. Thus, this function allows us to represent a range of social scenarios.

For the cost of contribution to the public good, $c(a)$, we use a simple quadratic function that represents accelerating marginal costs:

$$c(a_i) = a_i^2. \tag{2.19}$$

The external and internalized punishment functions, $\varepsilon_p(a_i, \alpha)$ and $\varepsilon_{\tau_i}(a_i, \alpha)$, respectively, both increase when the deviation of a focal individual's contribution from the group norm $|a_i - \alpha|$ increases. However, the behavioral and evolutionary stability conditions above also depend on the curvatures or second-order derivatives of these functions. For external punishment, $\varepsilon_p(a_i, \alpha)$, we use the following functional form:

$$\varepsilon_p(a_i, \alpha) = p(a_i - \alpha)^2.$$

This function captures the notion of graduated punishment (Ostrom, 1990) as applied to our setting, where small deviations from the norm encounter relatively small punishments, but the punishment escalates with larger deviations. The size of the punishment pool, p, modulates the amount of punishment. For internalized punishment function, we investigate two forms with different curvatures. The first form is analogous to the external punishment function above and is accelerating in terms of the deviation of contribution from the norm:

$$\varepsilon_{\tau_i}^{I}(a_i, \alpha) = \tau_i(a_i - \alpha)^2. \tag{2.20}$$

The second form captures the notion that as an individual's deviation from a norm grows, he or she might not experience infinitely increasing discomfort. Rather, individuals who are already far from a norm may feel relatively small additional discomfort for the same additional deviation compared to individuals who are closely adhering to the norm. This suggests an internalized discomfort function that plateaus at large deviations from the norm. We use the following function to represent this case:

$$\varepsilon_{\tau_i}^{II} = \tau_i \ln\left(1 + (a_i - \alpha)^2\right). \tag{2.21}$$

This function behaves similar to $\varepsilon_{\tau_i}^{I}$ at small deviations from the norm, but saturates at large deviations.

Analysis and Numerical Results

External Punishment Only

First, we assume that only external punishment is possible. Figure 2.1 depicts the evolved contribution and punishment levels and the evolved norm using the benefit and cost functions in Equations (2.18) and (2.19) with no synergistic interactions or $v = 0$. It shows that an evolutionarily stable (ES) social norm α^* and external punishment level p^* need a threshold level of relatedness within social groups. Below this threshold value of relatedness, a positive contribution to the punishment pool is not evolutionarily stable, which reflects the second-order dilemma that costly punishment poses. As relatedness crosses a threshold value, the ES punishment level increases from zero, positive selection arises on the social norm α, and an ES α^* and contribution level evolve. With increasing relatedness, punishment contributions increase while the contribution norm decreases. This is due to the fact that more relatedness allows for higher p and more punishment-induced cooperation. However, higher p increases the marginal cost of maintaining a social norm at a particular level, and this cost is increasingly paid by relatives. Despite decreasing the social norm α^*, increasing relatedness increases the equilibrium contribution to the public good (a^*) and the net fitness w. Thus, as relatedness increases, groups become more cooperative because they punish smaller deviations from the norm more harshly.

Figure 2.2 further shows that the critical value of relatedness needed to sustain a normative equilibrium decreases with increasing complementarity of contributions to the public goods. This is intuitively due to the fact that when contributions are complementary, a coordinated increase in the contribution level (actions) of

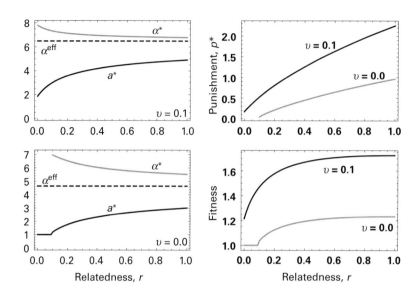

Figure 2.1 The evolutionarily stable (ES) contribution norm α^*, punishment investment p^*, and fitness as a function of relatedness for no complementarity, $v = 0.0$, and positive complementarity, $v = 0.1$ given $n = 10$ individuals. The norm and investment levels in the first column are relative to the baseline investment (ES investment and norm with no punishment). The fitness values in the lower-right panel are relative to the baseline fitness (ES investment and norm with no punishment). For low values of relatedness and $v = 0.0$, no positive punishment investment is evolutionarily stable; so all norms are neutral, since they do not get enforced.

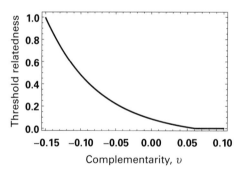

Figure 2.2 The critical value of relatedness needed for a normative equilibrium to be ES as a function of the complementarity parameter v. As the public good contributions get more complementary (or individuals become more collaborative), lower relatedness is needed to sustain a public goods contribution norm.

multiple individuals generates more benefits for everyone than if contribution levels are substitutes. Therefore, external punishment, which has the effect of increasing everyone's contribution, has a higher benefit in situations where actions are complementary.

Internalized Punishment Only

Now, suppose that there is no external punishment, $p = 0$, but internal punishment τ can evolve. We show in this case that norm internalization can only be evolutionarily stable if the internalized punishment function becomes saturating at some point. When there is no external punishment, the ESS condition for the contribution norm α, Equation (2.8), reduces to

$$\left[n \frac{\partial b}{\partial a} - c'(a^*) \right] \frac{\partial a^*}{\partial \alpha} = 0 ,$$

where $\frac{\partial a^*}{\partial \alpha}$ is given by Equation (2.13). This means that the first-order ESS condition for α can only be satisfied if either $\frac{\partial^2 \varepsilon_{\tau_i}}{\partial a_i \partial \alpha} = 0$ or the term in the square brackets vanish. The latter implies that the behavioral equilibrium maximizes total payoff of the group (Akçay and Van Cleve, 2012), which in turn means that the LHS of the ESS condition for τ, Equation (2.16), equals $\frac{1}{n-1}$. Satisfying the ESS condition for τ then requires either $\rho = 1$ or $r = 1$ (or both). The clonal condition with $r = 1$ is an uninteresting case, since in clonal groups no conflict over the public good exists. The other option, $\rho = 1$ from Equation (2.17), implies that $\Omega_{0\tau} = 0$, but this contradicts the stability condition for the behavioral equilibrium (Equation (2.2)). This means that for a social norm to exist with purely internalized punishment,

$$\frac{\partial a^*}{\partial \alpha} = 0;$$

i.e., the internalized enforcement of the norm should be such that at some point, increasing the group norm does not yield higher contribution. In other words, a necessary condition for an ESS with an internalized punishment only is that at some point increasing the social norm does not elicit higher contributions from group members. The intuition behind this result is that traits that raise the group norm α are not costly without norm enforcement (internal or external) but in the presence of norm internalization, they elicit higher contributions from group-mates. Thus, as long as $\frac{\partial a^*}{\partial \alpha}$ is positive, selection will act to increase the group norm. Only when norm internalizers stop responding to higher α values can we satisfy the ESS conditions.

Using Equation (2.13) in the absence of external punishment, we can see that $\frac{\partial a^*}{\partial \alpha} = 0$ requires

$$\frac{\partial^2 \varepsilon_{\tau_i}}{\partial a_i \partial \alpha} = 0.$$

This condition means that the internal punishment needs to decelerate, or start to saturate, at high deviations from the social norm. In other words, if individuals are already showing a big enough deviation from the norm, increasing the norm will not necessarily impose a high enough marginal discomfort on them to make them increase their contribution. In colloquial terms, these individuals would be "giving up" on trying to keep up their contribution level with the norm. Without external punishment, the contribution norm will evolve precisely to the level at which individuals are giving up

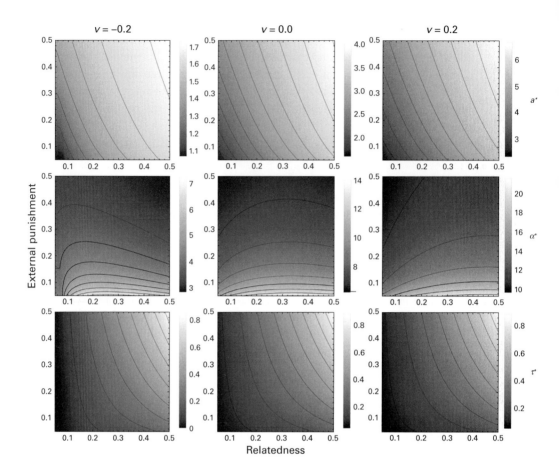

Figure 2.3 The ES contribution a^* (top row), ES norm α^* (middle row), and ES internalization τ^* (bottom row) for an accelerating internal punishment function ($\varepsilon^I_{\tau_i}$, in Equation (2.20)). Columns represent benefit functions varying from antisynergistic to synergistic going left to right. The ES contribution level a^* and norm α^* are shown relative to their values with no punishment (external or internal). The norm decreases with p and increases with r monotonically under all benefit functions, while norm internalization increases with both. The resulting contributions to the public good increase with both external punishment and relatedness despite the fact that the contribution norm might decline with external punishment.

on following it. This can lead to relatively high levels of cooperation, as all individuals will respond with the same increase in contribution level with an increase in the norm (until they stop responding), meaning that the marginal cost to the focal individual contributing more will be offset by the equivalent contributions from group-mates.

Internalization in the Presence of External Punishment

Finally, assume that there is a fixed level of external punishment p and that the level of social norm internalization τ can evolve. Figures 2.3 and 2.4 depict the ES

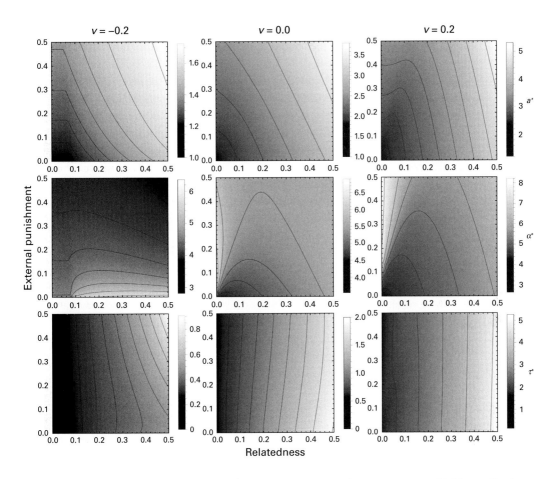

Figure 2.4 The ES contribution a^* (top row), ES norm α^* (middle row), and ES internalization τ^* (bottom row) as a function of relatedness and exogenously fixed external punishment, with decelerating internal punishment ($\varepsilon_{\tau_i}^{II}$, Equation (2.21)). The ES contribution level a^* and norm α^* are shown relative to their values with no punishment (external or internal). In the dark region for $v = -0.2$, the ES $\tau^* = 0$, and the social norm is purely maintained by external punishment. For antisynergistic benefits, increasing external punishment decreases the contribution norm, while for additive and synergistic benefits, it increases the norm. In all cases, norm internalization increases with relatedness but is relatively unresponsive to external punishment. The ES contribution generally increases with both external punishment and relatedness, but remains at lower levels compared to the accelerating function in Figure 2.3.

contribution norm and internalization level as a function of relatedness and external punishment under different internal punishment functions. For both accelerating and decelerating internal punishment functions, higher relatedness results in higher internalization. For accelerating internal punishment (Figure 2.3), a minimum level of relatedness has to be present before internalization, $\tau > 0$, can evolve (these low relatedness values are not plotted in Figure 2.3). For this case, the ES contribution norm decreases with increased external punishment: mirroring the case with external punishment only, increases in external punishment increase the cost of maintaining a high social norm, which generates selection for a lower contribution norm. Increases in the degree of complementary or synergy of the benefit of contribution (v) increase the ES norm and contribution level but have little effect on the level of internalization.

However, Figure 2.4 shows that decelerating internal punishment produces more complex patterns. When the contributions to the public goods are antisynergistic ($v < 0$), the patterns mirror those with accelerating internal punishment: the ES contribution norm decreases with increases in external punishment and a threshold relatedness is required for the ES τ to be positive. In contrast, when the benefit function is additive or synergistic (middle and right columns of Figure 2.4 where $v = 0$ and $v > 0$, respectively), the ES contribution norm increases with increasing external punishment but changes non-monotonically with changes in relatedness: α^* decreases at low relatedness and increases at higher relatedness. Furthermore, the level of the ES contribution norm and resulting ES contributions (relative to their levels with no punishment at all) are significantly lower for the decelerating internal punishment function as compared to the accelerating one.

These results, together with the analytical result above for the case of no external punishment, suggest a trade-off. While decelerating internal punishment functions might be easier to evolve in the absence of external enforcement, they might be less efficient in maintaining high cooperation than accelerating internal punishment, once the latter is stabilized by external enforcement.

Discussion

Not only have humans evolved a capacity to follow social norms, they can also internalize those norms so that they are intrinsically motivated to comply with group- or population-level standards of behavior. Such internalization is frequently crucial to the proper functioning of human groups and society. We have presented a model for how the capacity for internalization might evolve through evolution of social preferences. In particular, our model explores how external enforcement in the form of punishing deviations from the norm affects the evolution of norm internalization in the form of a preference for following the norm or a subjective discomfort from failing to follow the norm.

A main result from our model is that the interaction between external punishment of norm deviations and the evolution of norm internalization crucially depends on the shape of the function that describes the discomfort or internal punishment experienced by individuals. We distinguish between two functional forms for this internal punishment function: (i) accelerating functions representing cases where the marginal discomfort from deviation keeps increasing the farther an individual is from the norm, and (ii) decelerating functions, where at large enough deviations from the norm, the marginal discomfort starts decreasing and may vanish (i.e., individuals do not feel *additional* discomfort from deviating more). When internal punishment is an accelerating function, the evolution of internalization requires the presence of external punishment, while with a decelerating internal punishment function, internalization can evolve even if there is no external punishment. Accelerating internal punishment is not stable on its own because it causes a focal individual to always respond to an increase in the group norm by increasing their contributions. That means that group-mates of this focal individual could increase the value of the norm α and elicit more contribution from the focal individual. In the absence of external punishment, there is little cost to an individual increasing his or her part of the contribution norm α, but there is a significant benefit since everyone will contribute more. This causes the social norm to increase beyond optimal, and in response, the level of internalization τ is selected to decrease. This process eventually drives τ down to zero even as the norm α evolves to large values that are not enforced. This runaway increase in the norm and decrease in the internalization is thwarted when there is external punishment, since increasing the norm is now also directly costly to a focal individual due to the increased punishment. The runaway dynamics are also short-circuited when the internal punishment function starts leveling off, because as the norm keeps increasing, even an internalizing focal individual stops feeling the necessary additional discomfort to try to keep up. That caps how much the contribution level of a focal individual will increase in response to group-mates increasing the norm and stops the norm from increasing too much. This contrast between the two functions continues in the presence of external punishment. For accelerating functions, higher external punishment increases the strength of internalization (Figure 2.3), while with decelerating functions, internalization is most responsive to relatedness between group members, as opposed to external punishment (Figure 2.4).

These results suggest that internal punishment that eventually levels off might be more robust for evolving norm internalization on its own and that evolved norm psychology should have a mechanism for capping the subjective discomfort from norm violations. When this happens, individuals do not experience large marginal punishment or reward from marginal changes to their behavior or the norm. This bears some resemblance to the notion of goal disengagement (Wrosch et al., 2003) in psychology, which describes situations where individuals stop pursuing unattainable goals. In our model, the disengagement is "local" in the sense that individuals do not completely give up on prosocial behavior, but prove unwilling to go beyond a certain level of contribution. In a heterogeneous population (e.g., with individuals with different endowments), the point at which individuals disengage would vary, which might put pressure on group norms to diversify, leading to socially stratified social norms.

Further, our results show that presence of external punishment can drive the evolution of internalization. This effect is most pronounced for accelerating internal punishment functions, where external punishment keeps the social norm from increasing too much due to the fitness costs stemming from such punishment. With the social norm constrained by external punishment, internalization can evolve to take over some of the enforcement function, and reduce the amount of material punishment at equilibrium. In contrast, with decelerating internal punishment, internalization is much less sensitive to external punishment. In both cases, norm internalization becomes stronger with increased relatedness in a population. This is not surprising, since internalization is a trait directly benefiting group-mates, and it is expected to be favored when relatedness between group members is high. Under decelerating costs of internal punishment, but not under accelerating costs, the level of internalization also increases with the degree to which contributions to the public good are synergistic. However, since accelerating costs produce much stronger discomfort at the same level of internalization τ, they still produce greater stable contribution levels and norms. Thus, if external punishment is available through some social institution, accelerating internal punishment may produce more powerful norms with greater levels of cooperation.

There are only a handful of previous models of how the capacity to internalize social norms might have evolved. In an influential paper, Gintis (2003) developed a gene–culture coevolution model where a genetically transmitted allele allows for the social acquisition and internalization of a norm from parents and others. He found that such an allele will fix in a population if it allows the acquisition of individually beneficial norms, and that altruistic norms can "hitchhike" on this internalization capacity to also establish. In our model, internalization does not hitchhike on another beneficial trait; instead, it evolves because it allows individuals to reduce the punishment they experience and it generates positive behavioral responses. Another difference is that we consider both norms and the internalization to be inherited by the same mechanism (genetic or cultural), whereas Gintis models the coevolution of traits transmitted genetically and culturally. Considering gene–culture coevolution in a setting like ours will be an interesting future direction. More recently, Gavrilets and Richerson (2017) provide agent-based simulations of a setup closely related to ours, where individuals can contribute to a public good in a group as well as to the effort to punish deviations. Similar to our work in this chapter, they model internalization as an intrinsic reward for contributing to the public good and punishment, with a linear reward function and a fixed contribution norm. They find that internalization can evolve more easily in games where groups compete only indirectly (i.e., group success depends only on its own public good production; "us-vs-nature" games), compared to when groups compete directly (where group success depends on other groups' production; "us-vs-them" games). In the latter case, selection already favors high amounts of investment and therefore internalization is not required to sustain cooperation. Interestingly, Gavrilets and Richerson (2017) find that under many parameter regimes, substantial genetic variation in norm internalization is maintained, including dimorphisms between high norm internalizers and non-internalizers. Our analysis did not explicitly look for diversifying or balancing selection that could produce such polymorphisms, which is another interesting question for future research.

Conclusions

The phenomenon of internalization in humans is complex and is affected by many different processes at the individual, group, and societal levels. Here, we focused on the selective pressures that come into play for the coevolution of social norms and the intrinsic motivations to adhere to them in a public goods game setting. We find that the evolution of such intrinsic motivations is predicated on how the underlying mechanism processes deviations from the social norm and encodes the discomfort, guilt, or internal punishment from falling short. We find that internalization requires external punishment when the internal punishment is accelerating whereas functions that eventually level off in the deviation from the norm can support social norms through purely intrinsic preferences. Crucially, norm internalization can create a new conflict by generating an incentive to keep increasing the contribution norm as in the accelerating case or remove conflicts by causing coordinated increases in contributions in the decelerating function case. These conflicts or their resolution will reciprocally interact with the dynamics of institutions, social networks, and other group-level processes. Understanding these interactions remains an important goal of social evolutionary theory.

References

Akçay E., Van Cleve J., Feldman M. W., and Roughgarden J. (2009) A theory for the evolution of other-regard integrating proximate and ultimate perspectives. *Proceedings of the National Academy of Science USA*, 106(45): 19061–19066.

Akçay E., and Van Cleve J. (2012) Behavioral responses in structured populations pave the way to group optimality. *The American Naturalist*, 179(2): 257–269.

Alger I., and Weibull J. W. (2013) Homo Moralis – Preference evolution under incomplete information and assortative matching. *Econometrica*, 81(6): 2269–2302.

Axelrod R. (1986) An evolutionary approach to norms. *American Political Science Review*, 80 (4): 1095–1111.

Battigalli P., and Dufwenberg M. (2007) Guilt in games. *American Economic Review* 97(2): 170–176.

Bicchieri C. (2005) *The Grammar of Society: The Nature and Dynamics of Social Norms*. Cambridge University Press.

Bicchieri C. (2010) Norms, preferences, and conditional behavior. *Politics, Philosophy & Economics*, 9(3): 297–313.

Binmore K. (1998) *Game Theory and the Social Contract, Volume Two: Just Playing*. Cambridge, MA: MIT Press.

Bisin A., and Verdier T. (2001) The economics of cultural transmission and the dynamics of preferences. *Journal of Economic Theory*, 97(2): 298–319.

Boyd R., and Richerson P. J. (1992) Punishment allows the evolution of cooperation (or anything else) in sizable groups. *Ethology and Sociobiology*, 13(3): 171–195.

Chudek M., and Henrich J. (2011) Culture–gene coevolution, norm-psychology and the emergence of human prosociality. *Trends in Cognitive Sciences*, 15(5): 218–226.

Dekel E., Ely J. C., and Yılankaya O. (2007) Evolution of preferences. *Review of Economic Studies*, 74: 685–704.

Dufwenberg M., Gächter S., and Hennig-Schmidt H. (2011) The framing of games and the psychology of play. *Games and Economic Behavior*, 73(2): 459–478.

Ely J. (2001) Nash equilibrium and the evolution of preferences. *Journal of Economic Theory*, 97(2): 255–272.

Fehr E., and Gachter S. (2000) Cooperation and punishment in public goods experiments. *American Economic Review*, 90(4): 980–994.

Fehr E., and Schurtenberger I. (2018) Normative foundations of human cooperation. *Nature Human Behaviour*, 2(7): 458.

Gavrilets S., and Richerson P. J. (2017) Collective action and the evolution of social norm internalization. *Proceedings of the National Academy of Sciences*, 114(23): 6068–6073.

Gintis H. (2000) Strong reciprocity and human sociality. *Journal of Theoretical Biology*, 206 (2): 169–179.

Gintis H. (2003) The hitchhiker's guide to altruism: Gene-culture coevolution and the internalization of norms. *Journal of Theoretical Biology*, 220(4): 407–418.

Gintis H. (2010) Social norms as choreography. *Politics, Philosophy, & Economics*, 9(3): 251–264.

Gintis H., and Helbing D. (2015) Homo Socialis: An analytical core for sociological theory. *Review of Behavioral Economics*, 2(1–2): 1–59.

Güth W. (1995) An evolutionary approach to explaining cooperative behavior by reciprocal incentives. *International Journal of Game Theory*, 24(4): 323–344.

Heckathorn D. D. (1989) Collective action and the second-order free-rider problem. *Rationality and Society*, 1(1): 78–100.

Henrich J., Ensminger J., McElreath R. et al. (2010) Markets, religion, community size, and the evolution of fairness and punishment. *Science*, 327(5972): 1480–1484.

Kimbrough E. O., and Vostroknutov A. (2016) Norms make preferences social. *Journal of the European Economic Association*, 14(3): 608–638.

Morsky B., and Akçay E. (2019) Evolution of social norms and correlated equilibria. *Proceedings of the National Academy of Sciences*, 116(18): 8834–8839.

Ostrom E. (1990) *Governing the Commons: The Evolution of Institutions for Collective Action*. Cambridge: Cambridge University Press.

Ostrom E. (2000) Collective action and the evolution of social norms. *Journal of Economic Perspectives*, 14(3): 137–158.

Ostrom E., Walker J., and Gardner R. (1992) Covenants with and without a sword: Self-governance is possible. *American Political Science Review*, 86(2): 404–417.

Spitzer M., Fischbacher U., Herrnberger B., Grön G., and Fehr E. (2007) The neural signature of social norm compliance. *Neuron*, 56(1): 185–196.

Ty A., Mitchell D. G. V., and Finger E. (2017) Making amends: Neural systems supporting donation decisions prompting guilt and restitution. *Personality and Individual Differences*, 107: 28–36.

Van Cleve J., and Akçay E. (2014) Pathways to social evolution: Reciprocity, relatedness, and synergy. *Evolution*, 68: 2245–2258.

Wrosch C., Scheier M. F., Carver C. S., and Schulz R. (2003) The importance of goal disengagement in adaptive self-regulation: When giving up is beneficial. *Self and Identity*, 2(1): 1–20.

3 Reputation
A Fundamental Route to Human Cooperation

Junhui Wu, Daniel Balliet, and Paul A. M. Van Lange

Introduction

Humans are a remarkably social species. They form and live in groups and recurrently have to decide whether to cooperate or compete with others within and among groups. Cooperation has been essential for group survival and prosperity across human history. In hunter-gatherer societies, people need to form alliances in hunting to alleviate the risks from predator attacks. Likewise, modern societies require groups of people to cooperate in large ventures. Yet, social situations often involve a conflict between one's short-term personal interest and the long-term collective interest (i.e., social dilemmas; Dawes, 1980; Van Lange et al., 2013). In such mixed-motive situations, what is good for an individual may often harm the collective, and this makes people tempted to free ride and harvest the benefits from others' cooperation. Indeed, many societal problems and global issues (e.g., traffic problems, environmental pollution, and resource depletion) involve such conflicts of interests. Solving these problems often requires individuals to cooperate by paying a personal cost to benefit another person or the group.

Over the past several decades, many theories have been proposed to explain why humans perform costly behaviors that benefit others (for relevant reviews, see Nowak, 2006; Kurzban et al., 2015), and such behaviors are often measured with economic games in experimental studies (see Box 3.1 for explanations of games). The well-known kin selection theory proposed by Hamilton (1964) argues that natural selection favors cooperation directed at one's genetic relatives, as such behavior enhances the reproductive success of others who share one's genes (i.e., inclusive fitness). Although this theory can explain cooperation among genetic relatives, a large proportion of social interactions involves genetically unrelated strangers. When people encounter strangers in social interactions, they can choose to always cooperate (i.e., unconditional cooperation), only cooperate in certain situations (i.e., conditional cooperation), or never cooperate (i.e., unconditional defection). Unconditional cooperation is often costly, because unconditional cooperators may be exploited by free riders (e.g., Kuhlman and Marshello, 1975). In addition, those who never cooperate may be eventually excluded from the group. Thus, conditional cooperation is a favorable strategy because it minimizes the cost of exclusion and maximizes potential direct and indirect benefits from social interactions. (See Chapter 4 for a discussion of kin selection; Chapters 8–10 also discuss the role of kin selection in shaping social behavior.)

Box 3.1 Commonly Used Games of Cooperation and Conflict

Based on game theory (Camerer, 2003), researchers have proposed a number of simple games that capture different facets of the conflict between personal and collective interest, such as the public goods game (PGG), the prisoner's dilemma game (PDG), and the trust game (TG). Participants in these games make decisions about whether to share money with anonymous strangers. Decisions in these games are highly correlated and reflect one's tendency to provide a benefit to others at a personal cost (i.e., cooperation; Yamagishi et al., 2013; Peysakhovich et al., 2014).

Public goods game (PGG): In a public goods game involving n persons ($n \geq 2$), each person is initially endowed with E monetary units (MUs) and contributes x ($0 \leq x \leq E$) MUs to the group account. The remaining ($E - x$) MUs are kept in their private account. The total contribution is then multiplied by a factor k ($1 < k < n$) and divided equally among all members. Thus, each person can maximize their outcome when they do not contribute but still harvest the benefits generated from others' contribution. Yet, the collective outcome is maximized when everyone contributes E MUs. The number of MUs a person contributes to the group account in this game is the measure of cooperation.

Prisoner's dilemma game (PDG): The prisoner's dilemma game is a two-person game that is essentially a public goods game for $n = 2$ with dichotomous decisions. In this game, each person has one of two options: cooperate (C) or defect (D). Each person can maximize their own outcome by defecting with the other, regardless of the other's behavior. Yet, only mutual cooperation can maximize the collective outcome. Thus, cooperation in this game is measured as dichotomous choices (cooperate vs. defect).

Trust game (TG): The trust game is a two-person game involving a trustor and a trustee. The trustor is initially endowed with E monetary units (MUs) and can freely send x ($0 \leq x \leq E$) MUs to the trustee, who then receives the tripled amount ($3x$ MUs). Afterward, the trustee can return back y ($0 \leq y \leq 3x$) MUs to the trustor. In this game, x reveals the trustor's trust in the trustee, and $y/3x$ reveals the trustee's level of trustworthiness (or reciprocity). The measures of trust and trustworthiness in this game reveal both persons' levels of cooperation.

Veto game (VG): The veto game is an n-person game in which two or more individuals compete over a single individual who can provide a needed resource. The proposers can increase their offers and the player in the veto position decides whether or not to take them. Thus, the individual with veto power has control and, therefore, the ability to obtain the most resources as the other players outbid one another for access.

Dictator game (DG): The dictator game is a two-person game in which one individual is given an endowment to split between themselves and a partner. The partner has no active choice in this game. Thus, the dictator's choice measures altruism, as it reveals how much persons will donate when there is no extrinsic motivation to do so. This game is usually used in conjunction with the ultimatum game to determine how the partner's pressure impacts the dictator's decision.

Box 3.1 (*cont.*)

Ultimatum game (UG): The ultimatum game is identical to the dictator game except that after the first player makes a split of the endowment, the second player can determine whether to accept or reject it. If they accept, the endowment is split as proposed; if they reject, neither player gets anything. This game is argued to measure fairness and punishment preferences, and can be compared to the results of the DG to determine how these preferences impact decision-making.

Box 3.2 Agent-Based Modeling (ABM)

The agent-based modeling (ABM) approach uses computer programs to simulate individuals as autonomous agents that vary in attributes or decision rules. These agents interact with each other over a life span. The resulting payoff affects their reproductive success (i.e., the number of offspring that share similar traits or decision rules) in the next generation. This process is repeated for many generations and can help us address which decision rules can be evolutionarily stable in a population. (For reviews, see Smith and Conrey, 2007; Rand and Nowak, 2013.) As a case in point, computer simulations demonstrate that in a population characterized by repeated encounters, the simple rule of tit-for-tat (i.e., start with cooperating with the partner, and then follow the partner's previous behavior) wins over many other decision rules, as it can evolve and become more common in a population (Axelrod, 1984).

When two individuals interact repeatedly, it is possible for mutual cooperation to develop via direct reciprocity, the principle that people tend to help those who had previously helped them (Trivers, 1971). Thus, direct reciprocity is an ideal candidate that maintains cooperation in small-scale societies where people tend to know each other and are aware that they will interact over an extended period of time. Yet, the transition from small-scale societies to large-scale modern societies brings about new challenges for social interactions. In particular, modern societies tend to have larger groups within which repeated interactions are rare and there is more anonymity among group members (Durand, 1977; Kerr, 1989). This makes it less likely that one's cooperation is directly reciprocated by one's interaction partner. Fortunately, cooperation among strangers in societies with high relational mobility is more often explained by indirect reciprocity, the principle that people tend to help others who had previously helped someone else (also known as downstream reciprocity; see Nowak and Sigmund, 2005). An agent-based modeling (ABM) approach is often used to test the decision rules or behavioral strategies that are evolutionarily stable. (See Box 3.2 for details about ABM.) Computer simulations using this approach reveal that in a population where agents interact with one another and choose whether to pay a cost c for another to receive a benefit b ($b > c$), cooperation can be evolutionarily stable through indirect reciprocity (i.e., agents choose to help those who have helped someone else; Nowak and Sigmund, 1998a).

The key element of indirect reciprocity is knowledge about others' past behavior or reputation. Although people may directly observe others' previous behavior, an increase in group size makes direct observation less likely to occur. Notably, language enables humans to communicate and share socially diagnostic information about others in social interactions (i.e., gossip; Dunbar, 2004), and thus has important implications for the optimal fitness-relevant social strategies to select reliable coalition partners and manage one's own reputation. Thus, gossip and reputation spreading allow indirect reciprocity to occur and thus promote cooperation even when direct reciprocity is not possible.

To understand how reputation and indirect reciprocity can deter free riding and promote cooperation within and between groups, we first need to figure out the origins and functions of reputation in social interactions (reputation as an "outcome"). We propose that humans may have developed psychological adaptations to infer others' reputations from various situational cues so as to avoid cheaters. Meanwhile, we need to know how people treat others based on these others' reputation and adjust their own behavior when their reputation is at stake (reputation as an "input"). It is equally important to examine the relative ability of reputation, compared to material sanctions (e.g., reward and punishment), in promoting cooperation. Below, we review research that addresses these questions, discuss the broader implications of reputation and indirect reciprocity, and propose several important directions for future research.

Reputation Formation in Social Interactions

Origins and Functions of Reputation

Reputation is an important aspect of human social life. It is not a person's actual behavior or personality, but instead reflects others' beliefs about one's behavior or personality. Put in another way, reputation can be considered as a set of socially transmitted subjective beliefs or evaluations that a group or community has about one of its members (Emler, 1990; Bromley, 1993; Sperber and Baumard, 2012). Indeed, people belong to different groups and social networks, and care about the impressions they make on others, especially those whom they are closely connected with and regard as influential in their lives (Bromley, 1993). They also form impressions about others' reputation based on either direct experience (i.e., firsthand reputation) or gossip from third parties (i.e., secondhand reputation; Bromley, 1993; Anderson and Shirako, 2008). Empirical research suggests that people actively engage in prosocial gossip about absent third parties to help others identify cooperators and avoid being exploited by cheaters (Feinberg et al., 2012). Notably, even five-year-old preschoolers reliably engage in such prosocial gossip to guide their peers toward cooperative partners (Engelmann et al., 2016). In addition to gossip in daily conversations, reputation in the modern world can spread more quickly through mass media and the Internet, both of which many people have access to.

Reputation serves important functions in social interactions and thus can be utilized to navigate the social world. First, having a good reputation pays off in future interactions when one needs help, resources, or coalition partners to solve challenging tasks (Nowak and Sigmund, 2005). Second, knowing others' reputation helps people to select trustworthy partners and avoid potential cheaters or free riders (Dunbar, 2004). In addition, people are motivated to protect their good reputation and conceal their bad reputation, because reputation is important for individuals to navigate the social world and to integrate into group life. For example, recent research using multiple methods (i.e., cross-national data from the World Values Survey combined with decisions in hypothetical scenarios and real experiments) reveals that people tend to make substantial sacrifices to protect their moral reputations and to avoid reputation damage (Vonasch et al., 2017). Thus, reputation is an ideal candidate that enhances cooperation and drives the direction and consequences of future social interactions.

Reputation Assessment in Indirect Reciprocity

Humans recurrently encounter the adaptive problem to find reliable partners and avoid cheaters in social interactions. As a solution, natural selection may have equipped humans with specialized psychological mechanisms to infer others' reputation based on cues obtained from direct interactions, observations, or gossip with other third parties (Anderson and Shirako, 2008). While information from direct interactions and observations is more accurate, gossip may often be biased. Thus, in partner-choice settings, people may prefer to trust firsthand observation over gossip, and this preference already exists in seven-year-old children (Haux et al., 2017). Despite the potential bias in gossip, humans have also developed psychological adaptations to assess gossip veracity based on various situational cues in social exchange of information (Hess and Hagen, 2006). Importantly, gossip strongly affects how people behave toward others even when they can directly observe others' behavior (Sommerfeld et al., 2007).

Knowledge about others' reputation influences how one should treat others in future interactions (Nowak and Sigmund, 2005). Thus, how people assess others' reputation may provide insights into future decision processes and whether cooperation can be sustained in the population. Computer simulations that introduce binary reputations (i.e., good or bad) have derived four most prevailing reputation assessment rules, which are image scoring, standing strategy, stern judging, and shunning (see Table 3.1). These rules all assume that people would cooperate with those with a good reputation and defect with those with a bad reputation. Here, we illustrate these rules in a typical donation game where a donor decides whether to help a recipient by paying a personal cost c to bring a benefit b to the recipient ($b > c$). These rules vary in whether the donor's action (first-order information) and/or the recipient's reputation as a donor in the previous interaction (second-order information) are taken into account.

One simple first-order rule is image scoring (IS), which assigns a bad reputation to donors whenever they fail to help someone in need (Nowak and Sigmund, 1998a, 1998b). Yet, image scoring is too simple and has its weaknesses. In particular, people employing this rule would assign a good reputation to and thus help those who have helped others, but assign a bad reputation to and thus defect with those who refuse to help others. Accordingly, defecting with those who refuse to help others, even though

Table 3.1 Four key reputation assessment rules in indirect reciprocity

Donor's action Recipient's reputation in last interaction	C Good	C Bad	D Good	D Bad
Image scoring (IS)	Good	Good	Bad	Bad
Standing strategy (ST)	Good	Good	Bad	Good
Stern judging (SJ)	Good	Bad	Bad	Good
Shunning (SH)	Good	Bad	Bad	Bad

Note: C, cooperate; D, defect.

it is presumably justified, brings oneself a bad reputation and may elicit further defections. This problem can be solved by considering both the donor's action and the recipient's reputation. One such second-order rule is standing strategy (ST), which is a tolerant rule that further distinguishes between unjustified and justified defectors: assign a bad reputation to donors who refuse to help a good recipient, and a good reputation to donors who refuse to help a bad recipient (Leimar and Hammerstein, 2001; Panchanathan and Boyd, 2003). Thus, standing strategy outperforms image scoring because it enables justified defectors to receive benefits from others who employ the same rule. Another second-order rule is stern judging (SJ), which resembles standing strategy except that it distinguishes between unjustified and justified cooperators. According to stern judging, helping a good recipient or refusing to help a bad recipient leads to a good reputation, whereas helping a bad recipient or refusing to help a good recipient leads to a bad reputation (Pacheco et al., 2006). Different from stern judging, shunning (SH) is a simple but strict rule specifying that all donors get a bad reputation when interacting with a bad recipient, regardless of their behavior. With shunning, donors get a good reputation only when cooperating with a good recipient (Nowak and Sigmund, 2005).

Computer simulations demonstrate that these four rules can stabilize cooperation in some situations, but their robustness is sensitive to information availability. For example, image scoring can only stabilize cooperation in medium-sized groups where group members are able to keep track of each other's reputation (Nowak and Sigmund, 1998a). Ohtsuki and Iwasa (2007) studied the evolutionary dynamics of different behavioral strategies (i.e., unconditional cooperation, unconditional defection, and conditional cooperation) for each rule and found that only standing strategy, stern judging, and shunning can sustain cooperation. Other research suggests that stern judging is most robust in promoting cooperation, as it involves a straightforward rule that a donor's decision only depends on the recipient's previous reputation (Pacheco et al., 2006). Despite the coexistence of multiple reputation assessment rules that support the evolution of cooperation, whether evolution favors more tolerant or strict rules may depend on the specific context. To illustrate this issue, Yamamoto and colleagues (2017) used agent-based simulations to find the indispensable rules for cooperative societies to emerge by observing what happens when one rule is absent from the population in environments with or without errors. They found that cooperation cannot emerge without shunning and image scoring regardless of errors. Yet, in an environment with errors, there is an alternation from strict rules to tolerant rules, and standing strategy is also critical to stabilizing cooperation.

While these rules help us understand how people assess others' reputation, real-world situations are not simply driven by binary moral judgments. Besides, the same behavior may be judged as different in different contexts. People can assess others' reputation based on cues reflecting the intention or the cost of others' behavior. First, people who behave in ways that provide benefits to others tend to receive more reputational benefits. For example, in a computerized experiment, students who donate to UNICEF receive more money from others in a subsequent game, and also receive more votes in the election of the students' council (Milinski et al., 2002). Similarly, people who punish others' selfish behaviors to maintain cooperative norms are rated as more trustworthy, and receive more money in dyadic trust games (Barclay, 2006; Jordan et al., 2016). Second, other-regarding behaviors that are more costly may bring more reputational benefits. This is because costly behaviors or traits are often difficult to forge and thus can reliably reflect one's quality (Zahavi and Zahavi, 1997). In support of this argument, research shows that people tend to assign more prosocial reputation to lower-class donors than higher-class ones when they donate either the same amount or the same percentage of annual income to charity organizations, as such behavior is arguably more costly for lower-class donors (Yuan et al., 2018). Similarly, people who pay a higher (vs. lower) cost to punish others' unfair behavior are perceived as more fair, friendly, and generous, and are also transferred more money in a subsequent trust game (Nelissen, 2008).

Notably, individuals' reputation may not always correspond with their personality traits or behavioral history. The degree of discrepancy between one's reputation and one's personality or past behavior may be driven by one's social connections or unintended errors in social interactions (Tazelaar et al., 2004; Anderson and Shirako, 2008). For example, previous work suggests that more socially connected people tend to develop a reputation closely related to their past (un)cooperative behavior, whereas the behavior of less connected people has less impact on their reputation (Anderson and Shirako, 2008). Similarly, field research has found that one's cooperative reputation positively relates to the number of people one has cooperated with, but does not relate to the number of cooperative acts (Macfarlan et al., 2013). That is, the more people one has cooperated with, the more likely that this person will develop a cooperative reputation, but the number of times one cooperates with the same person does not promote one's cooperative reputation. This finding is interesting, as it suggests that gossip and reputation spreading in groups and social networks are powerful for reputation building.

Reputation, Indirect Reciprocity, and Cooperation

As noted earlier, indirect reciprocity occurs when a cooperative actor is reciprocated by another person who is not the previous partner. This reputation-based indirect reciprocity is a fundamental pathway that drives the evolution of cooperation. The phenomena of indirect reciprocity and reputation-based cooperation have been documented in both lab and field research, as well as research on children in the early stages of development.

Indirect Reciprocity in the Lab, Field, and in Early Development

People naturally prefer to interact with others who are trustworthy, helpful, and cooperative, and are willing to provide social benefits to these people. Prior experiments using economic games have consistently demonstrated that people do engage in indirect reciprocity. That is, being generous to others pays off in the long run as it builds up a good reputation that is later rewarded by other third parties. When participants play the donation game repeatedly as donors and recipients with different partners while their history of giving behavior is displayed to each other, they donate more frequently to receivers who had been generous to others in previous interactions (Wedekind and Milinski, 2000; Wedekind and Braithwaite, 2002). Notably, even some non-human species are sensitive to others' past behavior and reputation when they act as observers. For example, chimpanzees tend to develop a preference for gesturing to humans who share food with others over those who refuse to share food (Russell et al., 2008; Subiaul et al., 2008). In addition, clients of cleaner fish tend to spend more time next to "cooperative" cleaners that remove ectoparasites from clients than next to those who feed on client mucus (Bshary and Grutter, 2006; see Chapter 9 for further discussion).

Field research has also revealed that indirect reciprocity occurs. For example, people are more willing to offer free services on online platforms to service requesters whose profiles reveal that they have offered services to others in the past (van Apeldoorn and Schram, 2016). Similarly, customers in hair salons tend to offer more tips to hairdressers who participate in charity fundraising than those who do not (Khadjavi, 2016). A recent large-scale online experiment involving thousands of Airbnb users also found that others' reputation robustly predicts one's trust decisions (Abrahao et al., 2017). In this experiment, Airbnb users decided whether to invest 100 credits in the receivers they were ostensibly paired with in a trust game, while the receivers' profiles varied in similarity (i.e., the degree of match between participants and receivers' demographic attributes) and two reputation-relevant features (i.e., the average ratings and number of reviews on Airbnb). Indeed, Airbnb users trusted receivers with a better reputation even though they were dissimilar, and this pattern of results was also found in the real-world data of one million actual hospitality interactions among Airbnb users (Abrahão et al., 2017).

While the tendency to engage in indirect reciprocity is widely observed in human adults, whether this fundamental ability to condition one's cooperation on partners' reputations is acquired in early development has received less research attention. In recent years, researchers have started to examine whether young children also display the behavioral tendency to engage in indirect reciprocity. A field study observing preschoolers' natural interactions found that five- to six-year-olds performed more prosocial and affiliative behaviors toward a peer who had behaved prosocially toward others, compared to a peer who had shown no prosocial behaviors (Kato-Shimizu et al., 2013). In a similar experiment that presented 3.5-year-old children with a story in which a protagonist doll decided to distribute resources to other dolls, children guided the protagonist to share more resources with the dolls that had given to others, compared to the dolls that had not (Olson and Spelke, 2008, study 3). In addition, when distributing a small odd number of resources (i.e., wooden biscuits), 4.5-year-olds, but not 3-year-olds, gave more resources to a puppet who had helped another puppet than to a puppet who had violently hindered another puppet (Kenward and Dahl, 2011). Interestingly, in an eye-tracking experiment that presented ten-month-old infants with stimuli of fair or unfair donors being rewarded later by a third party, these infants spent a longer time looking at the event that the unfair donor was rewarded, suggesting that ten-month-old infants seem to expect third parties to act positively toward fair, rather than unfair, donors (Meristo and Surian, 2013). Thus, infants can generate evaluations of agents based on the fairness of their distributive actions and form expectations about how these agents should be treated by others. Taken together, these results among infants and young children demonstrate that the tendency to evaluate others' behavior and condition one's own behavior on these evaluations is universal and already shows up in early development.

Efficient Reputation Management in Social Interactions

People also have to face the adaptive problem of advertising themselves as reliable and trustworthy partners in social interactions. As discussed above, one's good or bad reputation can lead to fitness benefits (e.g., more access to resources, coalition or sexual partners) or fitness costs (e.g., social exclusions in competition with others). Thus, evolution may have selected for specialized mechanisms for individuals to be selectively more cooperative or less likely to engage in inappropriate behaviors (e.g., cheating) when the situation contains cues that relate to reputational consequences and potential indirect benefits or costs. One such cue is publicity or social visibility. Numerous studies reveal that people tend to contribute more to the public good, act more prosocially, and display more altruistic punishment of norm violators in public (vs. anonymous) situations in which their behavior can be observed by others (e.g., Kurzban et al., 2007; Ariely et al., 2009; Van Vugt and Hardy, 2010). Even a pair of eye images that creates subtle feelings of being watched can deter free riding and promote cooperation (e.g., Bateson et al., 2006; Nettle et al., 2012; Powell et al., 2012), although such effects are small and transient (Sparks and Barclay, 2013; see also Manesi et al., 2016).

In addition to cues of direct observation, people often can talk about absent others (i.e., gossip) in daily-life conversations and this is an important process for people to share and spread others' reputation. Thus, cues of gossip by others should serve as a robust social institution that can effectively maintain cooperative norms. In support of this, experimental studies find that people contribute more resources to the public good when their group members can gossip about them with their future partners, compared to situations with no such gossip (Feinberg et al., 2014; Wu et al., 2016a). Notably, such strategic cooperation in response to potential gossip by others is suggested to be mainly driven by one's enhanced concern for reputation rather than rational calculations of one's material benefits from different courses of action (Wu et al., 2015, 2016b).

Interestingly, even young children tend to strategically manage their reputation when there are cues of reputation in the environment. For example, experimental studies demonstrate that five-year-old children tend to share more stickers with others and also steal less when another child can observe their behavior (Engelmann et al., 2012), and that they share more stickers with an anonymous recipient when the observer can (vs. cannot) reciprocate them later or when the observer was an in-group (vs. out-group) member (Engelmann et al., 2013). In addition, five- to six-year-olds are also more resistant to cheating when they believe that someone is watching them, compared to when nobody is watching them (Piazza et al., 2011).

It is also important to note that the link between reputation and cooperation can be relatively strong or weak depending on potential personal or contextual factors. One possibility is that reputation may provide a stronger motivation to be cooperative for individuals with certain personality traits. For example, pro-self individuals who display a relative stable preference in prioritizing their own personal interest tend to be strategically more cooperative when their reputation is at stake (Simpson and Willer, 2008; Feinberg et al., 2012). In addition, prevention-focused individuals who tend to minimize negative outcomes and prevent losses are generally more concerned about their reputation (Cavazza et al., 2015), and also make more donations when they are exposed to cues of watching eyes (Pfattheicher, 2015). Another possibility is that cues of reputation may only elicit more cooperation when they lead to future consequences in terms of indirect benefits or costs (Barclay, 2012). Empirical evidence supporting this shows that people behave more generously when the recipient can gossip with their future partner, compared to when the recipient can gossip with an irrelevant person they will not meet or in a situation with no potential gossip (Wu et al., 2015).

Reputation versus Monetary Sanctions

In addition to reputation systems, monetary sanctions (i.e., punishment and reward) can also promote cooperation as they alter the costs and benefits of (non)cooperation (e.g., Fehr and Gächter, 2002; for a meta-analysis, see Balliet et al., 2011). Yet, monetary sanctions have their own weaknesses. For example, punishment is often costly for groups and organizations to implement and often harms both individual and group welfare, making it less efficient in promoting cooperation (Dreber et al., 2008; Egas and Riedl, 2008; Wu et al., 2009). In addition, punishing free riders or norm violators may induce retaliations from these people, leading to second-order free rider problems and the breakdown of cooperation, whereas reputation and partner choice seem to stabilize cooperation without such problems (Panchanathan and Boyd, 2004;

Nikiforakis, 2008; Barclay and Raihani, 2016). Besides, when there is punishment on free riders, people are less likely to believe that others are internally motivated to cooperate, and both trust and cooperation decrease when the existing punishment system is removed (Mulder et al., 2006). Indeed, a recent review of field experiments on solutions for cooperation problems shows that interventions that alter the costs and benefits of cooperation (e.g., material rewards) have mixed effects, whereas social interventions (e.g., reputation) are consistently highly effective in promoting cooperation (Kraft-Todd et al., 2015).

Some recent evidence suggests that reputation systems are more efficient than monetary sanctions (e.g., reward, punishment) at solving cooperation problems and maintaining group norms. For example, Wu and colleagues (2016a) conducted an online experiment involving real-time interactions to compare the relative ability of gossip (i.e., reputation sharing) and punishment to promote and maintain cooperation. Participants made their contribution decisions across four rounds of a four-person public goods game (PGG), and then played a two-round trust game (TG) as a trustor and a trustee, respectively. They interacted with different partners in each round of the two games. Participants were assigned to one of four conditions in which they could or could not (a) punish others by reducing others' earnings at a personal cost (cost-to-fine ratio of 1:3) or (b) gossip about others by sending notes to these others' partners in the next round after each round of the PGG. This research revealed that the gossip option elicited significantly more cooperation and personal gains, whereas punishment weakly affected cooperation and decreased personal gains. Notably, the initial gossip option made people more likely to trust others and be trustworthy themselves in the subsequent TG even when gossip was no longer possible.

Similarly, a recent field experiment conducted in a foraging society shows that social image concerns bring about the highest level of cooperation and economic gains (Grimalda et al., 2016). In this experiment, residents from eight villages played a one-shot prisoner's dilemma game (PDG) deciding whether to give their endowment to their partner. They were assigned to (a) a baseline condition, or one of four other conditions involving (b) an observer from the same village, (c) an observer from a different ethnic group, (d) options to reduce the other's earnings assuming two possible decisions, or (e) a local observer who could observe both their PDG and punishment choices. Both cooperation and payoffs were the highest when a local observer could observe their giving decisions. Although punishment also raised cooperation, it yielded lower gains when it was implemented alone or combined with social image concerns. Taken together, these findings from real-time interactions among online participants and residents in a foraging society consistently suggest that the desire to establish a positive reputation is a more effective solution than punishment in promoting cooperation.

In addition, reputation can promote large-scale cooperation in real-world social dilemmas. This was clearly demonstrated in a field experiment on sign-ups for a voluntary demand program to prevent blackouts during high electricity demand (Yoeli et al., 2013). This experiment showed that compared to an anonymous situation, making residents' names and unit numbers public on the sign-up sheet tripled the participation rate in this program. Moreover, the positive effect of observability was over four times larger than offering a monetary reward of $25. Taken together, gossip and reputation sharing is a cost-effective solution that promotes and maintains cooperation (for similar arguments, see Feinberg et al., 2012). Of course, future research may test whether reputation systems are still more effective when considering the potential cost (e.g., the psychological cost of gossiping) of implementing such systems.

Broader Implications and Future Directions

As noted earlier, although punishment can effectively promote cooperation, it is often costly to implement in groups and organizations, and negatively affects both individual and group payoff (Dreber et al., 2008; Egas and Riedl, 2008). Also, prior to punishing free riders or norm violators, people may prefer to monitor each other's behavior through gossip and reputation sharing, which are cheaper and easier to implement. Thus, the informal reputation systems may offer a better solution for real-world social dilemmas that arise in groups, organizations, and the society at large. Here, we highlight several research questions about reputation that are either understudied or currently a source of theoretical debate.

Does Negative Reputation Loom Larger Than Equally Positive Reputation?

People have a natural propensity to promote a good reputation and avoid reputation damage. Compared to the benefits brought by a good reputation, the detrimental effect of a bad reputation may be larger. More importantly, losing a good reputation can be easier than gaining it (Yaniv and Kleinberger, 2000). Thus, people may be more motivated to avoid getting a bad reputation than to gain a good reputation. This tendency may further elicit greater temptation to artificially manipulate one's reputation to be good when it is not. Although behavioral research on reputation often links it to complete knowledge of others' previous behavior, reputation in real-life situations may involve errors and thus can be biased. Biased reputations occur when cooperators are believed to be cheaters (i.e., biased negative reputation) or when cheaters are believed to be cooperators (i.e., biased positive reputation). Given that it is often more costly to choose cheaters as partners than to avoid cooperators in social interactions, people may use stricter standards when evaluating others' reputations in error-prone environments. This would imply that negative and positive reputation in real-life situations may have asymmetric effects on how individuals evaluate and behave toward others.

Related to this, a recent web-based experiment shows that compared to a neutral situation, people are more likely to consider others as untrustworthy and thus cooperate less with them when informed about their low reputation (i.e., being rated as one out of five stars) than consider them as more trustworthy or cooperate more with them based on their high reputation (i.e., being rated as five out of five stars; Capraro et al., 2016). This finding suggests that there might be a negativity bias (i.e., considering negative information as more reliable than positive information) in evaluating others' reputations. Indeed, such negativity bias is manifested in a range of domains, such as sensation, memory, impression formation, and moral judgment. (For a review, see Rozin and Royzman, 2001.) Future research can test whether and why this negativity bias occurs in evaluating and responding to others' reputations, as well as the efficiency of positively and negatively valenced reputation systems in promoting and maintaining cooperation.

Are Reputation Systems More or Less Efficient in Larger Groups?

Indirect reciprocity has been proposed to promote cooperation in large groups of genetically unrelated strangers (Yamagishi and Kiyonari, 2000; Nowak and Sigmund, 2005). While this argument implies that reputation can efficiently promote cooperation when groups grow larger, there is currently some debate on whether this is the case. One plausible idea is that larger groups contain a higher level of anonymity, and may be more vulnerable to free riding, making cooperation hard to evolve in larger groups. This idea has received support from some agent-based computer simulations. These simulations sought to analyze the evolution of cooperation in large communities while taking into account the effect of reputation and the general conclusion is that the evolution of indirect reciprocity becomes more difficult as group size increases (see Suzuki and Akiyama, 2005, 2007; dos Santos and Wedekind, 2015). However, other perspectives – such as the competitive altruism, biological markets, and costly signaling theories – suggest that reputation system can be more effective in promoting cooperation in larger groups. According to these perspectives, larger groups contain more competition to find reliable partners and to advertise oneself as a trustworthy partner (Van Vugt et al., 2007; Barclay, 2013). To be selected as coalition partners by others, one may need to display costly cooperative behaviors that signal one's genuine concern about others' welfare (Zahavi and Zahavi, 1997). Thus, people should be more sensitive to cues of reputation in larger, compared to smaller, groups, and would display a larger increase in cooperation when their behavior has reputational consequences in terms of future indirect benefits. Future research needs to resolve this debate on reputation in varied-size groups by testing the competing predictions from different perspectives.

Can Reputation Transcend Group Boundaries to Promote Cooperation?

People belong to groups in which they interact with other members over an extended period of time. The bounded generalized reciprocity perspective states that humans have an evolved tendency to cooperate with in-group members, because this tendency enables them to gain a positive reputation within their group and avoid the costs of social exclusion from their group (Yamagishi et al., 1999). One assumption underlying this logic is that reputation only matters in social interactions within one's group. In support of this claim, experimental studies using minimal group paradigms demonstrate that people express more in-group favoritism (i.e., behave more cooperatively toward in-group, compared to out-group, members) when their partner knows their group membership (Yamagishi and Mifune, 2008) or when they are exposed to reputational cues of eye images (Mifune et al., 2010). A recent meta-analysis also suggests that in-group favoritism only occurs when there is common (vs. unilateral) knowledge about each other's group membership (Balliet et al., 2014). These findings are complemented by developmental evidence that even five-year-olds share more resources with an anonymous recipient when an in-group member is observing them (Engelmann et al., 2013).

Clearly, both theory and evidence suggest that the strategy people use to manage their reputation depends on whether their actions contain reputational consequences within their groups. Yet, these findings do not necessarily mean that out-groups are not important for one's reputation or future interactions. Indeed, some evidence also suggests that a good reputation gained within and outside one's social group can make selfish individuals more cooperative in public good situations (Semmann et al., 2005). Similarly, people are more cooperative with both in-group and out-group members when their reputation is revealed to their partners in future interactions (Romano et al., 2017a). A recent large-scale cross-national study used natural groups (i.e., people of different nationalities) to test how people behave toward others with the same or different nationality. This study revealed that common (vs. unilateral) knowledge promoted cooperation with both in-group and out-group members (Romano et al., 2017b).

Although the above findings on the reputation–cooperation link within and outside one's group may seemingly conflict with each other, they share the same assumption that cooperation brings about a good reputation that may have future consequences. For one thing, people anticipate future interactions within their own group, and so behave more cooperatively with their in-group members to guarantee a good reputation. For another, when someone can observe and even gossip about their behavior with their future partners, people may invariably increase their cooperation with their current partner, regardless of this partner's group membership. Notably, the rapid development of social media and the Internet makes group boundaries less strict, and people can easily belong to different groups that compose densely connected social networks within which reputation can spread more quickly. Thus, reputation may transcend group boundaries to promote cooperation and the way people manage their reputation may be affected by dynamic social network properties, such as degree of separation between network members (Apicella et al., 2012), network density (Gallo and Yan, 2015), and the possibility to update network connections (Gross and Blasius, 2008; Rand et al., 2011).

Are There Cultural Variations in Indirect Reciprocity and Reputation-Based Cooperation?

Despite the different reputation assessment rules derived from computer simulations, existing research on reputation often assumes that allocating benefits to others would establish a good reputation (i.e., image scoring), without taking into account the recipient's past behavior or reputation (i.e., second-order information; Wedekind and Milinski, 2000). To what extent do people cooperate with others who have engaged in cooperation (vs. free riding) or punishing non-cooperators (vs. no punishment behavior) when their own reputation is at stake? Previous research suggests that people do actively seek information about their partners' past cooperation, and condition their cooperative decisions on this information to enhance their own reputation (Swakman et al., 2016). Moreover, people who punish free riders tend to receive more help from others (dos Santos et al., 2013). Second-order information reflects the social context of one's (un)cooperative behavior and allows others to judge one's real intentions. Yet, this information may not be treated equally across different cultures. Indeed, cross-cultural evidence suggests that Westerners tend to be context-independent, whereas Asians tend to be context-dependent when perceiving a salient object (Nisbett and Miyamoto, 2005). This variation in perception style may also exist in interpersonal context, such that Westerners and Asians may treat first-order and second-order information about others' behavior differently while deciding whether to cooperate with others. Thus, future research could examine whether Asians are more likely than Westerners to take the social context (i.e., second-order information) into account when assessing others' reputation or making their cooperative decisions toward others to enhance their own reputation.

Conclusions

Social interactions do not occur in a vacuum. They often take place in groups and social networks where people can monitor and spread each other's reputation. Despite the temptation to act selfishly when interacting with strangers, there is a never-ending conflict between the desire to act selfishly and the need to gain a good reputation (or avoid losing the good reputation one already has). While one's selfish behavior guarantees immediate material benefits, it may harm one's reputation and can lead to a long-term loss. Thus, reputation is a key element of indirect reciprocity that provides a fundamental route to human cooperation. In this chapter, we have discussed how reputation is formed and assessed in social interactions, reviewed empirical research that documents the phenomena of indirect reciprocity and reputation-based cooperation as well as evidence about the greater power of reputation over monetary sanctions in solving cooperation problems. Future research would benefit by investigating the negativity bias in reputation systems, the efficiency of reputation in varied-size groups, whether reputation transcends group boundaries to promote cooperation, and potential cultural variations. Taken together, we emphasize that reputation monitoring and spreading is a strong candidate to promote trust and cooperation, thereby reducing the possibility of social conflict, in a cost-effective manner, perhaps more so among people who are inclined to act selfishly.

References

Abrahao B., Parigi P., Gupta A., and Cook K. S. (2017) Reputation offsets trust judgments based on social biases among Airbnb users. *Proceedings of the National Academy of Sciences USA*, 114: 9848–9853.

Anderson C., and Shirako A. (2008) Are individuals' reputations related to their history of behavior? *Journal of Personality and Social Psychology*, 94: 320–333.

Apicella C. L., Marlowe F. W., Fowler J. H., and Christakis N. A. (2012) Social networks and cooperation in hunter-gatherers. *Nature*, 481: 497–501.

Ariely D., Bracha A., and Meier S. (2009) Doing good or doing well? Image motivation and monetary incentives in behaving prosocially. *American Economic Review*, 99: 544–555.

Axelrod R. (1984) *The Evolution of Cooperation*. New York: Basic Books.

Balliet D., Mulder L. B., and Van Lange P. A. M. (2011) Reward, punishment, and cooperation: A meta-analysis. *Psychological Bulletin*, 137: 594–615.

Balliet D., Wu J., and De Dreu C. K. W. (2014) Ingroup favoritism in cooperation: A meta-analysis. *Psychological Bulletin*, 140: 1556–1581.

Barclay P. (2006) Reputational benefits for altruistic punishment. *Evolution and Human Behavior*, 27: 325–344.

Barclay P. (2012) Harnessing the power of reputation: Strengths and limits for promoting cooperative behaviors. *Evolutionary Psychology*, 10: 868–883.

Barclay P. (2013) Strategies for cooperation in biological markets, especially for humans. *Evolution and Human Behavior*, 34: 164–175.

Barclay P., and Raihani N. (2016) Partner choice versus punishment in human prisoner's dilemmas. *Evolution and Human Behavior*, 37: 263–271.

Bateson M., Nettle D., and Roberts G. (2006) Cues of being watched enhance cooperation in a real-world setting. *Biology Letters*, 2: 412–414.

Bromley D. B. (1993) *Reputation, Image and Impression Management*. Oxford: John Wiley & Sons.

Bshary R., and Grutter A. S. (2006) Image scoring and cooperation in a cleaner fish mutualism. *Nature*, 441: 975–978.

Camerer C. (2003) *Behavioral Game Theory: Experiments in Strategic Interaction*. Princeton, NJ: Princeton University Press.

Capraro V., Giardini F., Vilone D., and Paolucci M. (2016) Partner selection supported by opaque reputation promotes cooperative behavior. *Judgment and Decision Making*, 11: 589–600.

Cavazza N., Guidetti M., and Pagliaro S. (2015) Who cares for reputation? Individual differences and concern for reputation. *Current Psychology*, 34: 164–176.

Dawes R. M. (1980) Social dilemmas. *Annual Review of Psychology*, 31: 169–193.

dos Santos M., Rankin D. J., and Wedekind C. (2013) Human cooperation based on punishment reputation. *Evolution*, 67: 2446–2450.

dos Santos M., and Wedekind C. (2015) Reputation based on punishment rather than generosity allows for evolution of cooperation in sizable groups. *Evolution and Human Behavior*, 36: 59–64.

Dreber A., Rand D. G., Fudenberg D., and Nowak M. A. (2008) Winners don't punish. *Nature*, 452: 348–351.

Dunbar R. I. M. (2004) Gossip in evolutionary perspective. *Review of General Psychology*, 8: 100–110.

Durand J. D. (1977) Historical estimates of world population: An evaluation. *Population and Development Review*, 3: 253–296.

Egas M., and Riedl A. (2008) The economics of altruistic punishment and the maintenance of cooperation. *Proceedings of the Royal Society B: Biological Sciences*, 275: 871–878.

Emler N. (1990) A social psychology of reputation. *European Review of Social Psychology*, 1: 171–193.

Engelmann J. M., Herrmann E., and Tomasello M. (2012) Five-year olds, but not chimpanzees, attempt to manage their reputations. *PLoS ONE*, 7: e48433.

Engelmann J. M., Herrmann E., and Tomasello M. (2016) Preschoolers affect others' reputations through prosocial gossip. *British Journal of Developmental Psychology*, 34: 447–460.

Engelmann J. M., Over H., Herrmann E., and Tomasello M. (2013) Young children care more about their reputation with ingroup members and potential reciprocators. *Developmental Science*, 16: 952–958.

Fehr E., and Gächter S. (2002) Altruistic punishment in humans. *Nature*, 415: 137–140.

Feinberg M., Cheng J. T., and Willer R. (2012) Gossip as an effective and low-cost form of punishment. *Behavioral and Brain Sciences*, 35: 25.

Feinberg M., Willer R., and Schultz M. (2014) Gossip and ostracism promote cooperation in groups. *Psychological Science*, 25: 656–664.

Feinberg M., Willer R., Stellar J., and Keltner D. (2012) The virtues of gossip: Reputational information sharing as prosocial behavior. *Journal of Personality and Social Psychology*, 102: 1015–1030.

Gallo E., and Yan C. (2015) The effects of reputational and social knowledge on cooperation. *Proceedings of the National Academy of Sciences USA*, 112: 3647–3652.

Grimalda G., Pondorfer A., and Tracer D. P. (2016) Social image concerns promote cooperation more than altruistic punishment. *Nature Communications*, 7: 12288.

Gross T., and Blasius B. (2008) Adaptive coevolutionary networks: A review. *Journal of the Royal Society Interface*, 5: 259–271.

Hamilton W. D. (1964) The genetical evolution of social behaviour. II. *Journal of Theoretical Biology*, 7: 17–52.

Haux L., Engelmann J. M., Herrmann E., and Tomasello M. (2017) Do young children preferentially trust gossip or firsthand observation in choosing a collaborative partner? *Social Development*, 26: 466–474.

Hess N. H., and Hagen E. H. (2006) Psychological adaptations for assessing gossip veracity. *Human Nature*, 17: 337–354.

Jordan J. J., Hoffman M., Bloom P., and Rand D. G. (2016) Third-party punishment as a costly signal of trustworthiness. *Nature*, 530: 473–476.

Kato-Shimizu M., Onishi K., Kanazawa T., and Hinobayashi T. (2013) Preschool children's behavioral tendency toward social indirect reciprocity. *PLoS ONE*, 8: e70915.

Kenward B., and Dahl M. (2011) Preschoolers distribute scarce resources according to the moral valence of recipients' previous actions. *Developmental Psychology*, 47: 1054–1064.

Kerr N. L. (1989) Illusions of efficacy: The effects of group size on perceived efficacy in social dilemmas. *Journal of Experimental Social Psychology*, 25: 287–313.

Khadjavi M. (2016) Indirect reciprocity and charitable giving – Evidence from a field experiment. *Management Science*, 63: 3708–3717.

Kraft-Todd G., Yoeli E., Bhanot S., and Rand D. (2015) Promoting cooperation in the field. *Current Opinion in Behavioral Sciences*, 3: 96–101.

Kuhlman D. M., and Marshello A. F. (1975) Individual differences in game motivation as moderators of preprogrammed strategy effects in prisoner's dilemma. *Journal of Personality and Social Psychology*, 32: 922–931.

Kurzban R., Burton-Chellew M. N., and West S. A. (2015) The evolution of altruism in humans. *Annual Review of Psychology*, 66: 575–599.

Kurzban R., DeScioli P., and O'Brien E. (2007) Audience effects on moralistic punishment. *Evolution and Human Behavior*, 28: 75–84.

Leimar O., and Hammerstein P. (2001) Evolution of cooperation through indirect reciprocity. *Proceedings of the Royal Society B: Biological Sciences*, 268: 745–753.

Macfarlan S. J., Quinlan R., and Remiker M. (2013) Cooperative behaviour and prosocial reputation dynamics in a Dominican village. *Proceedings of the Royal Society B: Biological Sciences*, 280: 20130557.

Manesi Z., Van Lange P. A. M., and Pollet T. V. (2016) Eyes wide open: Only eyes that pay attention promote prosocial behavior. *Evolutionary Psychology*, 14: 1474704916640780.

Meristo M., and Surian L. (2013) Do infants detect indirect reciprocity? *Cognition*, 129: 102–113.

Mifune N., Hashimoto H., and Yamagishi T. (2010) Altruism toward in-group members as a reputation mechanism. *Evolution and Human Behavior*, 31: 109–117.

Milinski M., Semmann D., and Krambeck H.-J. (2002) Donors to charity gain in both indirect reciprocity and political reputation. *Proceedings of the Royal Society B: Biological Sciences*, 269: 881–883.

Mulder L. B., Van Dijk E., De Cremer D., and Wilke H. A. M. (2006) Undermining trust and cooperation: The paradox of sanctioning systems in social dilemmas. *Journal of Experimental Social Psychology*, 42: 147–162.

Nelissen R. M. A. (2008) The price you pay: Cost-dependent reputation effects of altruistic punishment. *Evolution and Human Behavior*, 29: 242–248.

Nettle D., Nott K., and Bateson M. (2012) "Cycle thieves, we are watching you": Impact of a simple signage intervention against bicycle theft. *PLoS ONE*, 7: 8–12.

Nikiforakis N. (2008) Punishment and counter-punishment in public good games: Can we really govern ourselves? *Journal of Public Economics*, 92: 91–112.

Nisbett R. E., and Miyamoto Y. (2005) The influence of culture: Holistic versus analytic perception. *Trends in Cognitive Sciences*, 9: 467–473.

Nowak M. A. (2006) Five rules for the evolution of cooperation. *Science*, 314: 1560–1563.

Nowak M. A., and Sigmund K. (1998a) Evolution of indirect reciprocity by image scoring. *Nature*, 393: 573–577.

Nowak M. A., and Sigmund K. (1998b) The dynamics of indirect reciprocity. *Journal of Theoretical Biology*, 194: 561–574.

Nowak M. A., and Sigmund K. (2005) Evolution of indirect reciprocity by image scoring. *Nature*, 437: 1291–11298.

Ohtsuki H., and Iwasa Y. (2007) Global analyses of evolutionary dynamics and exhaustive search for social norms that maintain cooperation by reputation. *Journal of Theoretical Biology*, 244: 518–531.

Olson K. R., and Spelke E. S. (2008) Foundations of cooperation in young children. *Cognition*, 108: 222–231.

Pacheco J. M., Santos F. C., and Chalub F. A. C. (2006) Stern-judging: A simple, successful norm which promotes cooperation under indirect reciprocity. *PLoS Computational Biology*, 2: e178.

Panchanathan K., and Boyd R. (2003) A tale of two defectors: The importance of standing for evolution of indirect reciprocity. *Journal of Theoretical Biology*, 224: 115–126.

Panchanathan K., and Boyd R. (2004) Indirect reciprocity can stabilize cooperation without the second-order free rider problem. *Nature*, 432: 499–502.

Peysakhovich A., Nowak M., and Rand D. G. (2014) Humans display a 'cooperative phenotype' that is domain general and temporally stable. *Nature Communications*, 5: 4939.

Pfattheicher S. (2015) A regulatory focus perspective on reputational concerns: The impact of prevention-focused self-regulation. *Motivation and Emotion*, 39: 932–942.

Piazza J., Bering J. M., and Ingram G. (2011) "Princess Alice is watching you": Children's belief in an invisible person inhibits cheating. *Journal of Experimental Child Psychology*, 109: 311–320.

Powell K. L., Roberts G., and Nettle D. (2012) Eye images increase charitable donations: Evidence from an opportunistic field experiment in a supermarket. *Ethology*, 118: 1096–1101.

Rand D. G., Arbesman S., and Christakis N. A. (2011) Dynamic social networks promote cooperation in experiments with humans. *Proceedings of the National Academy of Sciences USA*, 108: 19193–19198.

Rand D. G., and Nowak M. A. (2013) Human cooperation. *Trends in Cognitive Sciences*, 17: 413–425.

Romano A., Balliet D., and Wu J. (2017a) Unbounded indirect reciprocity: Is reputation-based cooperation bounded by group membership? *Journal of Experimental Social Psychology*, 71: 59–67.

Romano A., Balliet D., Yamagishi T., and Liu J. H. (2017b) Parochial trust and cooperation across 17 societies. *Proceedings of the National Academy of Sciences USA*, 114: 12702–12707.

Rozin P., and Royzman E. B. (2001) Negativity bias, negativity dominance, and contagion. *Personality and Social Psychology Review*, 5: 296–320.

Russell Y. I., Call J., and Dunbar R. I. (2008) Image scoring in great apes. *Behavioural Processes*, 78: 108–111.

Semmann D., Krambeck H., and Milinski M. (2005) Reputation is valuable within and outside one's own social group. *Behavioral Ecology and Sociobiology*, 57: 611–616.

Simpson B., and Willer R. (2008) Altruism and indirect reciprocity: The interaction of person and situation in prosocial behavior. *Social Psychology Quarterly*, 71: 37–52.

Smith E. R., and Conrey F. R. (2007) Agent-based modeling: A new approach for theory building in social psychology. *Personality and Social Psychology Review*, 11: 87–104.

Sommerfeld R. D., Krambeck H. J., Semmann, D., and Milinski M. (2007) Gossip as an alternative for direct observation in games of indirect reciprocity. *Proceedings of the National Academy of Sciences USA*, 104: 17435–17440.

Sparks A., and Barclay P. (2013) Eye images increase generosity, but not for long: The limited effect of a false cue. *Evolution and Human Behavior*, 34: 317–322.

Sperber D., and Baumard N. (2012) Moral reputation: An evolutionary and cognitive perspective. *Mind and Language*, 27: 495–518.

Subiaul F., Vonk J., Okamoto-Barth S., and Barth J. (2008) Do chimpanzees learn reputation by observation? Evidence from direct and indirect experience with generous and selfish strangers. *Animal Cognition*, 11: 611–623.

Suzuki S., and Akiyama E. (2005) Reputation and the evolution of cooperation in sizable groups. *Proceedings of the Royal Society B: Biological Sciences*, 272: 1373–1377.

Suzuki S., and Akiyama E. (2007) Evolution of indirect reciprocity in groups of various sizes and comparison with direct reciprocity. *Journal of Theoretical Biology*, 245: 539–552.

Swakman V., Molleman L., Ule A., and Egas M. (2016) Reputation-based cooperation: Empirical evidence for behavioral strategies. *Evolution and Human Behavior*, 37: 230–235.

Tazelaar M. J. A., Van Lange P. A. M., and Ouwerkerk J. W. (2004) How to cope with "noise" in social dilemmas: The benefits of communication. *Journal of Personality and Social Psychology*, 87: 845–859.

Trivers R. L. (1971) The evolution of reciprocal altruism. *Quarterly Review of Biology*, 46: 35–57.

van Apeldoorn J., and Schram A. (2016) Indirect reciprocity: A field experiment. *PloS ONE*, 11: e0152076.

Van Lange P. A. M., Joireman J., Parks C. D., and Van Dijk E. (2013) The psychology of social dilemmas: A review. *Organizational Behavior and Human Decision Processes*, 120: 125–141.

Van Vugt M., and Hardy C. L. (2010) Cooperation for reputation: Wasteful contributions as costly signals in public goods. *Group Processes and Intergroup Relations*, 13: 101–111.

Van Vugt M., Roberts G., and Hardy C. (2007) Competitive altruism: Development of reputation-based cooperation in groups. In Dunbar R., and Barrett L., eds., *Handbook of Evolutionary Psychology*. Oxford: Oxford University Press, pp. 531–540.

Vonasch A. J., Reynolds T., Winegard B. M., and Baumeister R. F. (2017) Death before dishonor: Incurring costs to protect moral reputation. *Social Psychological and Personality Science*, 9: 604–613.

Wedekind C., and Braithwaite V. A. (2002) The long-term benefits of human generosity in indirect reciprocity. *Current Biology*, 12: 1012–1015.

Wedekind C., and Milinski M. (2000) Cooperation through image scoring in humans. *Science*, 288: 850–852.

Wu J., Balliet D., and Van Lange P. A. M. (2015) When does gossip promote generosity? Indirect reciprocity under the shadow of the future. *Social Psychological and Personality Science*, 6: 923–930.

Wu J., Balliet D., and Van Lange P. A. M. (2016a) Gossip versus punishment: The efficiency of reputation to promote and maintain cooperation. *Scientific Reports*, 6: 23919.

Wu J., Balliet D., and Van Lange P. A. M. (2016b) Reputation management: Why and how gossip enhances generosity. *Evolution and Human Behavior*, 37: 193–201.

Wu J. J., Zhang B. Y., Zhou,Z. X., He Q. Q., Zheng X. D., Cressman R., and Tao Y. (2009) Costly punishment does not always increase cooperation. *Proceedings of the National Academy of Sciences USA*, 106: 17448–17451.

Yamagishi T., Jin N., and Kiyonari T. (1999) Bounded generalized reciprocity: Ingroup boasting and ingroup favoritism. In Lawler E. J. (Series ed.) and Thye S. R., Lawler E. J., Macy M. W., and Walker H. A. (Vol. eds.), *Advances in Group Processes*. Bingley: Emerald, pp. 161–197.

Yamagishi T., and Kiyonari T. (2000) The group as the container of generalized reciprocity. *Social Psychology Quarterly*, 63: 116–132.

Yamagishi T., and Mifune N. (2008) Does shared group membership promote altruism? Fear, greed, and reputation. *Rationality and Society*, 20: 5–30.

Yamagishi T., Mifune N., Li Y. et al. (2013) Is behavioral pro-sociality game-specific? Pro-social preference and expectations of pro-sociality. *Organizational Behavior and Human Decision Processes*, 120: 260–271.

Yamamoto H., Okada I., Uchida S., and Sasaki T. (2017) A norm knockout method on indirect reciprocity to reveal indispensable norms. *Scientific Reports*, 7: 44146.

Yaniv I., and Kleinberger E. (2000) Advice taking in decision making: Egocentric discounting and reputation formation. *Organizational Behavior and Human Decision Processes*, 83: 260–281.

Yoeli E., Hoffman M., Rand D. G., and Nowak M. A. (2013) Powering up with indirect reciprocity in a large-scale field experiment. *Proceedings of the National Academy of Sciences USA*, 110: 10424–10429.

Yuan M., Wu J., and Kou Y. (2018) Donors' social class and their prosocial reputation: Perceived authentic motivation as an underlying mechanism. *Social Psychology*, 49: 205–218.

Zahavi A., and Zahavi A. (1997) *The Handicap Principle: A Missing Piece of Darwin's Puzzle*. Oxford: Oxford University Press.

4 Finding the Right Balance
Cooperation and Conflict in Nature

Elizabeth A. Ostrowski

Introduction

Cooperation is the defining feature of societies; in these groups, members work together to achieve something that the individuals alone cannot. We marvel at cooperation in part because it requires communication and coordination, complex behaviors that speak directly to the creative, constructive power of natural selection. Nevertheless, societies can be disrupted by internal conflicts (Hurst et al., 1996; Chapman, 2006; Ratnieks et al., 2006; Burt and Trivers, 2009; Queller and Strassmann, 2018; Sachs et al., 2018). Conflict can be defined in many ways, but it amounts to an incentive to defect because actions that benefit the individual (e.g., do not pay taxes) run counter to those that benefit the group (everyone pays their taxes). In some cases, conflict results in a tragedy of the commons, where cooperation produces goods that are available to all, but some individuals deplete the public good without contributing to its production.

Evolutionary biologists have identified the key ingredients that promote cooperation and reduce conflict, and the foundation of social evolution theory is strong. Nevertheless, cheaters or defectors are observed in many societies, and their presence can select for counter-mechanisms (such as policing) that prevent their success or limit their damage. Given this potential for adaptations and counter-adaptations, the long-term evolutionary dynamics of social conflict, as well as the balance between cooperation and conflict – if such a balance exists, or is even possible – is less well understood.

An Overview of Social Evolution Theory

The problem of cooperation goes back to Darwin, who commented that, "Natural selection will never produce in a being anything injurious to itself, for natural selection acts solely by and for the good of each" (Darwin, 1861). In light of Darwin's words, it would be difficult to understand examples of altruism, defined exactly as Darwin states: where an individual does something harmful to themselves, apparently for the good of others. A solution to the problem of how natural selection could favor altruism was hinted at by several biologists, including J. B. S. Haldane

(Haldane, 1955), but the formal theory to support it was developed by William Hamilton in 1964 (Hamilton, 1964a, 1964b).

Hamilton showed that natural selection maximizes a quantity he termed "inclusive fitness," which incorporates the direct effect of an individual's genes on that individual's own fitness, as well as the indirect effect of these same genes in relatives (Hamilton, 1964a; Gardner et al., 2011). Thus, maximizing inclusive fitness might at times involve behaviors that reduce the individual's direct fitness but enhance its indirect fitness via its relatives. Moreover, if the benefit to relatives more than offsets the detriment to the individual, then the genes that encode such costly behaviors can increase in frequency by natural selection. The upshot of Hamilton's insight was that while selection can favor genetic variants that confer a net benefit across all individuals who harbor them, it can involve a detriment in some. More important, it points to a key feature of societies where we expect cooperation to succeed: those consisting of relatives.

Decades of work have confirmed that relatedness is a key feature of many forms of altruistic behavior (Bourke, 2011a, 2011b). For instance, a combination of queen monogamy and relatedness is likely to be important in explaining an extreme form of reproductive division of labor known as eusociality. Eusociality occurs in bees, wasps, ants, termites, and naked mole rats. In these organisms, individuals form reproductive and non-reproductive castes. Though the workers do not reproduce themselves, they can gain indirect fitness benefits by caring for their siblings. These advantages may be especially strong in haplodiploid insects, where diploid females form from fertilized eggs, and haploid males form from unfertilized eggs. In species where the queen mates only once, workers will be more closely related to their sisters (relatedness $r = 0.75$) than to their own offspring ($r = 0.5$). This higher relatedness may strengthen selection favoring abstaining from self-reproduction in order to raise siblings (i.e., those produced by the queen). (See Chapter 10 for additional discussion of eusocial insect societies.)

If relatedness is a key principle governing the evolution of cooperation, how can we explain cooperation in societies where individuals are not highly related? For example, mutualism involves cooperation between members of different species, such as between legumes and the rhizobia that colonize them. The rhizobia provide the plants with fixed nitrogen, and the plant provides the bacteria with carbon. Similarly, mitochondria and nuclear genomes cooperate to carry out cellular functions. Unlike intraspecific cooperation, where an individual can gain an indirect fitness benefit from a costly behavior because it provides a benefit to relatives, here members of different species do not share a common gene pool and thus are unrelated. (See Chapter 9 for additional discussion of interspecies mutualism.)

Queller (2000) distinguishes between these different types of cooperation in categorizing the major evolutionary transitions, events that brought about large increases in biological complexity. He describes "fraternal" transitions as those that involve reproductive division of labor (i.e., lower units giving up their opportunities for fitness by promoting the survival and reproduction of others); these transitions evolve by kin selection. In contrast, "egalitarian" transitions involve mergers between units that retain their independent functions and ability to replicate, and these units can therefore be unrelated. Examples of egalitarian transitions include cooperation between the nuclear and mitochondrial genomes within cells and many cases of mutualism.

How can fraternal transitions evolve? In the case of interspecies mutualism, the explanations are complicated and multifactorial, but one important aspect is partner fidelity, which aligns fitness interests, so that what is good for one partner is also good for the other (Leigh, 2010; Hillesland, 2018). Fitness alignment can be accomplished through vertical transmission, which leads to co-inheritance, creating strong positive correlations between fitness of one member and fitness of the other. For example, while nearly all multicellular organisms harbor bacterial infections of some kind, only in some cases has the host become reliant on the bacteria, and vice versa, resulting in a mutualism. Among the best-known examples are the endosymbiotic bacteria that infect aphids (Douglas, 1998). These bacteria, called *Buchnera*, are housed in specialized host cells called bacteriocytes. Despite having lost most of its genes, the *Buchnera* have retained or acquired genes that are necessary to produce key amino acids required by the aphid host. In exchange, the aphid provides the *Buchnera* with energy-rich nutrients, such as carbohydrates. Endosymbiotic bacteria like *Buchnera* that can no longer exist outside their hosts probably evolve such extreme dependence and mutualism in part through vertical transmission. Co-inheritance means that successful production of insect offspring is critical to *Buchnera* as well.

How might societies form where individuals are both unrelated (making kin selection ineffective) and where fitness is also not aligned? Robert Trivers formulated one of the major hypotheses for altruism that applies to this scenario: if individuals repeatedly interact, then reciprocal altruism can evolve (Trivers, 1971). Under reciprocal altruism, individuals cooperate with others who cooperate, and they reject or defect from cooperation when others do the same. Thus, individuals can undertake costly behaviors with the confidence that they will be the beneficiaries in the future.

One aspect in common to different mechanisms of cooperation is that they all involve some form of assortment, such that cooperators disproportionately benefit other cooperators (Fletcher and Doebeli, 2009; Rankin and Taborsky, 2009). If kin selection operates, then an individual harboring a gene for cooperative behavior helps its kin, and kinship means that the relative is likely to be a cooperator as well. Under reciprocal altruism, cooperators also help other cooperators. Assortment can be produced passively, if cooperators do not disperse far (Taylor 1992; Wilson et al., 1992; Lehmann et al., 2008; Cornwallis et al., 2009), or actively through recognition and preferential interactions with relatives (Buss, 1982; Bekoff, 1992; Hepper, 2005).

Defection from Cooperation

One of the biggest problems with cooperation is that, by definition, its benefits are contingent on participation of others in the group, and participation is typically costly.

Costs can be offset in part by the benefits of cooperation for the group, but selection can still favor individuals that behave selfishly by shirking costly behaviors, especially if they can still access the group-level benefits. These actors can be referred to as "defectors" or "cheaters" (Ghoul et al., 2014). In addition to cheaters and defectors, we can define "loners" as individuals who do not cooperate, but also forgo the benefits (Tarnita et al., 2015). Cheating is a type of selfishness, the latter of which encompasses behaviors that provide a benefit to the individual but bear a cost to the group. Individuals can also behave spitefully toward others (in other words, cause harm to both themselves and others), and spite can be selected if it provides relatives with a fitness benefit (Foster et al., 2000, 2001; Gardner and West, 2004; Pizzari and Foster, 2008).

Do defectors and cheaters arise? And if so, what might prevent their invasion and success? Previous work has demonstrated the existence of defectors or cheaters across wide-ranging systems. In some haplodiploid eusocial insects, for example, workers can behave selfishly by developing ovaries and laying their own unfertilized eggs, which become males (Ratnieks, 1988; Foster and Ratnieks, 2000; Ratnieks and Wenseleers, 2005). Similarly, in the stingless bees *Melipona*, individuals have an unusual degree of control over their own caste fate and, consistent with an opportunity to behave selfishly, as many as 20% of individuals adopt the queen fate – in turn, necessitating a large cull of the excess queens (Bourke and Ratnieks, 1999; Wenseleers et al., 2004; Ratnieks and Wenseleers, 2005). Some leaf-cutting ants have lineages that are "royal cheats," which show a developmental bias toward adopting the queen fate (Hughes and Boomsma, 2008). And some eusocial insects have parasitic lineages where workers can reproduce through parthenogenesis, which results in the production of female offspring (Jordan et al., 2008; Dobata et al., 2011).

Selfish behaviors also arise among genes within cells. While meiosis is typically "fair," meaning each of the two gene copies in a diploid will be present in half of all gametes, selfish variants can arise that gain a disproportionate representation (Sandler and Novitski, 1957; Hurst, 1998; Burt and Trivers, 2009; Werren, 2011; Bravo Núñez et al., 2018). "Meiotic drive" is an umbrella term that encompasses a variety of mechanisms by which a fair meiotic process is disrupted by a genetic variant, resulting in its preferential transmission to the next generation (Lindholm et al., 2016). One way involves exploitation of the asymmetric cell divisions that result in egg production, such that one allele is more likely to end up in the daughter cell that c goes on to become the egg instead of the ones that degenerate into polar bodies. One type of meiotic drive thus involves alleles that disproportionately end up in the egg as opposed to the polar bodies. Other routes to transmission bias occur *after* meiosis, typically when the gene products of one allele have trans-acting effects that enable them to kill gametes that lack them. This phenomenon can result in nearly 100% of the gametes carrying a drive allele, compared to the expected 50% when meiosis is fair.

One of the best-known examples of a selfish gene is the *t*-haplotype in mice, which is actually a complex of several linked genes and effectors. Following meiosis in heterozygous males (those that carry one copy of the standard allele and one copy of the selfish *t*-haplotype), sperm that fail to inherit the *t*-allele exhibit impaired mobility, which prevents them from fertilizing the egg. How can a gene have these effects on cells where it is not present? The *t*-haplotype consists of several tightly linked genes that sit next to each other on the chromosome and are co-inherited as a unit, one encoding a long-lasting poison and the other encoding a short-lived antidote (Bravo Núñez et al., 2018). Sperm that receive the *t*-haplotype will have their mobility partially rescued through the continued production of the antidote, whereas those that do not retain the *t*-haplotype also lose the gene for the antidote, and thus become susceptible to the effects of the long-lasting toxin. Thus, the *t*-haplotype is a selfish gene: it torpedoes the success of gametes that lack it. In doing so, however, it gains a competitive advantage over the other allele, displacing it in the population. As a consequence, *+/t* males transmit the *t* haplotype to >99% of offspring.

Should these traits be classified as selfish? Or should they simply be considered competitive? This issue arises in many discussions of what it means to "behave selfishly" and whether the term should be applied to behaviors that lack any conscious decision-making. Although the trait is competitive at the allele level (the alleles are competing against one another), competition at this level has negative impacts on the higher level of organization – the organism for whom half of its gametes are now hobbled by a toxin. Thus, competition between alleles can disrupt the performance of the organism that harbors them, leading to a conflict between lower and higher levels of organization. From this perspective, we can see that a gene that behaves selfishly within an organism is no different from an individual behaving selfishly within a society: what is good for competition at lower levels can nevertheless disrupt benefits conferred at a higher level of organization. (See Chapter 9, which discusses a toxin–antidote system in the microorganism *Wolbachia*.)

In addition to *t*-haplotypes, there are many other documented examples of selfish genes; these include the segregation distorter (*SD*) in *Drosophila* and cytoplasmic male sterility (CMS) in plants (Hurst et al., 1996; Touzet and Budar, 2004; Larracuente and Presgraves, 2012; Brand et al., 2015; Lindholm et al., 2016). CMS arises in plants that are hermaphroditic, i.e., those that produce both male and female gametes on each flower, and is fairly common (Chase, 2007). It is caused by maternally inherited variants, typically encoded by mitochondrial genes, that sterilize the male reproductive portions of flowers – an act that can enhance the transmission of the mitochondria through the seed, but can be detrimental to the nuclear genes, which are transmitted through both seed and pollen. Finally, transposable elements, which comprise upwards of 50–90% of the genome in humans and maize, respectively, are selfish insofar as they gain a replication advantage within the germline of hosts, yet their transposition into functional genes can be deleterious, leading to a cost in lineages that carry them (Werren, 2011). Together, these examples demonstrate the existence of conflict within and between genomes.

In addition to gene-level selfishness within cells, selfish behaviors arise among cells within multicellular organisms and between multicellular organisms and the societies they form (Buss, 1983, 1987; Frank, 2003). On one hand, many multicellular organisms are founded from a single cell that divides to form the other cells of the body. This process results in cells that are near clonal, and this high relatedness is thought to discourage intra-organismal conflicts and promote cell–cell cooperation (Grosberg and Strathmann, 1998). A major exception is the gametes, as the chromosomes from the individual's two parents have segregated into different daughter cells, which now differ genetically from one another. We can predict from kin selection theory that variation in relatedness among gamete cells might promote cooperative and competitive behaviors – and, as we saw in the previous section, we see selfish behaviors emerging at this stage (Pizzari and Foster, 2008; Ostrowski and Shaulsky, 2009).

In addition, there are some organisms where cells aggregate to become multicellular, rather than undergoing repeated mitotic divisions from a single starting cell, or else distinct individuals can fuse (De Tomaso et al., 2005; Litman, 2006; Lakkis et al., 2008; Rosengarten and Nicotra, 2011; Zhao et al., 2015; Araujo Casares and Faugeron, 2016; Chang et al., 2018). In these cases, there is the potential for genetic chimerism, where different cell lineages coexist within a single organism. Here again, we can predict that selection might favor selfish behaviors, and studies of organisms that form chimeras suggest that this intuition is correct. For example, in the social amoeba *Dictyostelium discoideum* (sometimes referred to as a cellular slime mold), single-celled amoebae aggregate when starved and differentiate to form a multicellular organism. The cells initially form a migratory slug, which is attracted to light and heat. Eventually the slug transforms into a multicellular fruiting body, where ~20% of the cells vacuolize and die, having assembled themselves into a tall, rigid stalk. The remainder of the cells form hardy spores and sit on top of the stalk, where they can be dispersed by passing animals to new environments. Stalk formation is thought to be altruistic, in that stalk cells die to benefit others in their group. Adoption of spore versus stalk cell fate is also analogous to the reproductive division of labor seen in other multicellular organisms (i.e., germline and soma) and in eusocial insects (queens and workers; Strassmann et al., 2000; see also Chapter 10).

Consistent with the other examples, however, this system of cooperative fruiting body formation is also susceptible to selfish behaviors. In *Dictyostelium*, some lineages contribute fewer cells to the stalk than others. This behavior constitutes cheating, insofar as these lineages obtain the benefits of sitting on top of a stalk without paying their fair share for its production. Indeed, previous work has shown that when two or more natural isolates cooperate to build a fruiting body, they do not necessarily contribute equally to the production of the stalk (Strassmann et al., 2000; Buttery et al., 2009). Genetics screens, where pools of mutagenized cells, each containing a unique insertion mutation, pass through multiple rounds of selection, can also enrich for strains that preferentially form spores and avoid forming stalk. One such screen identified more than 150 sites (both intra- and intergenic) that, when disrupted, caused the mutant to cheat the wild-type (Santorelli et al., 2008). The results of this screen indicate that there are many genomic sites that can be mutated to produce cheating behaviors and that this organism may lack intrinsic safeguards to prevent cheating from occurring. Combined with the observation that cheating strains occur in nature, it suggests that the potential for these behaviors to disrupt the society is strong.

Selfish cell lineages have been found in other organisms where genetic chimerism occurs, such as in some species of marine invertebrates. In the marine tunicate *Botryllus schlosseri*, for example, neighboring sessile individuals can fuse. Injection of cells from one individual into the vasculature of another also shows that some lineages act as germ cell parasites (Stoner and Weissman, 1996); that is, they disproportionately gain access to the germline and thereby achieve a fitness benefit in terms of reproduction. As in *Dictyostelium*, these lineages gain the advantages of the somatic tissues while disproportionately propagating their own alleles into the next generation. The existence of selfish cell lineages in organisms that can form chimeras through either fusion (as in *Botryllus*) or by aggregation (as in *Dictyostelium*) supports the idea that selfish cell lineages are a realistic threat to the integrity of multicellular organisms (Buss, 1982; Lakkis et al., 2008).

How Is Fairness Ensured?

Given the threats of defecting and cheating, how can cooperative societies be maintained? Are they maintained? There are several possibilities, which are not mutually exclusive. One important possibility is that societies occasionally or even frequently succumb to cheating – and thus we might see a continual cycle where societies form, cheating arises, and the society collapses. An alternative possibility is that cheating is problematic, yet tolerable at low levels. A third possibility, which is not mutually exclusive, is that societies evolve mechanisms to guard against cheating and to reduce its impacts.

Unfortunately, the outcomes of these different scenarios are only evident over evolutionary timescales, and so it can be difficult to observe firsthand these dynamics in the brief span of time over which most research takes place. Nevertheless, laboratory microcosm experiments can provide some proof-of-principle. For example, experiments with the bacterium *Pseudomonas fluorescens* have demonstrated the rise and fall of cooperative societies (Rainey and Rainey, 2003). In these experiments, static cultures of bacteria reproducibly give rise to mat-forming variants, which overproduce a sticky polymer that enables them to colonize the liquid–air interface and gain better access to oxygen (Rainey and Travisano, 1998). However, production of the polymer is energetically costly, and non-producers quickly invade. They spread rapidly and cause the mat to sink, leading to extinction of the population (Rainey and Rainey, 2003). The formation and destruction of the mat provides a compelling example of how cheats can invade and drive the extinction of cooperative societies.

Alternatively, we know several ways in which selfishness can be repressed (Frank, 2003). One way is through restricting societies to relatives. Note that limiting societies to related individuals does not by itself prevent cheats from arising. However, when a cheat cheats its relative, it essentially cheats itself – and under high relatedness, selection should thus be effective at weeding out cheating behaviors when they are detrimental to others in the group. Consistent with the hypothesized importance of restricting societies to relatives, many organisms that undergo fusion or aggregation (e.g., *Botryllus* and *Dictyostelium*, as discussed above) also have genetically encoded mechanisms of kin recognition, which allow fusion or chimerism to persist only when individuals are related (De Tomaso et al., 2005; Ostrowski et al., 2008; Hirose et al., 2011; Nicotra, 2019).

Cheating can also be mitigated is through policing. For example, workers that develop ovaries can be attacked and killed by other workers (Ratnieks, 1988; Foster and Ratnieks, 2000; Ratnieks and Wenseleers, 2005). In plant–rhizobial mutualisms, hosts can sanction nodules (which house the rhiozobia) that do not fix nitrogen by withholding carbon (Kiers et al., 2003). In pig-tailed macaques, some group members intervene to curtail or prevent aggressive encounters between other individuals. However, policing is rare in other types of macaques. As noted by Flack and colleagues (2005), the circumstances that promote or prevent successful policing are likely to be complex, in part because they involve learning, but also because they typically involve intervening in an aggressive encounter between two other individuals – an act that may be costly and dangerous. Indeed, the studies of the pig-tailed macaques indicate that there are social correlates of these bystander interventions (Flack et al., 2005). For example, powerful individuals are more likely to police, perhaps because those with more power are less likely to experience retaliation. Several other chapters in this book discuss the role of punishment and sanctions in enforcing human cooperative behavior: see Chapters 1–3.

An imposed inability to defect from cooperation is sometimes referred to as coercion. Similar to cooperation and cheating, behaviors that potentially conform to this definition are found across different levels of biological organization. Although the distinctions in terminology are not necessarily clear-cut, I argue we should think of policing as a form of punishment that occurs after an act has been initiated or committed, whereas coercion involves a system in place that prevents expression of these traits in the first place. (See also Box 9.1 in Chapter 9.)

What are examples of coercion at the cell and molecular levels? One possible example of coercion is the uniparental inheritance of organelles, which happens in many species. One hypothesis is that uniparental inheritance evolved to ensure high relatedness of organelles to each other within a cell, as well as across cells of a single multicellular organism (Birky, 1995; Harrison et al., 2014; Greiner et al., 2015). Of course, in the case of cytoplasmic male sterility, *de novo* chromosomal rearrangements in the mitochondria continuously give rise to novel selfish variants, but uniparental inheritance might nonetheless help to reduce the frequency and success of selfish variants by reducing genetic diversity in the organelle population. Analogous hypotheses have been proposed for uniparental transmission of other types of endosymbionts, where bottlenecks alongside vertical transmission ensure genetic homogeneity and reduce conflict (Frank, 1996). In this sense, conflict reduction is being imposed upon the endosymbiont by the transmission mechanisms of its host. In some cases, the ability for individuals to act as independent Darwinian entities may have been suppressed by taking away their DNA altogether; for example, butterflies produce a class of sperm that lacks DNA (Pizzari and Foster, 2008), and there are examples of organelles that have completely lost their own genomes through a slow process of gene transfer to nuclear genomes (León-Avila and Tovar, 2004; Hjort et al., 2010).

In the *Pseudomonas* example above, cheaters increased in frequency quickly enough that they caused the population to crash. However, if invasion is not too fast, then cooperators may be able to evolve counter-adaptations to resist or suppress cheating. For example, in *Dictyostelium*, the presence of a cheater selects for resistant strains (Khare et al., 2009), and another study has showed that repression can arise rapidly enough to prevent cheaters from taking over (Levin et al., 2015). Manhes and Velicer (2011) have shown a similar result for the cooperative bacterium *Myxococcus xanthus*, where the presence of a cheater selected for a strain that can "police" – that is, it prevented the cheater from cheating both itself and other strains. Nevertheless, while these studies show that counter-evolution of resistance is a plausible outcome of social conflict, many of these studies are proof-in-principle using laboratory microbes, and more studies of repression in natural systems are needed.

Recognizing, Unmasking, and Studying the History of Conflicts in the Biological World

One potential consequence of the suppression of selfish behaviors is the establishment of an arms race. In an arms race, selfish behaviors repeatedly arise and select for counter-adaptations to suppress them, which in turn selects for new ways to overcome this resistance – and so on, potentially without end (Hurst et al., 1996; Werren, 2011; Dobata, 2012; Ågren, 2013; Ostrowski et al., 2015; Queller and Strassmann, 2018; Geist et al., 2019).

An important consequence of successful suppression is that any outward evidence of the conflict may disappear. For this reason, conflict may be frequently cryptic, and we potentially underestimate its frequency and importance (Sandler and Novitski, 1957; Lindholm et al., 2016). How might we unearth evidence of cryptic conflict? Two major approaches are being used. First, where populations evolve different sets of selfish elements and suppressors, crosses and backcrosses can separate selfish elements from their suppressors by recombination, revealing their latent existence. Second, where genes that mediate conflicts can be identified, analyses of their sequence evolution can be a powerful approach to identifying conflict and understanding its impact on the evolutionary trajectory of a species (Ostrowski et al., 2015; Ghoul et al., 2017; Mank, 2017).

A cryptic selfish gene was recently uncovered in the well-known model organism, the nematode *Caenorhabditis elegans* (Ben-David et al., 2017). The gene, called *pha-1*, had been studied for many years for reasons unrelated to selfishness: when the gene is deleted, the mutant fails to form a pharynx and dies during embryonic development. For this reason, the gene was originally characterized as a master-regulator of pharynx development. However, its true role was revealed only years later, by accident, when researchers crossed individuals from two distinct populations: the standard lab strain N2 was crossed with a highly diverged strain of *C. elegans* from Hawaii called DL238. In the F2, 25% of individuals died at the embryonic stage, a result that suggests a genetic incompatibility between the N2 and DL238 lines. Eyal Ben-David and his colleagues hypothesized that this type of pattern could occur if mothers were transmitting a long-lived toxin molecule in the cytoplasm of their eggs, and a gene encoding an antidote was then required to detoxify the toxin. In investigating this potential explanation, two genes were uncovered that appear to act as the antidote and toxin, respectively – which turned out to be *pha-1* and a previously unknown gene called *sup-35*. Because *pha-1 and sup-35* are both fixed within populations of N2, all offspring receive the toxin molecule, but also inherit the antitoxin gene. Thus, the effects of the toxin are always suppressed, and there is no outward evidence of the underlying toxin–antitoxin interaction. Similarly, the DL328 population lacks the antidote *pha-1*, but it does not matter, because it lacks the toxin gene, too. Thus, the entire system only becomes apparent from several rounds of crossing N2 and DL328 individuals, at which point, some offspring have received the toxin but not the antidote, leading to death. This *pha-1/sup-35* example demonstrates clearly how populations might fix different selfish genes and their suppressors and how the act of suppression can obscure the conflict. As Ben-David and colleagues (2017) state, "Selfish elements conferring genetic incompatibilities may be more common than previously thought, and some of them may be hiding in plain sight."

The *pha-1* gene is not the only example of a gene that acts this way. Similar selfish genes, as well as their suppressors, have been identified in a variety of other organisms and are sometimes explored through the unmasking that occurs when divergent populations are crossed (Tao et al., 2001). For example, Phadnis and Orr (2009) showed that individuals of two subspecies of the fruit fly *Drosophila pseudoobscura* produced hybrid males that were mostly sterile. However, the hybrid males were later found to be weakly fertile when aged – and surprisingly, the offspring of these hybrid males were all female, indicating a drive gene on the X chromosome. Indeed, the authors ultimately found that the gene causing the hybrid sterility is in fact a cryptic segregation distorter. This finding not only underscores the power of crossing diverged populations to reveal latent selfish genes but also provides experimental evidence to support a long-standing yet controversial hypothesis that the dynamics of selfish genes, followed by their suppression, can drive speciation – referred to as "conflict speciation." Thus selfish genes may not only be critical to the dynamics of social evolution, but they may also be a driver of evolutionary diversification and speciation more generally (Sandler and Novitski, 1957; Frank, 1991; Werren, 2011; Crespi and Nosil, 2013; Ågren and Clark, 2018; Patten, 2018).

Sequence Evolution Provides a Glimpse into the History of Conflict

Interpopulation crosses can reveal the existence of cryptic selfish genes, but this approach is limited to model systems, where controlled crosses are feasible. Nevertheless, recent advances in genetics and genomics are offering new opportunities to dissect the genetic bases of social behaviors in non-model organisms (Sucgang et al., 2011; Ghoul et al., 2017; Mank, 2017). These technologies have provided a new tool in the arsenal to identify conflict – that is, by reconstructing the evolutionary history of these genes, and especially by comparing their history to those of other genes in the genome, it is possible to identify the footprints of natural selection and to unearth the dynamics of conflict that would otherwise remain hidden from sight.

Genetics and Genomics of Social Interactions

Several "omics" tools are available to researchers of social behavior. First, transcriptomic analysis allows researchers to identify all of the genes that are expressed in a given individual or tissue and is a powerful way to trace behavioral traits back to their genetic bases (Robinson et al., 2005). For example, genes that are preferentially expressed in queens versus workers can provide insights into how caste fate is determined in social insects (Morandin et al., 2016; Okada et al., 2017; Toth, 2017) and genetic screens can identify candidate genes that impact phenotypes of interest. In addition, once candidate genes that influence a given social behavior have been identified, analyses of sequence variation can be used to investigate the history of the genes and traits. These analyses allow researchers to infer, for example, when particular genetic variants arose in populations and the speed with which they displaced other genetic variants, with unusually rapid displacement being a strong indicator of natural selection.

One example of the application of genomics to social conflict involves social organization in *Solenopsis invicta*, a fire ant that originated in South America but invaded the southern United States (Keller and Ross, 1998; Ross and Keller, 1998). Researchers knew that different colonies had different social behaviors: one type forms only single queen colonies (called monogyny), whereas the other forms multiple queen colonies (called polygyny). The genetic basis of this divergent social organization was associated with the presence or absence of a genetic variant for *Gp-9* (for "general protein 9"). Queens in multi-queen colonies are always heterozygous at this locus (that is, *Bb* genotype), whereas queens in the single-queen colonies are all homozygous dominant (*BB*). Because *bb* individuals cannot survive, colonies are always either all *BB* or a mix of *Bb* and *BB*, which are then monogynous and polygynous, respectively. When *BB* queens were introduced into colonies, they were attacked and killed by workers, whereas *Bb* queens were tolerated. In addition, *BB* queens were attacked primarily by *Bb* and not *BB* workers (Keller and Ross, 1998; Ross and Keller, 1998). Based on these findings, the *Gp-9* locus is thought to be a rare example of a "greenbeard" gene. A greenbeard refers to a hypothetical scenario envisioned by Richard Dawkins in *The Selfish Gene*, which he used to illustrate his point that altruism is a form of genetic selfishness. Dawkins states that an ideal gene would detect copies of itself in others and behave altruistically only towards those that harbor it (Dawkins, 2006). He suggested this process might work if the gene could produce some outward sign of its presence – for instance, if it caused its bearers to have a green beard. The idea that a single gene could confer the ability to recognize and preferentially cooperate with individuals that carry it seemed preposterous at first glance. The *Gp-9* locus, where *b* alleles enable their bearers to recognize whether others have it or do not, and then behave cooperatively only with those that do, clearly fits the criteria, demonstrating that such a gene might exist.

Further "omics" analyses of *Gp-9* have helped to explain how a single gene can effect such complex behaviors. Researchers have since shown that individuals that carry the different alleles not only differ at that one gene but across a long stretch of the chromosome. In fact, individuals with *B* or *b* alleles harbor alternative genetic variants over a large non-recombining region of the chromosome, spanning 13 million basepairs of DNA, equal to 55% of the chromosome (Wang et al., 2013). Large stretches of chromosomes that lack recombination typically indicate one thing: a chromosomal inversion. In essence, the inversion ensures that the entire region, comprising 616 genes, is inherited as a single unit: a supergene. Thus, the answer to how a single gene could cause such diverse and sophisticated effects is that it cannot: it required a molecular event that created fitness alignment among a collection of genes, ensuring their co-inheritance and shared evolutionary fate. (Notably, the *pha-1* and *sup-35* are also carried on a chromosomal inversion, which may have similarly ensured that the two genes were not separated by recombination.)

A second example where sequence analysis has shed light on the existence of social conflict involves the loss of cooperative behavior in the bacterium *Pseudomonas aeruginosa*, an opportunistic pathogen that colonizes the lungs of cystic fibrosis patients (Andersen et al., 2015). These bacteria cooperatively scavenge iron by producing and secreting a molecule called pyoverdine, which binds extracellular iron and enables its uptake via a receptor. Pyoverdine is a public good: it is costly to make and available to all – and consistent with a public goods dilemma, strains that have lost pyoverdine production are moderately common (~12% of isolates). However, despite these findings, it remained unclear why pyoverdine loss frequently arose within patients: does its loss reflect the benefits of social cheating, or does the host environment somehow make iron-scavenging unnecessary? Andersen et al. (2015) took advantage of an existing strain collection, collected longitudinally from cystic fibrosis patients over periods ranging from years to decades, and sequenced several genes required for the synthesis of pyoverdine or its receptor. They showed that producers and non-producers frequently co-occur within patients. Second, when pyoverdine-producing strains are present, the receptor gene function is then maintained in isolates that have lost production, whereas inactivating mutations disproportionately accumulate if pyoverdine production has been lost. Put another way, the receptor is retained so long as pyoverdine is being produced by others – a key result that suggests that cheating provides the selective impetus for retention of the receptor. Here, knowledge of the genetic basis of the traits, as well as the ability to sequence the receptor and other genes in hundreds of isolates has been crucial in demonstrating a role for social conflict in mediating the within-host dynamics of this important bacterial pathogen.

DNA Sequence Analyses Uncover "Epidemics" of Selfish Genes

To illustrate how sequence analysis can uncover the historical dynamics of selfish genes that are hidden from view, we must consider the potential fates of selfish genes in populations. As described by Ostrowski et al. (2015), we can envision several different outcomes (see also Ingvarsson and Taylor, 2002; Van Dyken and Wade, 2012; Ghoul et al., 2017). First, alleles favoring selfishness may have an evolutionary advantage that allows them to invade and take over populations relatively unimpeded – and upon fixing, their impacts are no longer noticeable. An alternative possibility is that selfish behaviors are beneficial when rare, but as they increase in frequency, the victims correspondingly decline and the selective advantage is eroded. The expected result is a balanced polymorphism, where both selfish and cooperative behaviors can be maintained in populations, neither one able to fully displace the other. Finally, a third possibility – a non-adaptive scenario – is that these variants arise repeatedly in populations by mutation but never gain a foothold, because selection continually removes these alleles. Under this latter scenario, we should expect to find selfish behaviors in natural populations, but they should be at consistently low frequency and have little impact on the evolutionary trajectory of the species.

How can one distinguish among these different scenarios if they unfold over evolutionary timescales, far longer than the life span of one or even many researchers? The key is recognizing that these different dynamics leave distinctive footprints in the underlying gene sequences (Nielsen, 2005). The first scenario of an evolutionary takeover (referred to as a "selective sweep") perturbs patterns of genetic variation in recognizable ways: when a variant takes over a population quickly, its rise can wipe away preexisting variation at that locus as well as other loci in the vicinity. These selective sweeps can be detected through analysis of sequence variation at the focal locus and closely linked loci by comparing these sites to other sites in the genome. Selective sweeps also drive elevated divergence between species, which can be detected by sequencing a closely related species and comparing its genome to that of the focal organism. The second scenario, a balanced polymorphism, results in near opposite signatures as the first: long-term maintenance of alternative alleles, which allows sequence variation to accumulate, leads to elevations in polymorphism (Charlesworth, 2006). Finally, if mutations conferring selfish behaviors are pushed out of populations by natural selection, then we may see patterns of purifying selection in the underlying genes. These patterns include unusually low levels of sequence polymorphism within species and strong sequence conservation between species.

Ingvarsson and Taylor (2002) applied this general approach to elucidate the historical dynamics of selfish genes involved in cytoplasmic male sterility in the weedy plant *Silene vulgaris* (see above). Recall that cytoplasmic male sterility is caused when selfish mitochondrial variants destroy the male reproductive structures, providing the mitochondria with a transmission advantage. Ingvarsson and Taylor (2002) compared the level of polymorphism in maternally transmitted chloroplast genes to that of nuclear genes, which are transmitted both maternally and paternally (that is, through the ovule and pollen). The chloroplast genes, which should show similar patterns to the causal mitochondrial genes, showed unusually low levels of genetic variation and other patterns consistent with purifying selection. Collectively, these patterns point to what Ingvarsson and Taylor refer to as "epidemics" of selfish genes, where new variants repeatedly arise but are quickly extinguished by natural selection. Interestingly, in a related species, *Silene acaulis*, where cytoplasmic male sterility also occurs, there were multiple, divergent mitochondrial haplotypes that appeared to be maintained over long periods of time, implicating balancing selection in that species (Städler and Delph, 2002). Recently, a similar approach was used to examine the evolutionary dynamics at loci that mediate cheating behaviors in *Dictyostelium*, comparing these loci to other sites in the genome. This work similarly revealed excess polymorphism and haplotype structure, indicative of balancing selection (Ostrowski et al. 2015).

Where the molecular basis of selfish genes is known, the frequency of these genes throughout the species' geographic range can also be examined, allowing additional inferences about the history and dynamics of these genes in natural populations. One of the best-known examples of a selfish gene is the *SD* locus in *Drosophila* (Larracuente and Presgraves, 2012). Like the *t*-haplotype example, the *SD* locus is a complex of multiple, tightly linked genes that form a genetic coalition. The locus consists of the *Sd* gene (a truncated Ran-GTPAse activating protein) and the *Rsp* (Reponder) gene. The RSp^i allele confers insensitivity to the product of the *Sd* gene. Under a fair meiosis, heterozygous males should produce *SD* and *SD+* gametes in equal frequencies, but these males in fact produce only *SD* offspring. The molecular mechanism of the *SD* gene drive is interesting in its own right, but the important point here is that researchers have quantified how common the *SD* variant is worldwide (Brand et al., 2015). These analyses revealed that the selfish *SD* allele is ubiquitous – that is, it is present in most populations. However, its frequency is remarkably low, only ~1–5% within a given population, which is surprisingly low, given that the *SD* allele has a near absolute advantage over the wild-type *SD+* allele within heterozygous males. Sequence analysis of the *SD* chromosomal regions revealed that the *Sd* allele arose as a partial gene duplication in Africa and then spread worldwide, indicating a single origin of the selfish gene. However, distinct variants of *SD* chromosomes in different locations show unusually low sequence variability within populations and other signatures that point to their recent expansion, a finding that, likely reflects frequent turnover among *SD* chromosomes. As Brand and colleagues (2015) point out, any given selfish *SD* locus may be in an arms race with newly arisen suppressors, including insensitive variants of the responder locus. The upshot of this work is that selfish genes may be highly successful, in the sense of being ubiquitous, but they may also be limited in their ability to take over a population. Careful detective work, leveraging advances in genome sequencing, can reveal that apparent stability is an illusion, with ongoing, frequent epidemics of selfish genes.

Conclusions

Cooperation provides a foundation for the emergence of biological complexity. However, cooperation also opens the door to individuals who can take advantage of it, and evolutionary theory has been a superb guide to explaining why conflicts emerge. Recent studies of selfish behaviors at the molecular level have shown that counter-evolutionary changes to suppress and mitigate conflicts of interest can occur – and the take-home message of this work is that sociality may involve not only complex adaptations to promote communication and cooperation but also mechanisms to protect cooperation from being co-opted by selfishness. Finally, the counter-evolution of selfishness-suppression points to the possibility that conflict might frequently go unnoticed. For this reason, we might fail to appreciate the important role that conflict plays in shaping how cooperative societies work. As Strassmann and Queller state, cooperation and conflict are "the yin and yang of biological interactions" (Queller and Strassmann, 2018). The reference to yin and yang likely reflects a view of cooperation and conflict not so much as forces in opposition to one another, but as forces that are complementary, where each gives rise to the other in turn.

Finally, the ability to unearth evidence of conflict has also been propelled by recent innovations in the realm of high-throughput analyses of genotypes and phenotypes (Ghoul et al., 2017; Mank, 2017). By first identifying the genetic bases of cooperation- and conflict-related traits, analyses of molecular evolution can then reveal the historical and temporal dynamics of these behaviors over evolutionary timescales. For example, analyses of molecular evolution can point to whether selfish genes have a history of repeated turnover in populations, suggesting their rapid spread and fixation (see Didion et al., 2016, for example); whether they show patterns consistent with epidemics, which ignite repeatedly before being beaten back by suppression (e.g., Ingvarsson and Taylor, 2002); or whether selfish genes show patterns of endemicity, where they gain a foothold but rarely take over (Brand et al., 2015; Ostrowski et al., 2015). Analyses of molecular variation can also reveal repeated evolution of social traits via independent mutations (Andersen et al., 2015; Wielgoss et al., 2019), as well as identify which social traits are subject to the strongest selection (such as kin recognition loci – see Ostrowski, 2019), and thereby help to elucidate the pivotal role of natural selection in driving the evolution of sociality. While molecular approaches to the study of social conflict are still in their infancy, these studies nonetheless offer an unprecedented view of the dynamic nature of cooperation and conflict and their potential roles as drivers of biological diversity.

References

Ågren J. A. (2013) Selfish genes and plant speciation. *Evolutionary Biology*, 40(3): 439–449.

Ågren J. A., and Clark A. G. (2018) Selfish genetic elements. *PLoS Genetics*, 14(11): e1007700.

Andersen S. B., Marvig R. L., Molin S., Krogh Johansen H., and Griffin A. S. (2015) Long-term social dynamics drive loss of function in pathogenic bacteria. *Proceedings of the National Academy of Sciences USA*, 112(34): 10756–10761.

Araujo Casares F., and Faugeron S. (2016) Higher reproductive success for chimeras than solitary individuals in the kelp *Lessonia* spicata but no benefit for individual genotypes. *Evolutionary Ecology*, 30(5): 953–972.

Bekof M. (1992) Kin recognition and kin discrimination. *Trends in Ecology and Evolution*, 7(3): 100.

Ben-David E., Burga A., and Kruglyak L. (2017) A maternal-effect selfish genetic element in *Caenorhabditis elegans*. *Science*, 356(6342): 1051–1055.

Birky C. W. (1995) Uniparental inheritance of mitochondrial and chloroplast genes: mechanisms and evolution. *Proceedings of the National Academy of Sciences USA*, 92(25): 11331–11338.

Bourke A. F. G. (2011a) *Principles of Social Evolution*. Oxford: Oxford University Press.

Bourke A. F. G. (2011b) The validity and value of inclusive fitness theory. *Proceedings of the Royal Society B Biological Science*, 278(1723): 3313–3320.

Bourke A. F. G., and Ratnieks F. L. W. (1999) Kin conflict over caste determination in social Hymenoptera. *Behavioral Ecology and Sociobiology*, 46(5): 287–297.

Brand C. L., Larracuente A. M., and Presgraves D. C. (2015) Origin, evolution, and population genetics of the selfish *Segregation Distorter* gene duplication in European and African populations of *Drosophila melanogaster*. *Evolution*, 69(5): 1271–1283.

Bravo Núñez M. A., Nuckolls N. L., and Zanders S. E. (2018) Genetic villains: killer meiotic drivers. *Trends in Genetics*, 34(6): 424–433.

Burt A., and Trivers R. (2009) *Genes in Conflict: The Biology of Selfish Genetic Elements*. Cambridge MA: Harvard University Press.

Buss L. W. (1982) Somatic cell parasitism and the evolution of somatic tissue compatibility. *Proceedings of the National Academy of Sciences USA*, 79(17): 5337–5341.

Buss L. W. (1983) Evolution, development, and the units of selection. *Proceedings of the National Academy of Sciences USA*, 80(5): 1387–1391.

Buss L. W. (1987) *The Evolution of Individuality*. Princeton, NJ: Princeton University Press.

Buttery N. J., Rozen D. E., Wolf J. B., and Thompson C. R. L. (2009) Quantification of social behavior in *D. discoideum* reveals complex fixed and facultative strategies. *Current Biology*, 19(16): 1373–1377.

Chang E. S., Orive M. E., and Cartwright P. (2018) Nonclonal coloniality: Genetically chimeric colonies through fusion of sexually produced polyps in the hydrozoan *Ectopleura larynx*. *Evolution Letters*, 2(4): 442–455.

Chapman T. (2006) Evolutionary conflicts of interest between males and females. *Current Biology*, 16(17): R744–R754.

Charlesworth D. (2006) Balancing selection and its effects on sequences in nearby genome regions. *PLoS Genetics*, 2(4): e64.

Chase C. D. (2007) Cytoplasmic male sterility: A window to the world of plant mitochondrial–nuclear interactions. *Trends in Genetics*, 23(2): 81–90.

Cornwallis C. K., West S. A., and Griffin A. S. (2009) Routes to indirect fitness in cooperatively breeding vertebrates: Kin discrimination and limited dispersal. *Journal of Evolutionary Biology*, 22(12): 2445–2457.

Crespi B., and Nosil P. (2013) Conflictual speciation: Species formation via genomic conflict. *Trends in Ecology and Evolution*, 28(1): 48–57.

Darwin C. R. (1861) *On the Origin of Species by Means of Natural Selection, or the Preservation of Favoured Races in the Struggle for Life*, 3rd ed. London: John Murray.

Dawkins R. (2006) *The Selfish Gene: With a New Introduction by the Author*. Oxford: Oxford University Press.

De Tomaso A. W., Nyholm S. V., Palmeri K. J., Ishizuka K. J., Ludington W. B., Mitchel K., and Weissman, I. L. (2005) Isolation and characterization of a protochordate histocompatibility locus. *Nature*, 438(7067): 454–459.

Didion J. P., Morgan A. P., Yadgary L. et al. (2016) *R2d2* drives selfish sweeps in the house mouse. *Molecular Biology and Evolution*, 33(6): 1381–1395.

Dobata S. (2012) Arms race between selfishness and policing: Two-trait quantitative genetic model for caste fate conflict in eusocial Hymenoptera. *Evolution*, 66(12): 3754–3764.

Dobata S., Sasaki T., Mori H., and Hasegawa E. (2011) Persistence of the single lineage of transmissible "social cancer" in an asexual ant. *Molecular Ecology*, 20(3): 441–455.

Douglas A. E. (1998) Nutritional interactions in insect-microbial symbioses: Aphids and their symbiotic bacteria Buchnera. *Annual Review of Entomology*, 43: 17–37.

Flack J. C., de Waal F. B. M., and Krakauer D. C. (2005) Social structure, robustness, and policing cost in a cognitively sophisticated species. *The American Naturalist*, 165(5): E126–E139.

Fletcher J. A., and Doebeli M. (2009) A simple and general explanation for the evolution of altruism. *Proceedings of the Royal Society of London B Biological Science*, 276(1654): 13–19.

Foster K. R., and Ratnieks F. L. (2000) Facultative worker policing in a wasp. *Nature*, 407 (6805): 692–693.

Foster K. R., Ratnieks F. L. W., and Wenseleers T. (2000) Spite in social insects. *Trends in Ecology and Evolution*, 15(11): 469–470.

Foster K. R., Wenseleers T., and Ratnieks F. L. W. (2001) Spite: Hamilton's unproven theory. *Annales Zoologici Fennici*, 38(3/4): 229–238.

Frank S. A. (1991) Divergence of meiotic drive-suppression systems as an explanation for sex-biased hybrid sterility and inviability. *Evolution*, 45(2): 262–267.

Frank S. A. (1996) Host-symbiont conflict over the mixing of symbiotic lineages. *Proceedings of the Royal Society of London B Biological Science*, 263(1368): 339–344.

Frank S. A. (2003) Repression of competition and the evolution of cooperation. *Evolution*, 57 (4): 693–705.

Gardner A., and West S. A. (2004) Spite and the scale of competition. *Journal of Evolutionary Biology*, 17(6): 1195–1203.

Gardner A., West S. A., and Wild G. (2011) The genetical theory of kin selection. *Journal of Evolutionary Biology*, 24(5): 1020–1043.

Geist K. S., Strassmann J. E., and Queller D. C. (2019) Family quarrels in seeds and rapid adaptive evolution in *Arabidopsis*. *Proceedings of the National Academy of Sciences USA*, 116(19): 9463–9468.

Ghoul M., Andersen S. B., and West S. A. (2017) Sociomics: Using omic approaches to understand social evolution. *Trends in Genetics*, 33(6): 408–419.

Ghoul M., Griffin A. S., and West S. A. (2014) Toward an evolutionary definition of cheating. *Evolution*, 68(2): 318–331.

Greiner S., Sobanski J., and Bock R. (2015) Why are most organelle genomes transmitted maternally? *BioEssays*, 37(1): 80–94.

Grosberg R. K., and Strathmann R. R. (1998) One cell, two cell, red cell, blue cell: The persistence of a unicellular stage in multicellular life histories. *Trends in Ecology and Evolution*, 13(3): 112–116.

Haldane J. B. S. (1955) Population genetics. *New Biology*, 18: 34–51.

Hamilton W. D. (1964a) The genetical evolution of social behaviour. I. *Journal of Theoretical Biology*, 7(1): 1–16.

Hamilton W. D. (1964b) The genetical evolution of social behaviour. II. *Journal of Theoretical Biology*, 7(1): 17–52.

Harrison E., MacLean R. C., Koufopanou V., and Burt A. (2014) Sex drives intracellular conflict in yeast. *Journal of Evolutionary Biology*, 27(8): 1757–1763.

Hepper P. G. (2005) *Kin Recognition*. Cambridge: Cambridge University Press.

Hillesland K. L. (2018) Evolution on the bright side of life: Microorganisms and the evolution of mutualism. *Annals of the New York Academy of Sciences*, 1422(1): 88–103.

Hirose S., Benabentos R., Ho H.-I., Kuspa A., and Shaulsky G. (2011) Self-recognition in social amoebae is mediated by allelic pairs of *tiger* genes. *Science*, 333(6041): 467–470.

Hjort K., Goldberg A. V., Tsaousis A. D., Hirt R. P., and Embley T. M. (2010) Diversity and reductive evolution of mitochondria among microbial eukaryotes. *Philosophical Transactions of the Royal Society of London B Biological Sciences*, 365(1541): 713–727.

Hughes W. O. H., and Boomsma J. J. (2008) Genetic royal cheats in leaf-cutting ant societies. *Proceedings of the National Academy of Sciences USA*, 105(13): 5150–5153.

Hurst L. D. (1998) Selfish genes and meiotic drive. *Nature*, 391(6664): 223.

Hurst L. D., Atlan A., and Bengtsson B. O. (1996) Genetic conflicts. *The Quarterly Review of Biology*, 71(3): 317–364.

Ingvarsson P. K., and Taylor D. R. (2002) Genealogical evidence for epidemics of selfish genes. *Proceedings of the National Academy of Sciences USA*, 99(17): 11265–11269.

Jordan L. A., Allsopp M. H., Oldroyd B. P., Wossler T. C., and Beekman M. (2008) Cheating honeybee workers produce royal offspring. *Proceedings of the Royal Society B Biological Science*, 275(1632): 345–351.

Keller L., and Ross K. G. (1998) Selfish genes: A green beard in the red fire ant. *Nature*, 394 (6693): 573–575.

Khare A., Santorelli L. A., Strassmann J. E., Queller D. C., Kuspa A., and Shaulsky G. (2009) Cheater-resistance is not futile. *Nature*, 461(7266): 980–982.

Kiers E. T., Rousseau R. A., West S. A., and Denison R. F. (2003) Host sanctions and the legume-rhizobium mutualism. *Nature*, 425(6953): 78–81.

Lakkis F. G., Dellaporta S. L., and Buss L. W. (2008) Allorecognition and chimerism in an invertebrate model organism. *Organogenesis*, 4(4): 236–240.

Larracuente A. M., and Presgraves D. C. (2012) The selfish *Segregation Distorter* gene complex of *Drosophila melanogaster*. *Genetics*, 192(1): 33–53.

Lehmann L., Ravigné V., and Keller L. (2008) Population viscosity can promote the evolution of altruistic sterile helpers and eusociality. *Proceedings of the Royal Society B Biological Science*, 275(1645): 1887–1895.

Leigh E. G., Jr. (2010) The evolution of mutualism. *Journal of Evolutionary Biology*, 23(12): 2507–2528.

León-Avila G., and Tovar J. (2004) Mitosomes of *Entamoeba histolytica* are abundant mitochondrion-related remnant organelles that lack a detectable organellar genome. *Microbiology*, 150(Pt 5): 1245–1250.

Levin S. R., Brock D. A., Queller D. C., and Strassmann J. E. (2015) Concurrent coevolution of intra-organismal cheaters and resisters. *Journal of Evolutionary Biology*, 28(4): 756–765.

Lindholm A. K., Dyer K. A., Firman R. C. et al. (2016) The ecology and evolutionary dynamics of meiotic drive. *Trends in Ecology and Evolution*, 31(4): 315–326.

Litman G. W. (2006) How *Botryllus* chooses to fuse. *Immunity*, 25(1): 13–15.

Manhes P., and Velicer G. J. (2011) Experimental evolution of selfish policing in social bacteria. *Proceedings of the National Academy of Sciences USA*, 108(20): 8357–8362.

Mank J. E. (2017) Population genetics of sexual conflict in the genomic era. *Nature Reviews Genetics*, 18(12): 721–730.

Morandin C., Tin M. M. Y., Abril S. et al. (2016) Comparative transcriptomics reveals the conserved building blocks involved in parallel evolution of diverse phenotypic traits in ants. *Genome Biology*, 17: article number 43.

Nicotra M. L. (2019) Invertebrate allorecognition. *Current Biology*, 29(11): R463–R467.

Nielsen R. (2005) Molecular signatures of natural selection. *Annual Review of Genetics*, 39: 197–218.

Okada Y., Watanabe Y., Tin M. M. Y., Tsuji K., and Mikheyev A. S. (2017) Social dominance alters nutrition-related gene expression immediately: Transcriptomic evidence from a monomorphic queenless ant. *Molecular Ecology*, 26(11): 2922–2938.

Ostrowski E. A. (2019) Enforcing cooperation in the social amoebae. *Current Biology*, 29(11): R474–R484.

Ostrowski E. A., Katoh M., Shaulsky G., Queller D. C., and Strassmann J. E. (2008) Kin discrimination increases with genetic distance in a social amoeba. *PLoS Biology*, 6(11): e287.

Ostrowski E. A., and Shaulsky G. (2009) Learning to get along despite struggling to get by. *Genome Biology*, 10(5): 218.

Ostrowski E. A., Shen Y., Tian X. et al. (2015) Genomic signatures of cooperation and conflict in the social amoeba. *Current Biology*, 25(12): 1661–1665.

Patten M. M. (2018) Selfish X chromosomes and speciation. *Molecular Ecology*, 27: 3772–3782.

Phadnis N., and Orr H. A. (2009) A single gene causes both male sterility and segregation distortion in *Drosophila* hybrids. *Science*, 323(5912): 376–379.

Pizzari T., and Foster K. R. (2008) Sperm sociality: Cooperation, altruism, and spite. *PLoS Biology*, 6(5): e130.

Queller D. C. (2000) Relatedness and the fraternal major transitions. *Philosophical Transactions of the Royal Society of London B*, 355(1403): 1647–1655.

Queller D. C., and Strassmann J. E. (2018) Evolutionary conflict. *Annual Review of Ecology, Evolution, and Systematics*, 49(1): 73–93.

Rainey P. B., and Rainey K. (2003) Evolution of cooperation and conflict in experimental bacterial populations. *Nature*, 425(6953): 72–74.

Rainey P. B., and Travisano M. (1998) Adaptive radiation in a heterogeneous environment. *Nature*, 394(6688): 69–72.

Rankin D. J., and Taborsky M. (2009) Assortment and the evolution of generalized reciprocity. *Evolution*, 63(7): 1913–1922.

Ratnieks F. L. W. (1988) Reproductive harmony via mutual policing by workers in eusocial Hymenoptera. *The American Naturalist*, 132(2): 217–236.

Ratnieks F. L. W., Foster K. R., and Wenseleers T. (2006) Conflict resolution in insect societies. *Annual Review of Entomology*, 51: 581–608.

Ratnieks F. L. W., and Wenseleers T. (2005) Policing insect societies. *Science*, 307(5706): 54–56.

Robinson G. E., Grozinger C. M., and Whitfield C. W. (2005) Sociogenomics: Social life in molecular terms. *Nature Reviews Genetics*, 6(4): 257–270.

Rosengarten R. D., and Nicotra M. L. (2011) Model systems of invertebrate allorecognition. *Current Biology*, 21(2): R82–R92.

Ross K. G., and Keller L. (1998) Genetic control of social organization in an ant. *Proceedings of the National Academy of Sciences USA*, 95(24): 14232–14237.

Sachs J. L., Quides K. W., and Wendlandt C. E. (2018) Legumes versus rhizobia: A model for ongoing conflict in symbiosis. *The New Phytologist*, 219(4): 1199–1206.

Sandler L., and Novitski E. (1957) Meiotic drive as an evolutionary force. *The American Naturalist*, 91(857): 105–110.

Santorelli L. A., Thompson C. R. L., Villegas E. et al. (2008) Facultative cheater mutants reveal the genetic complexity of cooperation in social amoebae. *Nature*, 451(7182): 1107–1110.

Städler T., and Delph L. F. (2002) Ancient mitochondrial haplotypes and evidence for intragenic recombination in a gynodioecious plant. *Proceedings of the National Academy of Sciences USA*, 99(18): 11730–11735.

Stoner D. S., and Weissman I. L. (1996) Somatic and germ cell parasitism in a colonial ascidian: Possible role for a highly polymorphic allorecognition system. *Proceedings of the National Academy of Sciences USA*, 93(26): 15254–15259.

Strassmann J. E., Zhu Y., and Queller D. C. (2000) Altruism and social cheating in the social amoeba *Dictyostelium discoideum*. *Nature*, 408(6815): 965–967.

Sucgang R., Kuo A., Tian X. et al. (2011) Comparative genomics of the social amoebae *Dictyostelium discoideum* and *Dictyostelium purpureum*. *Genome Biology*, 12(2): R20.

Tao Y., Hartl D. L., and Laurie C. C. (2001) Sex-ratio segregation distortion associated with reproductive isolation in *Drosophila*. *Proceedings of the National Academy of Sciences USA*, 98(23): 13183–13188.

Tarnita C. E., Washburne A., Martinez-Garcia R., Sgro A. E., and Levin S. A. (2015) Fitness tradeoffs between spores and nonaggregating cells can explain the coexistence of diverse genotypes in cellular slime molds. *Proceedings of the National Academy of Sciences USA*, 112(9): 2776–2781.

Taylor P. D. (1992) Altruism in viscous populations – An inclusive fitness model. *Evolutionary Ecology*, 6(4): 352–356.

Toth A. L. (2017) To reproduce or work? Insect castes emerge from socially induced changes in nutrition-related genes. *Molecular Ecology*, 26(11): 2839–2841.

Touzet P., and Budar F. (2004) Unveiling the molecular arms race between two conflicting genomes in cytoplasmic male sterility? *Trends in Plant Science*, 9(12): 568–570.

Trivers R. L. (1971) The evolution of reciprocal altruism. *The Quarterly Review of Biology*, 46(1): 35–57.

Van Dyken J. D., and Wade M. J. (2012) Detecting the molecular signature of social conflict: Theory and a test with bacterial quorum sensing genes. *The American Naturalist*, 179(4): 436–450.

Wang J., Wurm Y., Nipitwattanaphon M., Riba-Grognuz O., Huang Y.-C., Shoemaker D., and Keller L. (2013) A Y-like social chromosome causes alternative colony organization in fire ants. *Nature*, 493(7434): 664–668.

Wenseleers T., Hart A. G., Ratnieks F. L. W., and Quezada-Euan J. J. G. (2004) Queen execution and caste conflict in the stingless bee *Melipona beecheii*. *Ethology*, 110(9): 725–736.

Werren J. H. (2011) Selfish genetic elements, genetic conflict, and evolutionary innovation. *Proceedings of the National Academy of Sciences USA*, 108 (Suppl. 2): 10863–10870.

Wielgoss S., Wolfensberger R., Sun L., Fiegna F., and Velicer G. J. (2019) Social genes are selection hotspots in kin groups of a soil microbe. *Science*, 363(6433): 1342–1345.

Wilson D. S., Pollock G. B., and Dugatkin L. A. (1992) Can altruism evolve in purely viscous populations? *Evolutionary Ecology*, 6(4): 331–341.

Zhao J., Gladieux P., Hutchison E., Bueche J., Hall C., Perraudeau F., and Glass N. L. (2015) Identification of allorecognition loci in *Neurospora crassa* by genomics and evolutionary approaches. *Molecular Biology and Evolution*, 32(9): 2417–2432.

Part II

Neural Mechanisms

5 Social Living and Rethinking the Concept of "Prosociality"

Heather K. Caldwell and H. Elliott Albers

Introduction

A quick PubMed search reveals that use of the term "prosocial" has dramatically risen in the scientific literature over the last twelve years. With only 67 occurrences in 2006 and upward of 593 in 2018, this represents an almost ninefold increase in the use of the term. But what is meant by the term "prosocial"? The *Oxford English Dictionary* (2018) defines "prosocial" as follows:

Of, relating to, or designating something, esp. behavior, which is positive, helpful, and intended to promote social acceptance and friendship; (*Social Psychology*) relating to or designating behavior which adheres, sometimes in a rigid or conventional manner, to the moral standards accepted by the established social group (contrasted with *asocial* or *antisocial* behaviors or responses).

The study of prosocial behavior has been an active area of research in social psychology that dates back to the beginnings of the last century. (For a review see Penner et al., 2005,) This large body of literature includes a diverse range of phenomena centering around the origins and tendencies of humans helping other humans, including traits such as empathy. In psychology the term "prosocial behavior" is typically used to indicate a behavior that provides benefit to another person. However, this same term, and all that it implies, has been increasingly applied to nonhuman vertebrate animal behavior and the neural mechanisms regulating these behaviors. It is within this latter context that the term prosocial has been used rather loosely with no clear definitions provided.

The term prosocial behavior was first used in behavioral neuroscience as a descriptor for parental, reproductive, and attachment behaviors (Insel, 1992). More recently, it has been used to describe a wide range of behaviors in many different species such as biparental, socially monogamous, or cooperative behaviors (i.e. interactions that are collectively considered affiliative behaviors; Hung et al., 2017; Stetzik et al., 2018). The use of the term in these contexts is not particularly well-aligned with the human literature and may be misleading.

The original research discussed in this review was supported by NSF IOS 1353859 to HKC and by NIH grants MH109302 and MH110212 to HEA, as well as funds from the Brains and Behavior Program at Georgia State University.

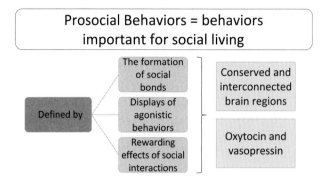

Figure 5.1 If prosocial behaviors in nonhuman mammals are inclusive of all behaviors that have some benefit to others or to the social group as a whole, then this term would apply to any social interactions that are rewarding, such as the formation of social bonds and the establishment of social status. Further, there is evidence that prosocial behaviors are regulated by conserved and interconnected brain regions, many of which are modulated by the oxytocin and vasopressin systems.

Of concern is the use of the term prosocial as a species descriptor, rather than as a behavioral outcome. For example, species that tend to be more competitive, solitary, or perhaps display fewer overt cooperative behaviors, as is observed in socially polygamous species, are sometimes labeled as being less prosocial or even asocial or antisocial. The reality is that most behaviors and/or species cannot be so simply categorized. Further, labeling species and/or behaviors as collectively prosocial does a disservice to the richness and nuance of social interactions. Thus, we would suggest that having too much fidelity to a narrow definition of the term prosocial is not particularly helpful.

The labeling of species as more or less "prosocial" has important implications for their use in biomedical research as well. If investigators buy too heavily into the idea that some preclinical models are better suited to investigate aspects of human behavior than others simply because of the models' perceived level of "prosociality," then other species that may be better models for other reasons could be marginalized. Therefore, we suggest a reevaluation of the use of the term "prosocial" and how it is considered in the context of neuroscience and biomedical research. Our suggestions will be framed in the context of the oxytocin/vasopressin family of neuropeptides, as they are key to the neural modulation of behaviors, and have fairly conserved roles as activators and inhibitors of behaviors that are important to social living in many mammalian species, particularly behaviors that are important for the maintenance of species-specific social structures. A conceptual overview of this chapter can be found in Figure 5.1.

Redefining the Concept of Prosocial

When investigators use the term "prosocial," what do they mean? Most often, this term seems to refer to animals that are deemed "motivated" to engage in social interactions or behaviors that are associated with a perceived "social goal." As behaviorists are well aware, it can be a dangerous thing to ascribe motive to an animal's behavior as we have a natural tendency to anthropomorphize. Rather, our job is to thoughtfully observe behavior and report those observations. One likely, though perhaps unintended, consequence of using the term "prosocial" as a stand-in for assumed social motivation is that it fails to recognize the complexity of species and their behaviors. If behaviors that are thought to have a positive social valence, such as cooperative/affiliative behaviors, are associated with an animal being socially motivated and "prosocial," then where does that leave behaviors that are oftentimes linked to negative social valence, such as those associated with conflict, i.e., aggression and territoriality? The reality is that social interactions are complex and for the most part animals engage in behaviors that are rewarding. Thus, it is important to recognize that even aggressive behaviors, including winning and perhaps losing, can also be rewarding (Meisel and Joppa, 1994; Martinez et al., 1995; Gil et al., 2013) and that competitive/conflict behaviors may not always be easily separated from what are often defined as prosocial behaviors. Many other chapters in this book raise similar points; for example, see Chapters 2, 4, 8, and 10.

For instance, in the nonhuman animal literature the formation of a pair bond is often referred to as a prosocial behavior, but a pair bond is not a specific behavior and it is not obvious how it improves another's welfare. Rather, the pair bond is the product of myriad behaviors that work together to help facilitate the formation of a selective social bond that is beneficial to both individuals. So, while the pair bond is characterized by a preference to spend time with a partner, it also includes aggressive behaviors associated with mate guarding, which does not intuitively fall into the category of prosocial behavior (Walum and Young, 2018). Thus, if we think of prosocial behaviors more broadly as behaviors that have some benefit to others or to the social group as a whole, then competition/conflict behaviors would also be included, within a given species, as competitive behaviors are just as important to the maintenance of social structures as are affiliative behaviors. By extension, if a display of aggressive or agonistic behaviors helps another individual, such as in the case of a mother defending her young, perhaps it too should be considered prosocial. A more holistic definition of prosocial behavior that might be more useful in the nonhuman literature is *any behavior that increases inclusive fitness as the result of an interaction with a conspecific*. This update to the definition of prosocial behavior focuses on the immediate outcome of the behavior, rather than on the unmeasurable motivational or cognitive state of the individual engaged in the behavior. This new, broader definition will also allow for better comparisons across multiple species, as there is a body of literature on behavior, often overlooked in discussions of prosociality in biomedical research, that has examined behavioral responses that increase positive outcomes for others and is focused on the result of a behavioral interaction or test rather than on the motivation. Examples of these include both positive outcomes (e.g., providing food benefits) or reducing negative outcomes to partners (e.g., freeing a trapped cage mate). These studies have defined prosocial behavior as behaviors that directly benefit another individual, whether intended or not (Sosnowski and Brosnan, 2019).

Hopefully it is apparent that there is little utility in labeling some species more or less prosocial compared to others. Within most species there can be fluid shifts from conflict to cooperation depending on the social context. These behavioral shifts also interact in complex ways with factors such as sex, developmental age, and time of year. (See Chapter 10 for the importance of these factors in social insects.) Based on this, investigators should be cautious when using the current, narrow definition of "prosociality" to rationalize the use of only a handful of preclinical models because of their presumed suitability to better interrogate aspects of human behavior. We would argue that, generally speaking, all mammalian species engage in prosocial behaviors and studying a diverse array of species is the best path to identifying the mechanisms that are conserved across species. It is also important to consider that simply because a preclinical model appears to have face validity for human behavior, that does not necessarily mean it has construct validity for human behavior. Rather, the behaviors of each species studied must really be considered more broadly in the context of that species' behavioral ecology.

What Defines the Neural Regulation of Prosocial Behaviors?

Now that we have a more comprehensive definition of prosocial behavior, what are the key neural features that are critical regulators of all behaviors that are important to social living? In particular, what brain regions and neurotransmitter systems contribute to these behaviors? As we do not provide a comprehensive review of these topics in this chapter, please refer to some of these papers for more detailed information: Caldwell, 2017; Carter, 2017; Johnson and Young, 2017; Ophir, 2017; Phelps et al., 2017; Walum and Young, 2018; Chapter 6 reviews brain systems in primates related to social behavior.

Conserved and Interconnected Brain Regions

While described in more detail elsewhere (Newman, 1999; Goodson and Kabelik, 2009; Albers, 2012; Caldwell and Albers, 2016; Johnson and Young, 2017), the social behavior neural network (SBNN) represents brain regions, i.e., "nodes," that are important to the neural regulation of social behaviors. Included as a part of the SBNN are the extended amygdala, the bed nucleus of the stria terminalis, the lateral septum, the periaqueductal gray, the medial preoptic area, the ventromedial hypothalamus, and the anterior hypothalamus. Characteristic of all of these "nodes" are their interconnectivity, the presence of steroid hormone receptors, and their importance to the neural modulation/regulation of more than one behavior. It is hypothesized that it is the differential patterns of activation and interaction among these nodes within a particular social context that shape behavioral output (Albers, 2015; Teles et al., 2015; Johnson and Young, 2017). Behaviors thought to be regulated by the SBNN include offensive and defensive aggression, social recognition memory, parental behavior, and social communication (Goodson and Thompson, 2010; O'Connell and Hofmann,

2011; Caldwell and Albers, 2016; Song and Albers, 2018). What is particularly compelling about the literature in support of the SBNN hypothesis is the cross-species conservation amongst vertebrates (O'Connell and Hofmann, 2011a, 2011b); this reinforces the idea that what we may learn in one species is likely to be of relevance to another.

More recently, a social salience network (SSN) has also been proposed that includes interconnected nodes that code for valence, or "wanting," in social contexts (Johnson et al., 2017). Included in this network are the main olfactory bulb and the anterior olfactory nucleus, the piriform cortex, the olfactory tubercle, the prefrontal cortex, the ventral tegmental area, the medial amygdala, the basolateral amygdala, the paraventricular nucleus of the hypothalamus, and the nucleus accumbens shell (Johnson et al., 2017). To date, this network has only really been examined in male prairie voles, so it is not yet clear how conserved it may be across species.

Navigating a social world is very complicated for animals and, as is suggested above, it is ever more apparent that there are numerous, conserved, neural networks that help to process social information and shape behavioral output. Fortunately, the ability of behavioral neuroscientists to evaluate the interconnectivity of brain regions has advanced dramatically in recent years. Technical advances, such as *in vivo* multi-electrode recordings, DREADDs, *in vivo* calcium imaging, and optogenetics, has made the testing of some of these hypotheses possible. It should be noted that the foundational work for these network hypotheses is rooted in experiments that systematically examined the contributions of individual neuroanatomical areas to the neural regulation of social behaviors. Through that thoughtful work, many of the aforementioned brain areas were identified as being important to the neural regulation of prosocial behaviors, ultimately putting investigators in a position to form hypotheses about the predicted output of these networks. Particularly exciting is the fact that some of the newer tools, while originally designed for use in laboratory mice, can be adapted for use in other species.

Two Evolutionarily Ancient Neuropeptides

The two neuropeptide systems that are integral to both the SBNN and the SSN are the oxytocin and arginine vasopressin systems. Both are highly conserved across vertebrate species and are consistently involved in the neural modulation of social behaviors (Albers, 2015; Caldwell and Albers, 2016; Bosch and Young, 2017; Caldwell, 2017; Ophir, 2017; Wilczynski et al., 2017; see also Chapter 7). Their receptors are also distributed throughout the SBNN and SSN, where they are thought to be critical for shaping network output. While extensive reviews of these two systems can be found elsewhere (Caldwell and Albers, 2016; Bosch and Young, 2017; Caldwell, 2017; Caldwell et al., 2017; Carter, 2017; Johnson and Young, 2017; Wilczynski et al., 2017; Freeman and Bales, 2018; Song and Albers, 2018), briefly, oxytocin and vasopressin are structurally similar, differing from one another by only two amino acids. Their receptors are also structurally similar, though differentially distributed throughout the brain. Because of the sequence and structure similarities of the receptors, there is also crosstalk between these systems, which further complicates their study (Song and Albers, 2018).

In the popular press, oxytocin is sometimes referred to as the "love" or "cuddle" hormone, as it is often associated with having "prosocial effects." However, given the nuances of this system and its complex interactions with the vasopressin system, thinking about oxytocin in this one-dimensional way obscures our understanding of its functions. It is also unlikely that postulates such as "exogenous oxytocin has prosocial effects" will help us better understand the neural mechanisms underlying social behaviors. Despite the term "prosocial" being regularly thrown around in the neurobiology of social behavior literature, it has not been used in a scientifically rigorous fashion to guide neurobiological studies. This is particularly obvious to those that study the oxytocin and vasopressin systems, as they are well aware that these systems often have very subtle effects that are sex-, species-, context-, and brain-region specific.

As mentioned above, the broad behavioral domains that are modulated by the oxytocin and vasopressin systems are fairly conserved across species and their behavioral effects can be moderated by many factors (Caldwell and Albers, 2016; Dumais and Veenema, 2016; Tamborski et al., 2016; Caldwell, 2017, 2018; Caldwell et al., 2017; Carter, 2017; Marlin and Froemke, 2017; Ophir, 2017; Perkeybile and Bales, 2017; Freeman and Bales, 2018). That said, the major mammalian behaviors that both of these systems help to regulate are those that are associated with social living. Specifically, behaviors that are thought to be regulated by the SBNN, i.e., offensive and defensive aggression, social recognition memory, parental behavior, and social communication. All of these behaviors are routinely associated with prosocial outcomes, using "prosocial" as defined in this chapter.

What Behaviors Define Social Living?

The Formation of Social Bonds

Social bonds allow for increased cooperativity within a group as well as provide stress buffering (Bales et al., 2017). Social bonds are also thought to be important for physical and psychological health in many mammalian species. While social bonding varies between species, much of what is understood about the role of the oxytocin and vasopressin systems with respect to cooperative prosocial behavior stems from work focused on understanding the formation of the aforementioned pair bond. The pair bond is a strong, selective bond with a primary sexual partner; not requiring sexual fidelity, just social fidelity. Often associated with pair bonds is selective aggression toward novel individuals; this highlights how aggressive/agonistic behaviors (discussed later in this chapter) often cannot be separated from other behaviors important for social living (Young et al., 2011; see also Chapter 10). Pair bonds are observed in prairie voles, as well as titi monkeys, marmosets, California mice, and humans. While somewhat rare in mammalian species, found in only 3–5% of species (Kleiman,

1977), the pair bond does represent an extreme of social bonding and has been particularly amenable for study, largely due to the fact that voles within the genus *Microtus* can be either socially monogamous, such as prairie voles (*Microtus ochrogaster*), or socially non-monogomous, such as montane (*Microtus montanus*) or meadow (*Microtus pennsylvanicus*) voles.

Particularly striking is that all evidence suggests that keys to the pair bond of any species are oxytocin and vasopressin. The prairie vole model, in particular, has been used to elucidate much of the neurochemistry. (Details can be found in Tabbaa et al., 2016, Walum and Young, 2018.) It should be noted that brain areas that are important to pair bond formation include those that are a part of the SBNN, the SSN, and the mesolimbic dopamine system. Recently, Bales and colleagues (2017) suggested that the titi monkey (*Callicebus cupreus*) may be particularly advantageous as a model of the pair bond, as titi monkeys may be more neurologically and socially similar to humans than some of the other species studied. Imaging studies in titi monkeys suggest that the brain areas involved with motivation are important for the formation and maintenance of the pair bond (Maninger et al., 2017). Also, the oxytocin receptor (OTR) and the vasopressin 1a receptor are expressed in brain areas important for the modulation of visual and sensory stimuli, including attention and learning and memory in this species (Bales et al., 2017). Further, a functional study found that intranasal administration of vasopressin in male titi monkeys affects approach behaviors (Jarcho et al., 2011). In humans, while studying the specific neurochemistry is not achievable at this time, identification of single nucleotide polymorphisms (Schneiderman et al., 2014; see also Chapter 7), quantification of plasma concentrations of oxytocin and vasopressin (Murphy et al., 1987; Gordon et al., 2008; Holt-Lunstad et al., 2008; Schneiderman et al., 2012), and imaging studies (Scheele et al., 2013; Grace et al., 2018; Sauer et al., 2019) all point to an important role for these neuropeptides in both the formation and maintenance of pair bonds.

Studies of pair bonding have greatly contributed to our understanding of the neurochemistry of social behavior and stimulated a great deal of research focused on oxytocin and vasopressin. That said, we know considerably less about the neurobiology of more casual social interactions, such as those between individuals that are not socially bonded or in individuals that are not even well known to each other. Since more casual social interactions occur frequently and are relevant for human social behavior, understanding their neural regulation is key to understanding social motivation.

Beyond pair bonding, oxytocin and vasopressin are important to other types of social bonds. For example, oxytocin facilitates approach behaviors associated with the mother–infant bond (Bosch and Neumann, 2012; Bridges, 2015). In wild chimpanzees, which are socially promiscuous, urinary oxytocin is elevated after the sharing of resources (Samuni et al., 2018), with sharing being used as a proxy for a non-sexual social bond. Even in humans there is evidence that oxytocin may be important for the modulation of non-sexual social bonds. Specifically, in humans, intranasal oxytocin has been found to increase social approach, particularly to strangers (Cohen, 2018; Cohen et al., 2018). There is also evidence that the epigenetic modulation of OTRs (e.g., methylation) can influence the effects of intranasal oxyto-cin on brain activity in a social interaction test, e.g., the prisoner's dilemma game (Chen et al., 2020; see Box 3.1 in Chapter 3 for description of games).

Displays of Agonistic Behaviors

Aggressive and agonistic behaviors are also important for the formation and mainten-ance of social bonds, and as mentioned earlier in this chapter, should probably be considered as prosocial behaviors. Within some species these behaviors aid in the formation of dominance hierarchies, which are known to contribute to social stability (Alexander, 1974). As there is always competition for resources both within social groups as well as between social groups, agonistic interactions – which include both aggressive and submissive behaviors – help individuals identify where they are in a hierarchy and avoid needless aggression and conflict over resources. It is of note that many of the behaviors that are characteristic of agonistic interactions are not "aggres-sive behaviors," but rather forms of social communication, such as scent marking, vocalizations, dominance displays, and so on, which can help lower the risk of injury (Albers et al., 2002; Fernald, 2014; Freeman et al., 2018). Similar to what has been described above, the oxytocin and vasopressin systems are important to the modula-tion of these behaviors in nuanced ways that are dependent upon a variety of factors.

Social forms of aggression and agonistic behaviors are most often considered within each sex, with intermale aggression and maternal aggression being the most common behaviors studied within this domain. That said, there are certain themes that have emerged with respect to how the oxytocin and vasopressin systems may be involved. The first theme is that an animal's prior social experience affects the way that the brain responds to oxytocin and vasopressin. For example, social status determines the ability of exogenously administered oxytocin to induce scent marking in male squirrel monkeys and in female Syrian hamsters (Winslow and Insel, 1991; Harmon et al., 2002b). Prior social experience also determines whether vasopressin can stimulate aggression in male hamsters (Ferris et al., 1997; Caldwell and Albers, 2004; Albers et al., 2006). In female rats, vasopressin can have differential effects on maternal aggression, depending on whether they are genetically predisposed to high or low anxiety, with increased vasopressin release in the central amygdala being posi-tively correlated with higher levels of offensive aggression only in high-anxiety dams (Bosch and Neumann, 2010).

The second theme is that oxytocin's and vasopressin's effects on aggression and agonistic behaviors depend greatly on their site of action. For instance, in male rats, pharmacological treatment with oxytocin may have anti-aggressive effects within the central amygdala (Calcagnoli et al., 2015), whereas vasopressin appears to facilitate aggressive behaviors in numerous other brain areas, including the ventrolateral hypo-thalamus (Delville et al., 1996), the lateral septum (Compaan et al., 1993; Everts et al., 1997), and the bed nucleus of the stria terminalis (Bester-Meredith and Marler, 2001). Similar site-specific effects are also observed in females (Ferris et al., 1992; Harmon et al., 2002a; Bosch et al., 2005; Consiglio et al., 2005; Bosch and Neumann, 2010;

Gutzler et al., 2010). These data, considered along with what was discussed in the section on social bonds, reinforce the idea that aggressive/agonistic behaviors are integral to social living, with oxytocin and vasopressin involved in the modulation of all behaviors important to animals' navigation of their social world.

Social Interactions Are Rewarding

Remarkably, our understanding of the parameters underlying the rewarding properties of social interactions and the neurobiological mechanisms responsible for social reward have rarely been studied; this is despite their known importance in psychological and physical health. Within this context it is also important to consider the underpinnings of how social interactions become less rewarding or even aversive. Recently, it has been proposed that the relationship between the reward value of social interactions and the frequency of choosing to socially interact are similar to what is observed in addiction models; as the reward value of both drugs and social interactions increase, the number of rewards obtained decreases (Maldonado et al., 1993; Doherty et al., 2013; Borland et al., 2018). Indeed, it has been proposed that there is an inverted-U relationship between the dose/duration/intensity of social interaction and the rewarding value of those interactions and that this relationship is initiated at lower concentrations in females than in males (Borland et al., 2019). Consistent with this hypothesis are the recent findings that females find the same amount of social interaction to be more rewarding than do males (Borland et al., 2019). The possibility of an inverted-U relationship between the intensity of social interactions and social reward is intuitively appealing, as it provides a single theoretical construct that can explain how emotional valence can change from positive to negative as the intensity of social interactions increases.

Oxytocin in particular is known to play a critical role in the regulation of social reward due to its interactions with the mesolimbic dopamine system. Inhibition of oxytocin receptors in two key brain sites within this system, the nucleus accumbens and the ventral tegmental area, inhibits the rewarding properties of social interactions in both males and females (Dolen et al., 2013; Song et al., 2016; Hung et al., 2017; Borland et al., 2019). As predicted by the inverted-U hypothesis, activation of oxytocin receptors can enhance or inhibit social reward depending on the amount of social interaction, and this may be done in a sex-dependent manner. These and other data suggest that in humans, as well as other animals, oxytocin can increase the positive valence of social stimuli but depending on the factors at play could also have the opposite effect, i.e., reduce the positive valence of social stimuli (Eckstein et al., 2014; Chen et al., 2017; Duque-Wilckens et al., 2017). Thus, previous suggestions that oxytocin has one-dimensional effects in that it uniformly increases the positive valence of social stimuli is likely inaccurate.

The vasopressin literature is scant in terms of this system's involvement in social reward. However, there is evidence of sex-specific effects of vasopressin on the rewarding properties of juvenile play behaviors in rats. Specifically, in juvenile females, social play increases the extracellular release of dopamine in the lateral septum, and this release can be blocked by treatment with a vasopressin 1a receptor antagonist; this same effect is not observed in juvenile males (Bredewold et al., 2018). Further, in humans there is evidence that vasopressin may increase the tendency to engage in cooperative behaviors, which has been suggested to be mediated by the dopamine system (Brunnlieb et al., 2016). Needless to say, there still remains much work to be done in the area of social reward, but there is a clear need to better understand which behaviors are rewarding, and which behaviors may be aversive, in both males and females. It is also highly likely that the oxytocin and vasopressin systems will emerge as important players in this facet of prosocial behavior as well.

Conclusions

The oxytocin and vasopressin systems are critical to the neurochemistry of social living in mammals. They help shape the integration of environmental information and behavioral output via their actions in overlapping neural networks. Within these networks they are involved in the neural modulation of social interactions, including the formation of social bonds, the establishment of social status, and the rewarding effects of social interactions. It is within this context that the definition of what it means to be "prosocial" should be considered. If foundational investigators apply the strictest definition of "prosocial" to their work on social living, then it seems likely that only a handful of behaviors could be included. However, if the term prosocial is used more broadly as a descriptor for all behaviors that benefit others or the social group as a whole, then behaviors that have to date been excluded, such as aggressive and/or agonistic behaviors, would also be recognized as being prosocial – as they too are critical for the formation and maintenance of social bonds. (Similar or complementary perspectives are provided in many of the other chapters in this book, including Chapters 2, 3, 8, and 10.) It is with this in mind that we challenge investigators to not be too narrow in their definitions of behaviors, or species, that are important for social living. Further, we suggest that to really advance our understanding of the "social brain" in both sexes, many behaviors and a variety of species should be studied, and that the nuances of the oxytocin and vasopressin systems be thoughtfully considered.

References

Albers H. E. (2012) The regulation of social recognition, social communication and aggression: Vasopressin in the social behavior neural network. *Hormones and Behavior,* 61(3): 283–292.

Albers H. E. (2015) Species, sex and individual differences in the vasotocin/vasopressin system: Relationship to neurochemical signaling in the social behavior neural network. *Frontiers in Neuroendocrinology,* 36: 49–71.

Albers H. E., Dean A., Karom M. C., Smith D., and Huhman K. L. (2006) Role of V1a vasopressin receptors in the control of aggression in Syrian hamsters. *Brain Research,* 1073–1074: 425–430.

Albers H. E., Huhman K. L., and Meisel R. L. (2002) Hormonal basis of social conflict and communication. In Pfaff D. W., Arnold A. P., Etgen A. M., Fahrbach S. E., and Rubin R. T., eds., *Hormones, Brain and Behavior*. Amsterdam: Academic Press, 393–433.

Alexander R. D. (1974) The evolution of social behavior. *Annual Review of Ecology and Systematics*, 5: 325–383.

Bales K. L., Arias Del Razo R., Conklin Q. A. et al. (2017) Titi Monkeys as a novel non-human primate model for the neurobiology of pair bonding. *Yale Journal of Biology and Medicine,* 90(3): 373–387.

Bester-Meredith J. K., and Marler C. A. (2001) Vasopressin and aggression in cross-fostered California mice (*Peromyscus californicus*) and white-footed mice (*Peromyscus leucopus*). *Hormones and Behavior*, 40(1): 51–64.

Borland J. M., Grantham K. N., Aiani L. M., Frantz K. J., and Albers H. E. (2018) Role of oxytocin in the ventral tegmental area in social reinforcement. *Psychoneuroendocrinology*, 95: 128–137.

Borland J. M., Rilling J. K., Frantz K. J., and H. E. Albers (2019) Sex-dependent regulation of social reward by oxytocin: An inverted U hypothesis. *Neuropsychopharmacology*, 44(1): 97–110.

Bosch O. J., Meddle S. L., Beiderbeck D. I., Douglas A. J., and Neumann I. D. (2005) Brain oxytocin correlates with maternal aggression: Link to anxiety. *Journal of Neuroscience*, 25 (29): 6807–6815.

Bosch O. J., and Neumann I. D. (2010) Vasopressin released within the central amygdala promotes maternal aggression. *European Journal of Neuroscience*, 31(5): 883–891.

Bosch O. J., and Neumann I. D. (2012) Both oxytocin and vasopressin are mediators of maternal care and aggression in rodents: From central release to sites of action. *Hormones and Behavior*, 61(3): 293–303.

Bosch O. J., and Young L. J. (2017) Oxytocin and social relationships: From attachment to bond disruption. In Hurlemann R., and Grinevich V., eds., *Behavioral Pharmacology of Neuropeptides: Oxytocin. Current Topics in Behavioral Neurosciences*, vol. 35. Cham: Springer, pp. 97–117.

Bredewold R., Nascimento N. F., Ro G. S., Cieslewski S. E., Reppucci C. J., and Veenema A. H. (2018) Involvement of dopamine, but not norepinephrine, in the sex-specific regulation of juvenile socially rewarding behavior by vasopressin. *Neuropsychopharmacology*, 43(10): 2109–2117.

Bridges R. S. (2015) Neuroendocrine regulation of maternal behavior. *Frontiers in Neuro-endocrinology*, 36: 178–196.

Brunnlieb C., Nave G., Camerer C. F., Schosser S., Vogt B., Munte T. F., and Heldmann M. (2016) Vasopressin increases human risky cooperative behavior. *Proceedings of the National Academy of Science USA*, 113(8): 2051–2056.

Calcagnoli F., Stubbendorff C., Meyer N., de Boer S. F., Althaus M., and Koolhaas J. M. (2015) Oxytocin microinjected into the central amygdaloid nuclei exerts anti-aggressive effects in male rats. *Neuropharmacology*, 90: 74–81.

Caldwell H. K. (2017) Oxytocin and vasopressin: Powerful regulators of social behavior. *Neuroscientist*, 23(4): 517–528.

Caldwell H. K. (2018) Oxytocin and sex differences in behavior. *Current Opinion in Behavioral Sciences*, 23: 13–20.

Caldwell H. K., and Albers H. E. (2004) Effect of photoperiod on vasopressin-induced aggression in Syrian hamsters. *Hormones and Behavior*, 46(4): 444–449.

Caldwell H. K., and Albers H. E. (2016) Oxytocin, vasopressin, and the motivational forces that drive social behaviors. *Current Topics in Behavioral Neuroscience*, 27: 51–103.

Caldwell H. K., Aulino E. A., Freeman A. R., Miller T. V., and Witchey S. K. (2017) Oxytocin and behavior: Lessons from knockout mice. *Developmental Neurobiology*, 77(2): 190–201.

Carter C. S. (2017) The oxytocin-vasopressin pathway in the context of love and fear. *Frontiers in Endocrinology*, 8: 356.

Chen X., Gautam P., Haroon E., and Rilling J. K. (2017) Within vs. between-subject effects of intranasal oxytocin on the neural response to cooperative and non-cooperative social interactions. *Psychoneuroendocrinology,* 78: 22–30.

Chen X., Nishitani S., Haroon E., Smith A. K., and Rilling J. K. (2020) OXTR methylation modulates exogenous oxytocin effects on human brain activity during social interaction. *Genes Brain and Behavior*, 19: e12555.

Cohen D., Perry A., Mayseless N., Kleinmintz O., and Shamay-Tsoory S. G. (2018) The role of oxytocin in implicit personal space regulation: An fMRI study. *Psychoneuroendocrinology*, 91: 206–215.

Cohen D., and Shamay-Tsoory S. G. (2018) Oxytocin regulates social approach. *Social Neuroscience*, 13(6): 680–687.

Compaan J. C., Buijs R. M., Pool C. W., de Ruiter A. J., and Koolhaas J. M. (1993) Differential lateral septal vasopressin innervation in aggressive and nonaggressive male mice. *Brain Research Bulletin* 30(1–2): 1–6.

Consigli A. R., Borsoi A., Pereira G. A., and Lucion A. B. (2005) Effects of oxytocin microinjected into the central amygdaloid nucleus and bed nucleus of stria terminalis on maternal aggressive behavior in rats. *Physiology and Behavior*, 85(3): 354–362.

Delville Y., Mansour K. M., and Ferris C. F. (1996) Testosterone facilitates aggression by modulating vasopressin receptors in the hypothalamus. *Physiology and Behavior*, 60(1): 25–29.

Doherty J. M., Cooke B. M., and Frantz K. J. (2013) A role for the prefrontal cortex in heroin-seeking after forced abstinence by adult male rats but not adolescents. *Neuropsychopharmacology*, 38(3): 446–454.

Dolen G., Darvishzadeh A., Huang K. W., and Malenka R. C. (2013) Social reward requires coordinated activity of nucleus accumbens oxytocin and serotonin. *Nature*, 501(7466): 179–184.

Dumais K. M., and Veenema A. H. (2016) Vasopressin and oxytocin receptor systems in the brain: Sex differences and sex-specific regulation of social behavior. *Frontiers in Neuroendocrinology*, 40: 1–23.

Duque-Wilckens N., Steinman M. Q., Busnelli M. et al. (2017) Oxytocin receptors in the anteromedial bed nucleus of the stria terminalis promote stress-induced social avoidance in female California Mice. *Biological Psychiatry*, 3(3): 203–213.

Eckstein M., Scheele D., Weber K., Stoffel-Wagner B., Maier W., and Hurlemann R. (2014) Oxytocin facilitates the sensation of social stress. *Human Brain Mapping,* 35(9): 4741–4750.

Everts H. G. J., De Ruiter A. J. H., and Koolhaas J. M. (1997) Differential lateral septal vasopressin in wild-type rats: Correlation with aggression. *Hormones and Behavior,* 31: 136–144.

Fernald R. D. (2014) Communication about social status. *Current Opinion in Neurobiology*, 28: 1–4.

Ferris C. F., Foote K. B., Meltser H. M., Plenby M. G., Smith K. L., and Insel T. R. (1992) Oxytocin in the amygdala facilitates maternal aggression. *Annals of the New York Academy of Science,* 652: 456–457.

Ferris C. F., Melloni R. Jr., Koppel G., Perry K. W., Fuller R. W., and Delville Y. (1997) Vasopressin/serotonin interactions in the anterior hypothalamus control aggressive behavior in golden hamsters. *Journal of Neuroscience,* 17(11): 4331–4340.

Freeman A. R., Hare J. F., Anderson W. G., and Caldwell H. K. (2018) Effects of arginine vasopressin on Richardson's ground squirrel social and vocal behavior. *Behavioral Neuroscience,* 132(1): 34–50.

Freeman S. M., and Bales K. L. (2018) Oxytocin, vasopressin, and primate behavior: Diversity and insight. *American Journal of Primatology,* 80(10): e22919.

Gil M., Nguyen N. T., McDonald M., and Albers H. E. (2013) Social reward: Interactions with social status, social communication, aggression, and associated neural activation in the ventral tegmental area. *European Journal of Neuroscience,* 38(2): 2308–2318.

Goodson J. L., and Kabelik D. (2009) Dynamic limbic networks and social diversity in vertebrates: From neural context to neuromodulatory patterning. *Frontiers in Neuroendocrinology,* 30(4): 429–441.

Goodson J. L., and Thompson R. R. (2010) Nanopeptide mechanisms of social cognition, behavior and species-specific social systems. *Current Opinion in Neurobiology,* 20(6): 784–794.

Gordon I., Zagoory-Sharon O., Schneiderman I., Leckman J. F., Weller A., and Feldman R. (2008) Oxytocin and cortisol in romantically unattached young adults: Associations with bonding and psychological distress. *Psychophysiology,* 45(3): 349–352.

Grace S. A., Rossell S. L., Heinrichs M., Kordsachia C., and Labuschagne I. (2018) Oxytocin and brain activity in humans: A systematic review and coordinate-based meta-analysis of functional MRI studies. *Psychoneuroendocrinology,* 96: 6–24.

Gutzler S. J., Karom M., Erwin W. D., and Albers H. E. (2010) Arginine-vasopressin and the regulation of aggression in female Syrian hamsters (*Mesocricetus auratus*). *European Journal of Neuroscience,* 31(9): 1655–1663.

Harmon A. C., Huhman K. L., Moore T. O., and Albers H. E. (2002a) Oxytocin inhibits aggression in female Syrian hamsters. *Journal of Neuroendocrinology,* 14(12): 963–969.

Harmon A. C., Moore T. O., Huhman K. L., and Albers H. E. (2002b) Social experience and social context alter the behavioral response to centrally administered oxytocin in female Syrian hamsters. *Neuroscience,* 109(4): 767–772.

Holt-Lunstad J., Birmingham W. A., and Light K. C. (2008) Influence of a "warm touch" support enhancement intervention among married couples on ambulatory blood pressure, oxytocin, alpha amylase, and cortisol. *Psychosomatic Medicine,* 70(9): 976–985.

Hung L. W., Neuner S., Polepalli J. S. et al. (2017) Gating of social reward by oxytocin in the ventral tegmental area. *Science,* 357(6358): 1406–1411.

Insel T. R. (1992) Oxytocin – A neuropeptide for affiliation: Evidence from behavioral, receptor autoradiographic, and comparative studies. *Psychoneuroendocrinology,* 17(1): 3–35.

Jarcho M. R., Mendoza S. P., Mason W. A., Yang X., and Bales K. L. (2011) Intranasal vasopressin affects pair bonding and peripheral gene expression in male *Callicebus cupreus*. *Genes Brain and Behavior,* 10(3): 375–383.

Johnson Z. V., Walum H., Xiao Y., Riefkohl P. C., and Young L. J. (2017) Oxytocin receptors modulate a social salience neural network in male prairie voles. *Hormones and Behavior,* 87: 16–24.

Johnson Z. V., and Young L. J. (2017) Oxytocin and vasopressin neural networks: Implications for social behavioral diversity and translational neuroscience. *Neuroscience Biobehavioral Reviews,* 76(Pt A): 87–98.

Kleiman D. G. (1977) Monogamy in mammals. *Quarterly Review of Biology*, 52: 39–69.

Maldonado R., Robledo P., Chover A. J., Caine S. B., and Koob G. F. (1993) D1 dopamine receptors in the nucleus accumbens modulate cocaine self-administration in the rat. *Pharmacology Biochemistry and Behavior,* 45(1): 239–242.

Maninger N., Hinde K., Mendoza S. P. et al. (2017) Pair bond formation leads to a sustained increase in global cerebral glucose metabolism in monogamous male titi monkeys (*Callicebus cupreus*). *Neuroscience*, 348: 302–312.

Marlin B. J., and Froemke R. C. (2017) Oxytocin modulation of neural circuits for social behavior. *Developmental Neurobiology*, 77(2): 169–189.

Martinez M., Guillen-Salazar F., Salvador A., and Simon V. M. (1995) Successful intermale aggression and conditioned place preference in mice. *Physiology and Behavior*, 58(2): 323–328.

Meisel R. L., and Joppa M. A. (1994) Conditioned place preference in female hamsters following aggressive or sexual encounters. *Physiology and Behavior*, 56(5): 1115–1118.

Murphy M. R., Seckl J. R., Burton S., Checkley S. A., and Lightman S. L. (1987) Changes in oxytocin and vasopressin secretion during sexual activity in men. *Journal of Clinical Endocrinology and Metabolism,* 65: 738–741.

Newman S. W. (1999) The medial extended amygdala in male reproductive behavior. A node in the mammalian social behavior network. *Annals of the New York Academy of Science*, 877: 242–257.

O'Connell L. A., and Hofmann H. A. (2011a) Genes, hormones, and circuits: An integrative approach to study the evolution of social behavior. *Frontiers in Neuroendocrinology*, 32(3): 320–335.

O'Connell L. A., and Hofmann H. A. (2011b) The vertebrate mesolimbic reward system and social behavior network: A comparative synthesis. *Journal of Comparative Neurology*, 519(18): 3599–3639.

Ophir A. G. (2017) Navigating monogamy: Nanopeptide sensitivity in a memory neural circuit may shape social behavior and mating decisions. *Frontiers in Neuroscience*, 11: 397.

Oxford English Dictionary (2018) "OED Online." From December 20, 2018, www.oed.com/view/Entry/152981.

Penner L. A., Dovidio J. F., Piliavin J. A., and Schroeder D. A. (2005) Prosocial behavior: Multilevel perspectives. *Annual Review of Psychology*, 56: 365–392.

Perkeybile A. M., and Bales K. L. (2017) Intergenerational transmission of sociality: The role of parents in shaping social behavior in monogamous and non-monogamous species. *Journal of Experimental Biology*, 220(Pt 1): 114–123.

Phelps S. M., Okhovat M., and Berrio A. (2017) Individual differences in social behavior and cortical vasopressin receptor: Genetics, epigenetics, and evolution. *Frontiers in Neuroscience*, 11: 537.

Samuni L., Preis A., Mielke A., Deschner T., Wittig R. M., and Crockford C. (2018) Social bonds facilitate cooperative resource sharing in wild chimpanzees. *Proceedings of the Royal Society B Biological Science*, 285: 20181643.

Sauer C., Montag C., Reuter M., and Kirsch P. (2019) Oxytocinergic modulation of brain activation to cues related to reproduction and attachment: Differences and commonalities during the perception of erotic and fearful social scenes. *International Journal of Psychophysiology*, 136: 87–96.

Scheele D., Wille A., Kendrick K. M. et al. (2013) Oxytocin enhances brain reward system responses in men viewing the face of their female partner. *Proceedings of the National Academy of Science USA*, 110(50): 20308–20313.

Schneiderman I., Kanat-Maymon Y., Ebstein R. P., and Feldman R. (2014) Cumulative risk on the oxytocin receptor gene (OXTR) underpins empathic communication difficulties at the first stages of romantic love. *Social Cognitive and Affective Neuroscience*, 9(10): 1524–1529.

Schneiderman I., Zagoory-Sharon O., Leckman J. F., and Feldman R. (2012) Oxytocin during the initial stages of romantic attachment: Relations to couples' interactive reciprocity. *Psychoneuroendocrinology*, 37(8): 1277–1285.

Song Z., and Albers H. E (2018) Cross-talk among oxytocin and arginine-vasopressin receptors: Relevance for basic and clinical studies of the brain and periphery. *Frontiers in Neuro-endocrinology*, 51: 14–24.

Song Z., Borland J. M., Larkin T. E., O'Malley M., and Albers H. E. (2016) Activation of oxytocin receptors, but not arginine-vasopressin V1a receptors, in the ventral tegmental area of male Syrian hamsters is essential for the reward-like properties of social interactions. *Psychoneuroendocrinology*, 74: 164–172.

Sosnowski M. J., and Brosnan S. F. (2019) Pro-social behavior. In Vonk J. and Shackelford T. K., eds., *Encyclopedia of Animal Cognition and Behavior*. New York: Springer, DOI: https://doi.org/10.1007/978-3-319-47829-6_1410-1.

Stetzik L., Payne R. E. 3rd, Roache L. E., Ickes J. R., and Cushing B. S. (2018) Maternal and paternal origin differentially affect prosocial behavior and neural mechanisms in prairie voles. *Behavioral Brain Research*, 360: 94–102.

Tabbaa M., Paedae B., Liu Y., and Wang Z. (2016) Neuropeptide regulation of social attachment: The Prairie Vole model. *Comprehensive Physiology,* 7(1): 81–104.

Tamborski S., Mintz E. M., and Caldwell H. K. (2016) Sex differences in the embryonic development of the central oxytocin system in mice. *Journal of Neuroendocrinology*, 28 (4), DOI: https://doi.org/10.1111/jne.12364.

Teles M. C., Almeida O., Lopes J. S., and Oliveira R. F. (2015) Social interactions elicit rapid shifts in functional connectivity in the social decision-making network of zebrafish. *Proceedings of the Royal Society B Biological Science*, 282: 20151099.

Walum H., and Young L. J. (2018) The neural mechanisms and circuitry of the pair bond. *Nature Reviews Neuroscience,* 19(11): 643-654.

Wilczynski W., Quispe M., Munoz M. I., and Penna M. (2017) Arginine vasotocin, the social neuropeptide of amphibians and reptiles. *Frontiers in Endocrinology*, 8: 186.

Winslow J. T., and Insel T. R. (1991) Social status in pairs of male squirrel monkeys determines the behavioral response to central oxytocin administration. *Journal of Neuroscience,* 11(7): 2032–2038.

Young K. A., Gobrogge K. L., Liu Y., and Wang Z. (2011) The neurobiology of pair bonding: Insights from a socially monogamous rodent. *Frontiers in Neuroendocrinology*, 32(1): 53–69.

6 The Role of the Temporal Lobe in Human Social Cognition

Katherine L. Bryant, Christina N. Rogers Flattery, and Matthias Schurz

Introduction

Humans exhibit an impressive array of social behaviors. We engage in complex cooperative behavior, employ flexibility in social responses, and navigate large social groups effectively. These behaviors are made possible by more fundamental cognitive abilities including facial recognition, communication, storing and accessing concepts about social entities, and processing emotions. All of these abilities have at least part of their neural basis in the temporal lobe (e.g., Deen et al., 2015), one of the major divisions of the cerebral cortex (see Box 6.1). The remarkable human facility for cooperation – and indeed, great conflict – suggests there may be equally remarkable features of the human brain.

For example, the capacity to process information on social entities is critical to cooperative behavior, and facial recognition is necessary for linking social information to relevant individuals of a social group (Hoffman and Haxby, 2000; Tsukiura et al., 2010). Effective communication is also critical for dynamic and interactive social responses, and language comprehension and components of its production are made possible by the temporal lobe (Pobric et al., 2007; Binder and Desai, 2011; Marchina et al., 2011). These and other temporal cortex functions illustrate that in order to have a complete picture of social cognition in humans, we need to examine the structure and function of the temporal lobe.

Cooperation, conflict, and other social responses require humans to identify individuals and keep track of their interactions with and among these individuals. To do this, we must be able to understand the perspective of others, deduce their motives, and adapt to changes in social relationships. Some regions of the temporal lobe appear to be highly specialized in humans to support these abilities. The fusiform face area, located in the ventral right temporal lobe in the fusiform gyrus, enables the recognition

Figure 6.1 used data provided in part by the Human Connectome Project, WU–Minn Consortium (Principal Investigators: David Van Essen and Kamil Ugurbil; 1U54MH091657) funded by the 16 NIH Institutes and Centers that support the NIH Blueprint for Neuroscience Research and by the McDonnell Center for Systems Neuroscience at Washington University. Versions of Figures 6.1 and 6.2 appeared in a previous publication (Bryant and Preuss, 2018) and are reproduced here with permission. The authors would like to acknowledge the Marie Skłodowska-Curie Individual Fellowship Grant (MSCA-IF 750026) and the Erwin Schroedinger Fellowship by the Austrian Science Fund (FWF-J4009-B27) for supporting this work.

Box 6.1 Definitions of Terms Used in This Chapter

Diffusion-weighted imaging, or DWI:	a type of neuroimaging method that measures the direction of diffusion of water molecules in the brain, commonly used to provide information about the location and orientation of white matter tracts.
Architectonics:	the study of the structure of the brain, typically on a microscopic level. This approach to understanding the brain focuses mainly on the spatial distribution and morphology of cells.
Cerebral cortex, or simply cortex:	the outer layer of the cerebrum, itself composed of up to six layers, depending on the region. The cortex is composed of gray matter and is made up of **cortical folds** termed **gyri** and **sulci**.
Gyrus (plural **gyri**):	a "ridge" or a "hill" in the folds of the brain.
Sulcus (plural **sulci**):	a "groove" or a "valley" in the folds of the brain.
Primary cortex, also called **sensory cortex** or **primary sensory cortex:**	the regions of the cerebral cortex responsible for the initial processing of sensory information. These include primary visual cortex, primary somatosensory (touch) cortex, primary auditory cortex, and more. Primary cortical areas also have associated **secondary** areas, which process higher-order features of sensory information.
Association cortex:	the regions of the cerebral cortex outside of the primary sensory areas. This includes secondary sensory areas, as well as cortical regions that integrate information from more than one sensory modality.
Unimodal brain regions:	regions that only process information from one type of sense. Primary cortex is unimodal.
Multimodal brain regions:	regions that process information from multiple senses, or sensory modalities. For example, Wernicke's area, involved in language comprehension, is active both when listening to spoken language (auditory information) or reading written language (visual information).
Dorsal and ventral pathways:	also called the **"where" and "what" pathways**, these are two components of the visual system. The dorsal, or "where," pathway is involved with processing an object's physical location. The ventral, or "what," pathway is involved with processing and object's identity.
Myelin:	a fatty substance that wraps around an **axon**, the part of the brain cell that conducts electrical impulses to send signals. Myelin insulates the axon to allow signals to travel more quickly.

Box 6.1 (*cont.*)

Theory of mind:	the ability to understand that others can have mental states – thoughts, emotions, knowledge, intentions – that can be different from one's own.
Sylvian fissure:	a fissure or sulcus that defines the boundary of the temporal lobe.
Hominids:	primates within the family Hominidae, which includes humans and great apes.
Fasciculus:	a coherent bundle of axons, also called a **tract**, within the white matter of the brain.

of individual faces (Grill-Spector et al., 2004; Kanwisher and Yovel, 2006). Another small region of cortex at the posterior limit of the temporal lobe, where it meets the parietal lobe, is involved in the ability to represent and reason about others' mental states and beliefs, a skill known as theory of mind (Saxe and Powell, 2006).

Other social cognitive processes are broader in scope, both functionally and structurally. Humans communicate in order to determine what social responses are appropriate in a context-sensitive manner, and we construct and connect information about individuals and concepts that are relevant to social relationships. These behaviors are supported by language comprehension and semantic processing, located in the temporal lobe. A large swath of cortex (see Box 6.1), dubbed the anterior temporal lobe (ATL), is involved in the processing and retrieval of human social knowledge (Olson et al., 2013) and semantic knowledge (Visser et al., 2012). The middle temporal gyrus, particularly the posterior portion, is critical for language comprehension (Hickok and Poeppel, 2007). Medial temporal cortex and subcortical structures within the temporal lobe also have key roles to play in human social cognition with regard to emotional processing (Adolphs, 2010).

Are there aspects of these brain regions, or the temporal lobe as a whole, that are unique to humans? Many insights on the neural substrates of human social cognition have been gleaned from comparisons with other primates. One of the goals of comparative neuroscience is to determine what features of human brains are conserved – that is, evolved early in our lineage as primates – and what are derived – evolved in the human lineage exclusively. Derived features of human neuroanatomy may be clearly unique, like the size of our brains relative to our closest primate relatives. In other cases, determining whether an aspect of brain structure or function is unique to humans or shared with other primates requires more in-depth analysis.

For several reasons – including their ease of upkeep and history of use in biomedical research – neuroscience research has relied heavily on rhesus macaques (*Macaca mulatta*) as a model for the human brain. Although humans and rhesus macaques share a common ancestry with Old World monkeys, our divergence from this species dates back to approximately 30 million years (Steiper and Seiffert, 2012). Thus, it is important to compare humans to species with whom we share a more recent evolutionary history. In recent years, advances in neuroimaging have begun to offer possibilities for noninvasive comparisons between human, chimpanzee (*Pan troglo-dytes*), and other nonhuman primate brains (Mars et al., 2014). Direct comparisons using these techniques have permitted new insights into the features of the human brain, including those within the temporal lobe, that are unique to humans, and those we share with other apes and/or with Old World monkeys.

Overview of Human Temporal Lobe Structure and Function in Relation to Other Primates

A thorough comparison of temporal lobe anatomy suggests that the human temporal lobe differs from that of other primates in ways ranging from gross structural changes to microstructural modifications. Human temporal lobes are larger, both relatively and absolutely. They have evolutionarily novel cortical folds and unique large-scale white matter connectivity patterns, and features of their local cortical circuits suggest differences in connectivity among neurons. In order to understand the organization of the temporal lobe in humans, we must first examine its features in context with its organization in other primates.

The temporal lobe is distinguished from the rest of the cerebral cortex by the Sylvian or lateral fissure. Species within several mammalian orders have evolved temporal expansions of cortex independently of primates, and these temporal lobes differ in their structural organization from one another (e.g., Hof and Van der Gucht, 2007; Lyras, 2009; Hecht et al., 2016). Unlike other mammalian orders, all members of the order Primates have a temporal lobe, and this temporal lobe probably evolved a single time early in the lineage (Allman, 1982). This leads us to two questions: first, how did the anterolateral expansion of cortex that is the primate temporal lobe come to be? and second, what are the unique structural features of the human temporal lobe that permit the unique social cognitive abilities of humans?

Primates are visual mammals. Early primates evolved large, forward-facing eyes for depth perception and greater visual acuity. Later in the evolution of primates, about 30 million years ago, the Old World monkeys (the group from which apes, and eventually humans, arose) evolved an extra cone for trichromatic color vision. In most mammals, the processing of visual information in the brain begins in the primary visual cortex, in the rear of the cerebrum. In primates, as processing of this infor-mation moves away from the primary visual cortex, it splits into two partly independ-ent pathways – a dorsal pathway, traveling toward the parietal cortex, where information about object locations and their movement in space is analyzed, and a ventral pathway, traveling through the temporal lobe, that specializes in finer-grained details, color processing, and object recognition (Mishkin et al., 1983; Livingstone and Hubel, 1988). Primates, in particular, excel in object recognition, and the expan-sion of cortex into a temporal lobe likely occurred to support these elaborate visual specializations (Allman, 1982; Kaas, 2006); moreover, primates with larger brains tend to have an increased size and number of neurons that subserve the ventral rather

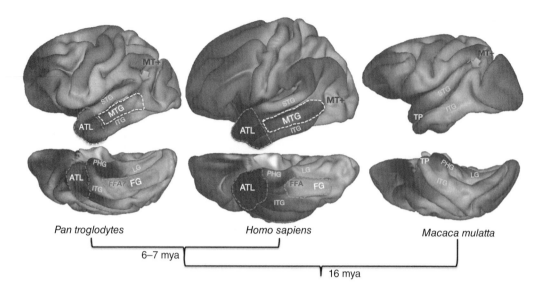

Figure 6.1 The organization of the human temporal lobe in evolutionary context. Divergence dates are from Steiper and Seiffert (2012). FG, Fusiform gyrus; MTG, Middle temporal gyrus; MT+, Visual motion area MT; ATL, Anterior temporal lobe. Dashed lines delineate evolutionarily novel cortical territories that have expanded in the hominid lineage. (A black and white version of this figure will appear in some formats. For the color version, please refer to the plate section.)

than dorsal pathway (Barton, 1998). Hominid (human and great ape) temporal lobes have further expanded to house territories that specialize in face perception and (at least in humans) in semantic processing (reviewed in Bryant and Preuss, 2018).

There are several key structural modifications to the temporal lobe that have occurred in the hominid lineage as compared to other primates. Hominid temporal lobes have developed two novel gyri: the middle temporal gyrus and the fusiform gyrus (Figure 6.1). Further, even when compared to other great apes, human temporal lobes have expanded disproportionately (Rilling and Seligman, 2002). This expansion has been focused on association cortices (Orban et al., 2004; Glasser et al., 2014), which are cortical areas that process higher-order unisensory and multisensory processing (for example, the integration of auditory and visual information). The dramatic change in the proportion of association cortex in the human brain suggests that human cognitive specializations are due to a specialization in multisensory integration. Lastly, among primates, humans are unique in the location of visual motion area MT, part of the dorsal visual pathway, and a useful landmark for delineating the posterior margin of the temporal lobe. In humans, unlike other hominids and Old World monkeys, this territory has been displaced posteriorly and inferiorly (Ungerleider and Desimone, 1986; Watson et al., 1993; Figure 6.1). Taken together, these observations indicate that both expansion and reorganization have occurred in the evolution of the human temporal lobe.

In comparison with other primates, humans have a large temporal lobe (Rilling and Seligman, 2002). The human brain is larger than would be expected for a primate of our size (Schoenemann, 1997; Rilling, 2006), and although apes have larger brains than most primates (except for the highly encephalized capuchin monkeys), human brains are also larger than expected for an ape of our size (Semendeferi and Damasio, 2000; Rilling and Seligman, 2002). In addition to expansion, modifications of temporal lobe organization have occurred as well. These modifications have important implications for the role of higher-order association cortices in humans. The primary auditory cortex in hominids is posteriorly displaced, and makes up a smaller proportion of cortical surface when compared to other primates (Hackett et al., 2001). In addition to visual motion area MT, other higher-order visual areas have been pushed posteriorly (Orban et al., 2004; Glasser and Van Essen, 2011), effectively expanding temporal association cortices in between auditory core and higher-order visual areas (Orban et al., 2004). The functional implication of the expansion of this area is still under investigation, but the critical role that posterior temporal lobe plays in human language (e.g., Turken and Dronkers, 2011), and integration of auditory and visual motion information (e.g., Kiefer et al., 2012), suggest that a specialization in conceptual processing in humans may have driven these structural changes.

A simple and useful way to compare human temporal lobe anatomy with other primates is to observe the gross morphological differences in the cortical surface. The outward and inward convolutions of the cortex are termed gyri and sulci, respectively, and are helpful landmarks for localizing functions. Many primates have a single longitudinal sulcus that travels along the lateral temporal lobe, dividing it into the superior and inferior temporal gyri. The superior temporal gyrus contains the primary auditory cortex and auditory association areas, while the inferior temporal gyrus houses the ventral visual pathway. Hominids (humans and great apes) share a morphological modification – the addition of an inferior temporal sulcus – which is absent in Old World monkeys, and gives rise to the middle temporal gyrus (Figure 6.1). On the ventral surface of the temporal lobe, hominids also possess a new cortical fold, the fusiform gyrus, also not found in macaques (Nasr et al., 2011; Figure 6.1). Beyond topological changes, we can also learn more about patterns of temporal expansion in humans by examining how the myelin content of cortex has been modified. Association cortices contain lower myelin content than primary cortical areas. In humans and chimpanzees, the lateral and anterior temporal cortex have lower cortical myelin density when compared with macaques (Glasser and Van Essen, 2011). These changes, combined with the expansion of temporal cortex in hominids, suggest that a key specialization of hominid brains, particularly humans, is the expansion of association areas in the temporal lobe (Preuss, 2011).

The organization of major fasciculi, which are the large coherent fiber bundles that form long-distance connections, has also been modified in the human and hominid lineages. Up until the early 2000s, most studies of fascicular anatomy were carried out using blunt dissection, but since the development of a specialized form of brain imaging based on MRI, termed diffusion-weighted imaging, there has been a resurgence of research on the organization of white matter in humans, macaques, and other species. In rhesus macaques, six fasciculi are generally recognized to connect to the temporal lobe; in humans, an additional fasciculus is recognized (Figure 6.2).

(A) Human

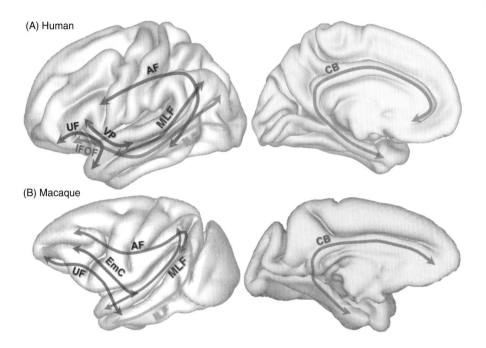

(B) Macaque

Figure 6.2 A simplified diagram of the major white matter tracts in humans and macaques that connect with and/or course through the temporal lobe. (A) Human fascicular trajectories are based on Catani and Thiebaut de Schotten (2008), Rilling et al. (2008), Saur et al. (2008), Frey et al. (2008), Makris et al. (2009), and Turken and Dronkers (2011). (B) Macaque fascicular trajectories are based on Schmahmann and Pandya (2009) and Schmahmann et al. (2007). AF, arcuate fasciculus; CB, cingulum bundle; EmC, extreme capsule; IFOF, inferior fronto-occipital fasciculus; ILF, inferior longitudinal fasciculus; MLF, middle longitudinal fasciculus; UF, uncinate fasciculus; VP, ventral pathway. Cortical reconstructions were generated using the Freesurfer image analysis suite, http://surfer.nmr.mgh.harvard.edu (Fischl, 2012). (A black and white version of this figure will appear in some formats. For the color version, please refer to the plate section.)

Fasciculi found in the temporal lobe of both species include the arcuate fasciculus, the middle longitudinal fasciculus, the inferior longitudinal fasciculus, the uncinate fasciculus, the ventral pathway or extreme/external capsule, and the cingulum bundle. In humans, the extreme/external capsule pathway is a component of a longer tract called the inferior fronto-occipital fasciculus, which reaches the occipital cortex as it traverses the temporal lobe, and recent neuroimaging studies indicate it exists in chimpanzees (Bryant et al., 2018; Mars et al., 2019). Most workers have argued that there is no true inferior fronto-occipital fasciculus in macaques (Schmahmann et al., 2007; Thiebaut de Schotten et al., 2012; Forkel et al., 2014; but see DeCramer et al., 2018).

The arcuate fasciculus has perhaps received the most attention among the major white matter bundles in humans, because of Geschwind's seminal work on its role in language and speech production (Geschwind, 1965, 1970). In humans, arcuate terminations are expansive, reaching all major lateral gyri in the temporal lobe (superior, middle, and inferior temporal gyri) connecting them to multiple frontal areas, including

Broca's area (Powell et al., 2006; Glasser and Rilling, 2008). From a comparative perspective, the arcuate has been studied the most thoroughly and shows dramatic differences between species. Rhesus macaques, for example, have very limited arcuate projections in the temporal lobe, limited to the posterior superior temporal gyrus (Schmahmann et al., 2007). In chimpanzees, the arcuate has more robust connections but these are still limited to the posterior superior temporal gyrus (Glasser and Rilling, 2008). Interestingly, chimpanzees and humans share lateralized arcuates; left arcuate fiber bundles are typically larger than the ones on the right (Rilling et al., 2011), consistent with other temporal asymmetries in chimpanzees (Gannon et al., 1998; Cantalupo et al., 2009; Hopkins et al., 2010) suggesting some features that contribute to language evolved prior to the split between humans and chimpanzees (Preuss, 2011).

Another fiber bundle with important evolutionary modifications is the inferior fronto-occipital fasciculus, which travels from occipital and parietal regions and passes through the middle and inferior temporal gyri, reaching the orbitofrontal and lateral prefrontal cortices through the extreme/external capsule, in close apposition with the uncinate fasciculus (Curran, 1909; Davis, 1921; Catani and Thiebaut de Schotten, 2008). The current lack of evidence for this fasciculus in macaques has led to the speculation that it is unique to the human brain (Catani, 2006; Thiebaut de Schotten et al., 2012). However, a study by Bryant et al. (2018) has demonstrated that a tract following this anatomical trajectory can be reconstructed with diffusion tracto-graphy in chimpanzees, suggesting that the inferior fronto-occipital fasciculus may not be unique to humans but might rather be a hominid specialization.

The ventral pathway is a fiber bundle that connects the prefrontal cortex with the temporal lobe in humans (Parker et al., 2005; Saur et al., 2008) and in other primates (Kaas et al., 1999; Romanski et al., 1999; Schmahmann et al., 2007), and plays an important role in human language comprehension and production. Complicating our understanding of its role in human evolution is the fact that this pathway has been variously described as the extreme capsule, the external capsule, as part of the uncinate fasciculus, or as synonymous with the prefrontal projections of the inferior fronto-occipital fasciculus in humans (for discussion, see Bryant and Preuss, 2018). More recent investigations using functional MRI and diffusion-weighted imaging in humans indicate this bundle is independent of the uncinate fasciculus and passes through the extreme capsule, connecting auditory association areas with ventral prefrontal cortex (Frey et al., 2008; Saur et al., 2008). One possible interpretation of the literature is that the ventral pathway is conserved among primates, and in hominids has expanded posteriorly to become the inferior fronto-occipital fasciculus.

The other major tracts that have connections in the temporal lobe are the inferior longitudinal fasciculus, the middle longitudinal fasciculus, the uncinate fasciculus, and the cingulum bundle. These fasciculi appear to be more conserved among primates; that is, clear homologous bundles exist in Old World monkeys and humans. The inferior longitudinal fasciculus is a ventral associative bundle that runs along the middle and inferior temporal gyri, lateral to the inferior fronto-occipital fasciculus in humans, connecting the occipital and temporal lobes, with short fibers reaching the amygdala and hippocampus (Catani et al., 2003; Catani and Thiebaut de Schotten,

2008). The inferior longitudinal fasciculus in chimpanzees has a similar organization to humans (Rilling et al., 2011; Bryant et al., 2018); in macaques, it connects the occipital and temporal cortex as it courses through the inferior temporal gyrus (Schmahmann et al., 2007). The middle longitudinal fasciculus connects the inferior parietal lobule, with some occipital branching, to the superior temporal gyrus in humans (Makris et al., 2009; De Witt Hamer et al., 2011), in chimpanzees (Bryant et al., 2018) and in macaques (Seltzer and Pandya, 1984; Schmahmann et al., 2007). The cingulum runs along the cingulate gyrus, reaching the orbitofrontal cortex and the anterior medial temporal lobe via the parahippocampal gyrus (Catani and Thiebaut de Schotten, 2008), and has been described in several primate species, including the marmoset and macaque (Beevor, 1891; Baleydier and Mauguiere, 1980; Goldman-Rakic et al., 1984; Schmahmann et al., 2007); in humans, it appears to be involved in visceral and affective processes (Bubb et al., 2018). The uncinate, which connects the anterior temporal lobe to the prefrontal cortex, has been described in both macaques (Petrides and Pandya, 2007; Schmahmann et al., 2007) and humans (Catani and Thiebaut de Schotten, 2008; Hau et al., 2017), and likely plays a role in autobiographical memory in humans (Levine et al., 2009).

Moving from large- to small-scale neuroanatomy, there exists a small but useful literature on the laminar and architectonic characteristics of the temporal lobe in nonhuman primates as compared with humans. Laminar refers to the organization of cortical cells into layers, and architectonics involves the study of the pattern of distribution of cell types and their columnar organization. One focus of this research is the degree to which the temporal lobes are asymmetric with respect to architectonic characteristics, and how this varies across species, which is relevant for understanding the evolution of lateralized functions like language and face recognition.

Neuropil refers to the space between neurons and glial cell bodies, containing axons, dendrites, and the processes of glial cells; a higher proportion of neuropil suggests a greater interconnectedness among neurons. In a survey of neuropil distribution across the cerebral cortex, chimpanzees were found to have a greater proportion of neuropil in primary auditory cortex than humans (Spocter et al., 2012). On a basic level, cortex is made up of cortical minicolumns, which are vertically oriented clusters of neurons that form a columnar shape. Neurons within minicolumns are interconnected with each other and receive common inputs and send out common outputs. In the mid-fusiform gyrus, corresponding to Brodmann area 37, chimpanzees show no laterality in minicolumn width, whereas humans have wider minicolumns in the left hemisphere as compared with the right. Humans also have overall wider minicolumns than chimpanzees in this region (Chance et al., 2013). The laminar pattern of the fusiform gyrus is generally similar across the two species, but humans do have larger and less dense cells, as well as greater depth of cortex, as compared with chimpanzees (Chance et al., 2013).

Differences in minicolumn width have been suggested to be related to coactivation patterns – wider minicolumns may be more important for segregating the processing of independent components while narrow minicolumns may specialize in holistic processing (Harasty et al., 2003). One area of special interest for neuroanatomists is the planum temporale, which constitutes part of Wernicke's area, a territory involved in speech comprehension (Geschwind, 1970; Shapleske et al., 1999). Humans have both wider minicolumns and greater neuropil in the left planum temporale as compared with the right, while chimpanzees and rhesus macaques show no asymmetry (Buxhoeveden et al., 2001). However, chimpanzees do show leftward asymmetry in terms of the volume (Gannon et al., 1998) and number of neurons in this region (Spocter et al., 2010). The asymmetrical patterns of minicolumn width in human fusiform gyrus and planum temporale may be related to human specializations for language and facial recognition, which are both lateralized functions; in other words, the left and right hemispheres are specialized to handle different aspects of these cognitive skills.

Cellular differences among primates are apparent in subcortical regions of the temporal lobe. A survey of subdivisions of the amygdala and their relative sizes among human, chimpanzee, gorilla, gibbon, orangutan, and long-tailed macaques revealed a relative expansion in the lateral nucleus in the human amygdala (Barger et al., 2012). The lateral nucleus contained the highest number of neurons in the human amygdala, whereas the basal nucleus had the most neurons in the other species. The lateral nucleus is the main recipient of cortical input, particularly from the temporal association cortex (Stefanacci and Amaral, 2002; LeDoux, 2007; Freese and Amaral, 2009), suggesting coordinated evolutionary change in these regions (Barger et al., 2012). With regard to the hippocampus, chimpanzees and humans similarly show a volumetric asymmetry, with a rightward bias in total hippocampal volume (Freeman et al., 2004). However, there are differences in the composition of the hippocampus, with the CA1 region and the subiculum taking up a greater proportion of the total neuron number in humans compared to chimpanzees (Rogers Flattery et al., 2020). Finally, the entorhinal cortex has an overall similar architecture and pattern of output to the hippocampus in humans and rhesus macaques, though the olfactory subfield may take up a lower proportion of the total entorhinal cortex in humans (Insausti et al., 2002).

Comparative neuroanatomical studies have demonstrated that the human temporal lobe has multiple structural specializations. Evolutionary changes to cortex include the expansion of the middle temporal gyrus, fusiform gyrus, and the anterior temporal lobe. The amount and organization of white matter has changed as well, with expansions and modifications to the arcuate fasciculus and the inferior fronto-occipital fasciculus. Laminar and architectonic changes have occurred in the fusiform gyrus and planum temporale. Additionally, subcortical changes are evident in the expansion of the lateral nucleus of the amygdala and the organization of the hippocampus. Together, this suggests that the evolutionary pressures that developed human cognitive and behavioral adaptations have caused the temporal lobe to be extensively modified. Next, we will explore in more detail the social cognitive functions in which these components are engaged.

Human Social Cognition Specializations in the Temporal Lobe

Cooperation, conflict, and the mediation of human interactions more generally depend on a complex of interrelated processes. Although these social responses do not rely uniquely on the temporal lobe, the temporal lobe does house critical cognitive processes, without which human social interaction would be impoverished, and in many cases, impossible. Take, for example, face perception, which is a specialization of the fusiform gyrus. Although individuals with prosopagnosia (the inability to recognize individual faces) can survive, if not thrive (Oliver Sacks is a notable example), most would agree that identifying conspecifics is an important part of daily social interaction. Beyond individual recognition, face perception – the ability to recognize the configuration of eyes, noses, and mouths – allows us to process signals for identifying emotional expressions. This is obviously an important part of social interaction, but also plays a key role in empathy and Theory of Mind.

Although it is well-established that decision-making is the domain of the prefrontal cortex, making informed decisions that are dynamically responsive to a changing social environment would not be possible without the emotional processing that occurs in the amygdala and medial temporal cortices. And although communication is not possible without a functioning left inferior frontal gyrus (Broca's area), it is also true that language comprehension, and coherent language production, requires the multisensory processing within the posterior temporal lobe (Wernicke's area). But perhaps even more directly relevant to human social behavior are the functions housed within the anterior temporal lobes. These territories appear to store and mobilize information about both concrete and abstract concepts, with a particular specialization in conceptual information that is relevant to individuals, including descriptors for the attitudes and behaviors of others (e.g., "honorable," or "tactless"; Zahn et al., 2007). They are therefore not only necessary for expressive language about social entities; they also permit the construction of webs of knowledge about social interactions that can inform decisions about whether to cooperate, avoid, or engage in conflict with others in one's social group.

Face Perception

Face perception is arguably one of the most specialized components of the human visual system. We express a great deal of information relevant to successful navigation of the social world through our faces. One basic question in social neuroscience asks whether social cognition is fundamentally similar to (i.e., it is accomplished through the same regions and basic processes) or separate from nonsocial cognition. A third possibility holds that the reality is somewhere in between, with social and nonsocial cognition accomplished through the same regions, performing different functions depending on co-activation patterns in the networks in which these regions are embedded. The specificity of response in certain regions of the temporal lobe to faces has provided some of the strongest evidence for the evolution of unique cortical territories that are devoted to specific cognitive tasks, or what can be termed modularity, in primate social cognition.

The fusiform gyrus contains the fusiform face area, which responds primarily to invariant aspects of faces (Kanwisher et al., 1997). The fusiform face area first receives input from the occipital face area, which processes components of faces, including the mouth, eyes, and nose (Pitcher et al., 2011). The processing of these components in a holistic fashion is termed "configural face processing" and is unique to this area. Configural face processing is asymmetrically dominant in the right hemisphere in humans (Kanwisher et al., 1997), though individual facial features are predominantly processed in the left hemisphere (Rossion et al., 2000). Studies involving neurosurgery patients demonstrate that stimulation of the fusiform face area leads to the spontaneous perception of a face, even in inanimate objects without face-like features – a phenomenon known as face pareidolia (Parvizi et al., 2012; Schalk et al., 2017).

In rhesus macaques, face processing occurs in the lateral temporal lobe inferior to the superior temporal sulcus (STS), an area called the inferotemporal cortex. Pioneering single-cell recording studies have revealed that cells in this territory respond more to faces than any other stimuli; some of these cells respond preferentially to frontal views and some respond to other orientations of the face, such as the profile (Desimone et al., 1984), and some cells in the anterior inferotemporal cortex respond to faces regardless of their orientation (Freiwald and Tsao, 2010). Rhesus macaques likely even experience face pareidolia (Taubert et al., 2017). Functional mapping of face-selective regions has also been done in the common marmoset, which is significant, considering that marmosets are New World monkeys that are more distantly related to humans than macaques. Marmosets appear to possess network of regionally specific, categorically selective face areas in inferotemporal areas, similar to macaques (Hung et al., 2015), suggesting that inferotemporal cortex processing of faces is a specialization of anthropoids (New and Old World monkeys), if not all primates. However, from an anatomical perspective, rhesus macaques and marmosets lack a fusiform gyrus, and the anterior portions of the inferotemporal cortex that process view-invariant faces in these species are unlikely to be structurally homologous to the fusiform face area of humans. In contrast, chimpanzees possess a small fusiform gyrus (Figure 6.1), and face processing in chimpanzees has been shown to activate this cortical region (Parr et al., 2009). Functionally, face perception in chimpanzees shares multiple features with humans, including configural processing (Parr and Taubert, 2011) that is right-lateralized (Dahl et al., 2013), and sensitivity to individual familiarity (Parr et al., 2011).

So far, it appears that great apes have one face recognition system, located in the fusiform gyrus, and that other primates have a different system for processing faces, which shares some functional similarities, in another region of the temporal cortex (the inferior temporal cortex). However, the situation is more complex. It turns out that humans, in addition to the fusiform gyrus, also recruit the superior temporal sulcus in the processing of faces. This sulcus is organized in functionally separable "patches" along the anterior-posterior axis (Deen et al., 2015), similar to rhesus macaques. Unlike the configural processing of the fusiform face area, these patches are devoted to processing changeable aspects of the face and other socially relevant stimuli (Hoffman and Haxby, 2000), including patches responsive to facial expression or movement, eye gaze (Engell and Haxby, 2007), body movement (Thompson et al., 2005), and voices (Belin et al., 2000). In rhesus macaques, the majority of face-responsive cortex is located along the superior temporal sulcus in at least five well-defined patches (Tsao et al., 2008). As in humans, these regions respond to changeable aspects of faces (Perrett et al., 1984), biological motion (Jastorff et al., 2012), and voice (Belin, 2006). And like humans, infant macaques show responsiveness for faces in the superior temporal sulcus early in development. A subregion of the superior temporal sulcus that responds to both facial movement and vocal sounds shows selectivity for faces by 4–6 weeks in human infants (Deen et al., 2017; Livingstone et al., 2017). In humans, this response is less selective than that of adults, suggesting this territory becomes increasingly specialized in development (Deen et al., 2017).

In both macaques and hominids, the ventral visual processing pathway is at least partially enabled by the inferior longitudinal fasciculus, which connects the occipital cortex with the inferior temporal gyrus (Catani et al., 2003; Schmahmann et al., 2007; Bryant et al., 2018). In humans, fMRI studies support the role of the inferior longitudinal fasciculus in face processing (ffytche et al., 2010), including a role in associating names with recognized faces (Fox et al., 2008) and emotions with faces (Pessoa et al., 2002). Sugiura and co-workers (2006) showed that the anterior temporal lobes are also important areas for recognition of personally familiar or known persons, which is based on relating semantic information to persons (see also Tsukiura et al., 2010).

The facial recognition systems of humans and rhesus macaques have important similarities, suggesting these features were shared by the common ancestor of Old World monkeys and great apes approximately 30 million years ago. The expansion of the ventral inferotemporal cortex to form the fusiform gyrus and the development of configural face processing in humans and chimpanzees suggest that these features developed more recently, during or after the emergence of great apes, but prior to the split between the ancestors of humans and chimpanzees. The enlargement of the fusiform gyrus in humans compared with chimpanzees, and our relatively larger social groups, suggests that the human fusiform may have been selected for enhanced facial recognition abilities; however, more research is needed on chimpanzee facial recognition to determine the relationship between the size and function of the fusiform gyrus in humans.

Theory of Mind

A meta-analysis on functional imaging in humans (Schurz et al., 2014) found consistent activation in the anterior temporal lobe for Theory of Mind (see Box 6.1) tasks containing social concept words (e.g., personality traits) and sequences of actions (false belief stories, cartoons showing everyday goal-directed actions). These findings were linked to storage and retrieval of social semantic scripts (e.g., Frith and Frith, 2003; Gallagher and Frith, 2003). With respect to action sequences, a script contains knowledge of the particular activities that will take place in a particular setting. Gallagher and Frith (2003, p. 77) give an illustrative example of such a script: "The activities associated with the restaurant script would include reading the menu, ordering a drink and getting the bill. If I catch the waiter's eye and make the gesture of writing on my left palm with my right forefinger he will usually bring me the bill.

He correctly interprets my actions on the basis of his knowledge of my likely goals." Furthermore, the anterior temporal lobes process not only scripts but social semantic concepts in general. Activity was also found (Zahn et al., 2007) for processing single words describing social concepts, for example when rating the similarity of word pairs that describe human social semantic concepts ("honorable" – "tactless") versus concepts of biological function ("nutritious" – "useful"). Consistent with the reviewed findings, Ross and Olson (2010) found in a within-subjects design that activity in anterior temporal lobes for processing social concept words overlaps with activation for watching animations of simple geometric shapes portraying human interactions. The authors suggested that the anterior temporal lobes contribute to Theory of Mind more broadly, both in terms of processing general conceptual knowledge and specific social concepts.

Moving posteriorly along the temporal lobe, the broad region around the posterior superior temporal sulcus and the junction between the temporal lobe and the parietal lobe (the temporo-parietal junction, or TPJ) can be divided into different subparts relevant for Theory of Mind. One part is located at the posterior end of the TPJ. In human functional imaging, this area is robustly active across a range of different mentalizing tasks (see Schurz et al., 2014), and was suggested to reflect a central process for Theory of Mind (e.g., Saxe and Kanwisher, 2003; Samson et al., 2004; Young et al., 2010). Furthermore, this area shows a distinctive pattern of whole brain connectivity compared to its neighboring regions (Yeo et al., 2011; Bzdok et al., 2012; Mars et al., 2014), which mainly links it to the default mode network involved in self-generated cognition detached from the material world (e.g., Andrews-Hanna, 2012). The default mode network is thought to be relevant for Theory of Mind, because there is no perceptual access to others' mental states (see Frith and Frith, 2003; Lieberman, 2007). It was speculated that this network relies on and adapts past experiences to imagine novel perspectives and events that go beyond the observable environment (Buckner and Carroll, 2007). Interestingly, also in more cognitive-oriented research, Theory of Mind was linked to a decoupling mechanism that distinguishes beliefs from reality (Frith and Frith, 2003; Gallagher and Frith, 2003; see also Leslie, 1987).

More anterior and ventral parts around the TPJ are mainly engaged for Theory of Mind if actions or biological cues are presented. In a meta-analysis of human imaging studies (Schurz et al., 2014), these areas were activated for mentalizing based on actions portrayed by moving shapes or cartoons, or for judging facial expressions of mental states ("reading mind in the eyes" tasks). More broadly, posterior parts of the STS were also found implicated in processing of other social stimuli, such as biological motion and faces (Deen et al., 2015). A main distinction between social processes activating posterior STS versus posterior TPJ is that the former relate to social information and mental states linked to observable stimuli (e.g., movement, facial expression), whereas the latter relate to mental states that are unobservable and require self-generated cognition and decoupling (e.g., false beliefs). In this vein, it has been proposed that Theory of Mind processing comes in two forms, overt versus covert, that map onto different neural networks, one that recruits the TPJ and the other that involves the pSTS (Gobbini et al., 2007).

The debate on whether Theory of Mind is unique to humans spans decades. Since Premack and Woodruff originally posed the question of whether chimpanzees possess Theory of Mind (Premack and Woodruff, 1978), psychologists and neuroscientists have endeavored to determine the important components of Theory of Mind and to what extent, if at all, they are experimentally demonstrable in our closest relatives. Evidence that chimpanzees do not use Theory of Mind has been found in studies finding that chimpanzees fail false belief tasks, indicating they cannot understand what others do not know (Krachun et al., 2010), that they lack development of triadic interactions (or joint attention; Povinelli and Eddy, 1996; Tomonaga et al., 2004), and that they cannot engage in complex perspective-taking (Karg et al., 2016). However, other experiments have found evidence that chimpanzees understand others' beliefs, at least in some simple test conditions (Hare et al., 2001), and although there is scant evidence for joint attention, they do demonstrate gaze-following (Povinelli and Eddy, 1996). Further evidence for components supporting Theory of Mind have been found in chimpanzees, including mirror self-recognition (Gallup et al., 1971; but not in macaques: Gallup, 1977), ability to engage in role-reversal (Povinelli et al., 1992a; again, not in macaques: Povinelli et al., 1992b), imitation of hierarchically structured tasks (Whiten, 1998), and deception (Hare et al., 2006). In a comprehensive review, Call and Tomasello (2008) conclude that chimpanzees possess both covert and overt forms of Theory of Mind, but with some important differences with humans, who are capable of more complex covert representations about others' beliefs and long-term goals.

Language and Conceptual Processing

Although it is not possible here to do a full review of human language abilities and their relationship to other species, it is important to touch on some of the relationships between temporal lobe anatomy and language in humans, as language is an inextricable component of human social life. The capacity for symbolic representation has been demonstrated in chimpanzees and bonobos, using either modified American Sign Language (Gardner and Gardner, 1969, 1975) or visual symbols (lexigrams; Savage-Rumbaugh et al., 1986). Unlike humans, chimpanzee symbol communication has little to no grammatical features (Terrace, 1979), or at least, has a more rudimentary form of grammar than human language. Although some researchers have argued that nonhuman ape symbol use centers on instrumental communication (Rivas, 2005), a careful reading of the literature demonstrates that chimpanzees and bonobos raised in enriched, naturalistic environments preferentially use symbolic representations for referential or social interactions (Gardner and Gardner, 1969, 1975; Fouts, 1973; Savage-Rumbaugh et al., 1986). Overall, it is reasonable to say that great ape abilities for symbol-based communication outshine those of Old World monkeys (Thomas Schoenemann, 1999; Fitch, 2005); this suggests that the expansion of association areas and terminations of the arcuate fasciculus may play a role in facilitating these abilities.

In humans, the arcuate fasciculus is a key structure facilitating language comprehension and production. The arcuate fasciculus is a large fiber bundle that connects Broca's area (Brodmann areas 44 and 45) to multiple temporal cortex areas, including the superior temporal gyrus, the superior temporal sulcus, and the middle temporal gyrus (encompassing Wernicke's area). Aphasias from arcuate lesions can result in conduction aphasia, which involves impairment of repetition (Geschwind, 1970; Damasio and Damasio, 1980; but see Bernal and Ardila, 2009). The role of the arcuate fasciculus in speech production is well-documented (e.g., Marchina et al., 2011; Yeatman et al., 2011), and is made possible by the so-called phonological pathway that connects the posterior superior temporal gyrus with the fronto-opercular cortex (Broca's area; Duffau et al., 2008; Glasser and Rilling, 2008). Although the majority of research has focused on the language functions of the left arcuate fasciculus (e.g., Nucifora et al., 2005; Glasser and Rilling, 2008; Rilling et al., 2008), the right arcuate may play a role in processing music, especially music with a vocal component (Halwani et al., 2011).

Other major fasciculi are also involved in language processing. The inferior fronto-occipital fasciculus and inferior longitudinal fasciculus are implicated in reading (Catani and Mesulam, 2008; Epelbaum et al., 2008), and the middle longitudinal fasciculus in language comprehension, particularly sentence comprehension (Menjot de Champfleur et al., 2013), possibly due to its connections between the angular gyrus and Wernicke's territory (Catani and ffytche, 2005). While the arcuate fasciculus is part of the dorsal pathway, the ventral pathway (white matter connecting the temporal cortex with prefrontal cortex) also plays a key role in language. Functionally, the ventral pathway processes sentences rather than individual words (Saur et al., 2008), and is involved in sound-to-word learning tasks (Wong et al., 2011), naming (Ueno et al., 2011), and the syntactic components of language (Weiller et al., 2011). In a recent review, Skeide and Friederici (2016) argue that the ventral pathway matures earlier than the dorsal pathway (the arcuate) and specializes in bottom-up processes, including phonological, morphosyntactic, and prosodic processing, along with lexical retrieval and lexical-semantic categorization.

In humans, semantic information required for affective cognition and the comprehension and manipulation of concepts occurs in large part in the anterior temporal lobe. Affective cognitive functions that have been localized to the anterior temporal lobe in humans include olfactory memory (Rausch et al., 1977; Eskenazi et al., 1986), taste recognition (Small et al., 1997), perception of emotional facial expressions (Cancelliere and Kertesz, 1990; Schmolck and Squire, 2001), generation of emotions in response to visual cues (Reiman et al., 1997), and emotional memory retrieval (Dolan et al., 2000). Semantic cognitive functions in the ATL include the production and comprehension of spoken and written words and pictures (Coccia et al., 2004; Pobric et al., 2007), coherent conceptual categorization of objects (Rogers et al., 2004; Lambon Ralph et al., 2010), comprehension of social concepts (Zahn et al., 2007, 2009; Ross and Olson, 2010), and a storage site for unique, socially relevant entities, such as familiar people and landmarks (Damasio et al., 2004; Frith, 2007; Kriegeskorte et al., 2007).

Because of the anterior temporal lobe's multimodal properties, it has been argued that this region, in humans, functions as a transmodal or amodal hub, in which abstract conceptual concepts are generated by the convergence of multiple sensory streams (Lambon Ralph et al., 2010; Pobric et al., 2010; Binder and Desai, 2011; Visser et al., 2012). This model of anterior temporal lobe function is supported by studies of patients with bilateral degeneration of the temporal poles who show under- and overgeneralization of conceptual categories (Lambon Ralph and Patterson, 2008). Other workers have focused the connections between the anterior temporal lobe, the amygdala, and the orbitofrontal cortices via the uncinate fasciculus, suggesting the chief contribution of this area is the storage and processing of social knowledge (Simmons and Martin, 2009; Simmons et al., 2010; Olson et al., 2013). Complex social cognitive processes that rely on affective as well as conceptual processing have been found to activate this area in human imaging studies, including ethical decision-making (Heekeren et al., 2003) and moral and social judgments (Moll et al., 2001, 2002). Together, these findings suggest the human anterior temporal lobe is involved in multiple interrelated and complex cognitive processes that interlink conceptual information, social cognition, and behavior.

Emotional Processing

The amygdala and nearby medial temporal cortical areas have an important function in emotional processing. The amygdaloid complex in primates (and other mammals) is a collection of 13 nuclei, broadly grouped into the basolateral, cortical-like, and centromedial divisions (Pabba, 2013). The spatial resolution of neuroimaging techniques presents a challenge for understanding at a fine-grained level the role of each nucleus in human socio-emotional cognition. Nevertheless, there is an enormous literature implicating the amygdala, as a whole, in various disparate functions. Overall, we can be reasonably certain that the amygdala processes complex information from social stimuli, including faces; processes codes for salience, relevance, or "impact" of a stimulus; and in addition to fear learning, also participates in reward learning (Adolphs, 2010).

In humans, a great deal of emotional information is conveyed by the face (as well as the voice and body motion); accordingly, the amygdala receives highly processed information from the temporal cortices (Adolphs, 2003). This is in part facilitated by the inferior longitudinal fasciculus and the inferior fronto-occipital fasciculus, which connect the amygdala to temporal cortical areas that house semantic and visual representations (ffytche et al., 2010), and the cingulum bundle, which connects the amygdala to limbic areas, linking emotional processing with memory and attention functions (Catani, 2006; Rudrauf et al., 2008). In primates, information from the superior temporal sulcus guides social behavior by influencing memory, decision-making, and attention based on the significance of the stimulus. The human amygdala is involved in emotion recognition, particularly but not exclusively fear (Adolphs et al., 1994, 1995; Breiter et al., 1996; Calder, 1996; Morris et al., 1996), and it contains cells that discriminate faces from inanimate objects (Fried et al., 1997), including ones that respond selectively to whole faces (Rutishauser et al., 2011).

One line of evidence for the role of the amygdala in human social cognition comes from neuropsychological patients with medial temporal lobe lesions. The most striking example is Patient SM, an individual with a rare condition leading to bilateral amygdala destruction (Adolphs et al., 1995; Adolphs, 2010). Patient SM showed impaired recognition of emotion in images of faces; however, when explicitly told to orient her attention to the eyes, emotional recognition was rescued. Indeed, in healthy controls, amygdala activity correlates with the degree to which participants shifted their gaze toward the eyes when looking at pictures of faces (Gamer and Büchel, 2009). Single-cell recordings in neurosurgical patients implicate the amygdala in the processing of faces (Fried et al., 1997). Moreover, stimulation of the amygdala can produce feelings of fear, but is also capable of producing positive emotions (Bijanki et al., 2014). The basolateral amygdala is typically the target of electrophysiological recordings in humans (Rutishauser et al., 2015).

The function of the amygdala in nonhuman primates appears to be similar to its function in humans. A greater variety of methodological possibilities in model species allows more insight into the individual amygdaloid nuclei. A role for the amygdala in social and emotional processing has been suspected since at least the monumental temporal lobectomy studies by Klüver and Bucy (1937, 1939). Monkeys with their temporal lobes removed showed an inability to access the "value" of social behaviors or use them in appropriate contexts. Subsequent experiments selectively lesioning the amygdala have resulted in monkeys showing less caution in approaching predators that they would normally fear (Machado et al., 2009) and less avoidance of human strangers (Mason et al., 2006). In macaques, as in humans, the amygdala is involved with fixations onto eyes in faces (Rutishauser et al., 2015). Amygdala size is correlated with social group size in both humans (Bickart et al., 2011) and macaques (Sallet et al., 2011).

The amygdala has been studied in a handful of other primate species. Notably, despite their close relatedness, bonobos have more neuropil (Issa et al., 2018) and greater serotonin innervation (Stimpson et al., 2016) in the amygdala than chimpanzees. Bonobos also have more gray matter in the right dorsal amygdala and a larger pathway linking the amygdala to anterior cingulate cortex (Rilling et al., 2012). This is intriguing, given the social behavioral differences between the two species: bonobos are less territorial (Sobolewski et al., 2012) and more tolerant of strangers than chimpanzees (Tan and Hare, 2013; Tan et al., 2017). When it comes to intragroup interactions, differences in social behavior between chimpanzees and bonobos appear to be more complex. Chimpanzees have been found to share food with group members more frequently, and more tolerantly, than bonobos (Jaeggi et al., 2010). Compared with chimpanzees, bonobo males cooperate with females at a higher rate (Surbeck et al., 2017), and bonobo females are more likely to share food with males (Jaeggi et al., 2013). Overall, one of the more notable differences is the greater likelihood of peaceful interactions between groups of bonobos (Jaeggi et al., 2016). As future studies explore the differences in amygdalar anatomy among humans, bonobos, and chimpanzees, they should have important implications for human social-behavioral evolution.

Conclusions

The temporal lobe is an evolutionary innovation that originally evolved as a response to the complex object-based visual processing necessary for the primate lifestyle. Early primates specialized in foraging for fruits and flowers on the fine terminal branches of trees, which requires keen diurnally adapted vision, and skills in object-guided reaching and grasping (Passingham and Wise, 2012; Kaas, 2013). This ecological niche required the ability to learn and remember the location of ephemerally ripe fruits and navigate in complex three-dimensional environments.

Humans inherited stereoscopic color vision from early primates, as well as many other specializations of the visual system that give us expertise in object recognition. But many features of the life history, ecological niche, and social behavior of humans have diverged a great deal from our early primate ancestors. Coupled with the dramatic expansion of our brains, it appears that the human temporal lobe has been co-opted from its original function as an expansion of the ventral visual pathway to enable a wide repertoire of social behaviors. Some of these modifications appear to have evolved in the hominid lineage, prior to our divergence from our most recent common ancestor with the great apes. These include configural face processing, Theory of Mind, and symbolic representation. Great apes, including humans, have complex social behaviors facilitated by these evolutionary specializations, including the ability for deception, cooperation within and between groups, and the capacity for inter-individual and intergroup aggression.

The human temporal lobe's unique contribution to human social behavior is derived from a combination of conserved features – object recognition, emotional processing systems with hominid specializations (Theory of Mind and configural face processing) with elaborate conceptual processing and language. Thus humans can not only identify and keep track of individuals and relevant entities, and assess their emotional salience and valence, but we can also communicate or withhold this information from conspecifics strategically. The recognition of others' emotions via facial expressions in the context of verbal communication and other information about social relationships can be used to deduce others' goals and trustworthiness. In partnership with prefrontal areas, humans can make decisions over long time periods based on the social, conceptual, and emotional information that we acquire and consolidate. In these ways, the temporal lobe is fundamental for social cognition and behavior that leads to conflict, cooperation, and other group interactions that are uniquely, humanly, complex.

References

Adolphs R. (2003) Is the human amygdala specialized for processing social information? *Annals of the New York Academy of Science*, 985: 326–340.
Adolphs R. (2010) What does the amygdala contribute to social cognition? *Annals of the New York Academy of Science*, 1191: 42–61.

Adolphs R., Tranel D., Damasio H., and Damasio A. (1994) Impaired recognition of emotion in facial expressions following bilateral damage to the human amygdala. *Nature*, 372: 669–672.

Adolphs R., Tranel D., Damasio H., and Damasio A. R. (1995) Fear and the human amygdala. *Journal of Neuroscience*, 15: 5879–5891.

Allman J. (1982) Reconstructing the evolution of the brain in primates through the use of comparative neurophysiological and neuroanatomical data. In Armstrong E., and Falk D., eds., *Primate Brain Evolution: Methods and Concepts*. Boston, MA: Springer US, pp. 13–28.

Andrews-Hanna J. R. (2012) The brain's default network and its adaptive role in internal mentation. *Neuroscientist*, 18: 251–270.

Baleydier C., and Mauguiere F. (1980) The duality of the cingulate gyrus in monkey. Neuroanatomical study and functional hypothesis. *Brain*, 103: 525–554.

Barger N., Stefanacci L., Schumann C. M. et al. (2012) Neuronal populations in the basolateral nuclei of the amygdala are differentially increased in humans compared with apes: A stereological study. *Journal of Comparative Neurology*, 520: 3035–3054.

Barton R. A. (1998) Visual specialization and brain evolution in primates. *Proceedings of the Royal Society B Biological Sciences*, 265: 1933–1937.

Beevor C. E. (1891) On the course of the fibres of the cingulum and the posterior parts of the corpus callosum and fornix in the marmoset monkey. *Proceedings of the Royal Society B Biological Sciences*, 182: 135–199.

Belin P. (2006) Voice processing in human and non-human primates. *Proceedings of the Royal Society B Biological Sciences*, 361: 2091–2107.

Belin P, Zatorre R. J., Lafaille P., Ahad P., and Pike B. (2000) Voice-selective areas in human auditory cortex. *Nature,* 403: 309–312.

Bernal B., and Ardila A. (2009) The role of the arcuate fasciculus in conduction aphasia. *Brain*, 132: 2309–2316.

Bickart K. C., Wright C. I., Dautoff R. J., Dickerson B. C., and Barrett L. F. (2011) Amygdala volume and social network size in humans. *Nature Neuroscience,* 14: 163–164.

Bijanki K. R., Kovach C. K., McCormick L. M. et al. (2014) Case report: Stimulation of the right amygdala induces transient changes in affective bias. *Brain Stimulation*, 7: 690–693.

Binder J. R., and Desai R. H. (2011) The neurobiology of semantic memory. *Trends in Cognitive Science*, 15: 527–536.

Breiter H. C., Etcoff N. L., Whalen P. J. et al. (1996) Response and habituation of the human amygdala during visual processing of facial expression. *Neuron*, 17: 875–887.

Bryant K. L., Glasser M. F., Li L. et al. (2019) Organization of extrastriate and temporal cortex in chimpanzees compared to humans and macaques. *Cortex*, 118: 223–243.

Bryant K. L., Li L., and Mars R. B. (2018) White matter projection maps in chimpanzees in comparison with humans and macaques. *Cortical Evolution Conference* 2018.

Bryant K. L., and Preuss T. M. (2018) A comparative perspective on the human temporal lobe. In Bruner E., Ogihara N., and Tanabe H. C., eds., *Digital Endocasts: From Skulls to Brains*. Tokyo: Springer Japan, pp. 239–258.

Bubb E. J., Metzler-Baddeley C., and Aggleton J. P. (2018) The cingulum bundle: Anatomy, function, and dysfunction. *Neuroscience and Biobehavioral Reviews*, 92: 104–127.

Buckner R. L., and Carroll D. C. (2007) Self-projection and the brain. *Trends in Cognitive Science*, 11: 49–57.

Buxhoeveden D. P., Switala A. E., Litaker M., Roy E., and Casanova M. F. (2001) Lateralization of minicolumns in human planum temporale is absent in nonhuman primate cortex. *Brain Behavior and Evolution*, 57: 349–358.

Bzdok D., Schilbach L., Vogeley K., Schneider K., Laird A. R., Langner R., and Eickhoff S. B. (2012) Parsing the neural correlates of moral cognition: ALE meta-analysis on morality, theory of mind, and empathy. *Brain Structure and Function*, 217: 783–796.

Calder A. J. (1996) Facial emotion recognition after bilateral amygdala damage: Differentially severe impairment of fear. *Cognitive Neuropsychology*, 13: 699–745.

Call J., and Tomasello M. (2008) Does the chimpanzee have a theory of mind? Thirty years later. *Trends in Cognitive Science*, 12: 187–192.

Cancelliere A. E., and Kertesz A. (1990) Lesion localization in acquired deficits of emotional expression and comprehension. *Brain and Cognition*, 13: 133–147.

Cantalupo C., Oliver J., Smith J., Nir T., Taglialatela J. P., and Hopkins W. D. (2009) The chimpanzee brain shows human-like perisylvian asymmetries in white matter. *European Journal of Neuroscience*, 30: 431–438.

Catani M. (2006) Diffusion tensor magnetic resonance imaging tractography in cognitive disorders. *Current Opinion in Neurology*, 19: 599–606.

Catani M., and ffytche D. H. (2005) The rises and falls of disconnection syndromes. *Brain*, 128: 2224–2239.

Catani M., Jones D. K., Donato R., and ffytche D. H. (2003) Occipito-temporal connections in the human brain. *Brain*, 126: 2093–2107.

Catani M., and Mesulam M. (2008) The arcuate fasciculus and the disconnection theme in language and aphasia: History and current state. *Cortex*, 44: 953–961.

Catani M., and Thiebaut de Schotten M. A (2008) diffusion tensor imaging tractography atlas for virtual in vivo dissections. *Cortex*, 44: 1105–1132.

Chance S. A., Sawyer E. K., Clover L. M., Wicinski B., Hof P. R., and Crow T. J. (2013) Hemispheric asymmetry in the fusiform gyrus distinguishes Homo sapiens from chimpanzees. *Brain Structure and Function*, 218: 1391–1405.

Coccia M., Bartolini M., Luzzi S. Provinciali L., and Ralph M. A. L. (2004) Semantic memory is an amodal, dynamic system: Evidence from the interaction of naming and object use in semantic dementia. *Cognitive Neuropsychology,* 21: 513–527.

Curran E. J. (1909) A new association fiber tract in the cerebrum with remarks on the fiber tract dissection method of studying the brain. *Journal of Comparative Neurology and Psychology,* 19: 645–656.

Dahl C. D., Rasch M. J., Tomonaga M., and Adachi I. (2013) Laterality effect for faces in chimpanzees (*Pan troglodytes*). *Journal of Neuroscience*, 33: 13344–13349.

Damasio H., and Damasio A. R. (1980) The anatomical basis of conduction aphasia. *Brain*, 103: 337–350.

Damasio H., Tranel D., Grabowski T., Adolphs R., and Damasio A. (2004) Neural systems behind word and concept retrieval. *Cognition*, 92: 179–229.

Davis L. E. (1921) An anatomic study of the inferior longitudinal fasciculus. *Archives of Neurology and Psychology*, 5: 370–381.

DeCramer, T., Swinnen, S., Van Loon, J., Janssen, P., & Theys, T. (2018) White matter tract anatomy in the rhesus monkey: a fiber dissection study. *Brain Structure and Function*, 223(8) 3681–3688.

Deen B., Koldewyn K., Kanwisher N., and Saxe R. (2015) Functional organization of social perception and cognition in the superior temporal sulcus. *Cerebral Cortex*, 25: 4596–4609.

Deen B., Richardson H., Dilks D. D. et al. (2017) Organization of high-level visual cortex in human infants. *Nature Communications,* 8: 13995.

Desimone R., Albright T. D., Gross C. G., and Bruce C. (1984) Stimulus-selective properties of inferior temporal neurons in the macaque. *Journal of Neuroscience,* 4: 2051–2062.

De Witt Hamer P. C., Moritz-Gasser S., Gatignol P., and Duffau H. (2011) Is the human left middle longitudinal fascicle essential for language? A brain electrostimulation study. *Human Brain Mapping*, 32: 962–973.

Dolan R. J., Lane R., Chua P., and Fletcher P. (2000) Dissociable temporal lobe activations during emotional episodic memory retrieval. *Neuroimage*, 11: 203–209.

Duffau H., Gatignol P., Mandonnet E., Capelle L., and Taillandier L. (2008) Intraoperative subcortical stimulation mapping of language pathways in a consecutive series of 115 patients with Grade II glioma in the left dominant hemisphere. *Journal of Neurosurgury*, 109(3): 461–471.

Engell A. D., and Haxby J. V. (2007) Facial expression and gaze-direction in human superior temporal sulcus. *Neuropsychologia*, 45: 3234–3241.

Epelbaum S., Pinel P., Gaillard R. et al. (2008) Pure alexia as a disconnection syndrome: New diffusion imaging evidence for an old concept. *Cortex*, 44: 962–974.

Eskenazi B., Cain W. S., Novelly R. A., and Mattson R. (1986) Odor perception in temporal lobe epilepsy patients with and without temporal lobectomy. *Neuropsychologia*, 24: 553–562.

ffytche D. H., Blom J. D., and Catani M. (2010) Disorders of visual perception. *Journal of Neurology, Neurosurgery, and Psychiatry*, 81: 1280–1287.

Fischl B. (2012) FreeSurfer. *Neuroimage,* 62: 774–781.

Fitch W. T. (2005) The evolution of language: A comparative review. *Biology and Philosophy*, 20: 193–203.

Forkel S. J., Thiebaut de Schotten M., Kawadler J. M., Dell'Acqua F., Danek A., and Catani M. (2014) The anatomy of fronto-occipital connections from early blunt dissections to contemporary tractography. *Cortex*, 56: 73–84.

Fouts R. S. (1973) Acquisition and testing of gestural signs in four young chimpanzees. *Science,* 180: 978–980.

Fox C. J., Iaria G., and Barton J. J. S. (2008) Disconnection in prosopagnosia and face processing. *Cortex*, 44: 996–1009.

Freese J. L., and Amaral D. G. (2009) Neuroanatomy of the primate amygdala. In Whalen P. J., and Phelps E. A., eds., *The Human Amygdala*. New York: Guilford Press, pp. 3–42.

Freeman H. D., Cantalupo C., and Hopkins W. D. (2004). Asymmetries in the hippocampus and amygdala of chimpanzees (Pan troglodytes). *Behavioral Neuroscience*, 118(6): 1460.

Freiwald W. A., and Tsao D. Y. (2010) Functional compartmentalization and viewpoint generalization within the macaque face-processing system. *Science*, 330: 845–851.

Frey S., Campbell J. S. W., Pike G. B., and Petrides M. (2008) Dissociating the human language pathways with high angular resolution diffusion fiber tractography. *Journal of Neuroscience*, 28: 11435–11444.

Fried I., MacDonald K. A., and Wilson C. L. (1997) Single neuron activity in human hippocampus and amygdala during recognition of faces and objects. *Neuron*, 18: 753–765.

Frith C. D. (2007) The social brain? *Proceedings of the Royal Society B Biological Science*, 362: 671–678.

Frith U., and Frith C. D. (2003) Development and neurophysiology of mentalizing. *Proceedings of the Royal Society B Biological Sciences*, 358: 459–473.

Gallagher H. L., and Frith C. D. (2003) Functional imaging of "theory of mind". *Trends in Cognitive Science*, 7: 77–83.

Gallup G. G. Jr. (1977) Absence of self-recognition in a monkey (*Macaca fascicularis*) following prolonged exposure to a mirror. *Developmental Psychobiology: The Journal of the International Society for Developmental Psychobiology*, 10: 281–284.

Gallup G. G., McClure M. K., Hill S. D., and Bundy R. A. (1971) Capacity for self-recognition in differentially reared chimpanzees. *Psychology Record*, 21: 69–74.

Gamer M., and Büchel C. (2009) Amygdala activation predicts gaze toward fearful eyes. *Journal of Neuroscience*, 29: 9123–9126.

Gannon P. J., Holloway R. L., Broadfield D. C., and Braun A. R. (1998) Asymmetry of chimpanzee planum temporale: Humanlike pattern of Wernicke's brain language area homolog. *Science*, 279: 220–222.

Gardner B. T., and Gardner R. A. (1975) Evidence for sentence constitutents in the early utterances of child and chimpanzee. *Journal of Experimental Psychology: General*, 104: 244.

Gardner R. A., and Gardner B. T. (1969) Teaching sign language to a chimpanzee. *Science*, 165: 664–672.

Geschwind N. (1965) Disconnexion syndromes in animals and man. II. *Brain*, 88: 585–644.

Geschwind N. (1970) The organization of language and the brain. *Science*, 170: 940–944.

Glasser M. F., Goyal M. S., Preuss T. M., Raichle M. E., and Van Essen D. C. (2014) Trends and properties of human cerebral cortex: Correlations with cortical myelin content. *Neuroimage*, 93(Pt 2): 165–175.

Glasser M. F., and Rilling J. K. (2008) DTI tractography of the human brain's language pathways. *Cerebral Cortex*, 18: 2471–2482.

Glasser M. F., and Van Essen D. C. (2011) Mapping human cortical areas in vivo based on myelin content as revealed by T1- and T2-weighted MRI. *Journal of Neuroscience*, 31: 11597–11616.

Gobbini M. I., Koralek A. C., Bryan R. E., Montgomery K. J., and Haxby J. V. (2007) Two takes on the social brain: A comparison of theory of mind tasks. *Journal of Cognitive Neuroscience*, 19: 1803–1814.

Goldman-Rakic P. S., Selemon L. D., and Schwartz M. L. (1984) Dual pathways connecting the dorsolateral prefrontal cortex with the hippocampal formation and parahippocampal cortex in the rhesus monkey. *Neuroscience*, 12: 719–743.

Grill-Spector K., Knouf N., and Kanwisher N. (2004) The fusiform face area subserves face perception, not generic within-category identification. *Nature Neuroscience*, 7: 555–562.

Hackett T. A., Preuss T. M., and Kaas J. H. (2001) Architectonic identification of the core region in auditory cortex of macaques, chimpanzees, and humans. *Journal of Comparative Neurology*, 441: 197–222.

Halwani G. F., Loui P., Rüber T., and Schlaug G. (2011) Effects of practice and experience on the arcuate fasciculus: Comparing singers, instrumentalists, and non-musicians. *Frontiers in Psychology*, 2: 156.

Harasty J., Seldon H. L., Chan P., Halliday G., and Harding A. (2003) The left human speech-processing cortex is thinner but longer than the right. *Laterality*, 8: 247–260.

Hare B., Call J., and Tomasello M. (2001) Do chimpanzees know what conspecifics know? *Animal Behaviour*, 61: 139–151.

Hare B., Call J., and Tomasello M. (2006) Chimpanzees deceive a human competitor by hiding. *Cognition*, 101: 495–514.

Hau J., Sarubbo S., Houde J. C. et al. (2017) Revisiting the human uncinate fasciculus, its subcomponents and asymmetries with stem-based tractography and microdissection validation. *Brain Structure and Function*, 222: 1645–1662.

Hecht E. E., Gutman D. A., Dunn W., Keifer O. P. Jr., Sakai S., Kent M., and Preuss T. (2016) Neuroanatomical variation in domestic dog breeds. Program No. 834.13/III15.

Heekeren H. R., Wartenburger I., Schmidt H., Schwintowski H.-P., and Villringer A. (2003) An fMRI study of simple ethical decision-making. *Neuroreport*, 14: 1215–1219.

Hickok G., and Poeppel D. (2007) The cortical organization of speech processing. *Nature Reviews Neuroscience*, 8: 393–402.

Hoffman E. A., and Haxby J. V. (2000) Distinct representations of eye gaze and identity in the distributed human neural system for face perception. *Nature Neuroscience*, 3: 80–84.

Hof P. R., and Van der Gucht E. (2007) Structure of the cerebral cortex of the humpback whale, *Megaptera novaeangliae* (Cetacea, Mysticeti, Balaenopteridae). *Anatomical Record,* 290: 1–31.

Hopkins W. D., Taglialatela J. P., Nir T., Schenker N. M., and Sherwood C. C. (2010) A voxel-based morphometry analysis of white matter asymmetries in chimpanzees (*Pan troglodytes*). *Brain Behavior and Evolution*, 76: 93–100.

Hung C.-C., Yen C. C., Ciuchta J. L., Papoti D., Bock N. A., Leopold D. A., and Silva A. C. (2015) Functional mapping of face-selective regions in the extrastriate visual cortex of the marmoset. *Journal of Neuroscience*, 35: 1160–1172.

Insausti R., Marcos P., Arroyo-Jiménez M. M., Blaizot X., and Martínez-Marcos A. (2002) Comparative aspects of the olfactory portion of the entorhinal cortex and its projection to the hippocampus in rodents, nonhuman primates, and the human brain. *Brain Research Bulletin*, 57: 557–560.

Issa H. A., Staes N., Diggs-Galligan S. et al. (2018) Comparison of bonobo and chimpanzee brain microstructure reveals differences in socio-emotional circuits. *Brain Structure and Function*, 224(1): 239–251.

Jaeggi A. V., Boose K. J., White F. J., and Gurven M. (2016) Obstacles and catalysts of cooperation in humans, bonobos, and chimpanzees: Behavioural reaction norms can help explain variation in sex roles, inequality, war and peace. *Behaviour*, 153: 1015–1051.

Jaeggi A. V., De Groot E., Stevens J. M. G., and Van Schaik C. P. (2013) Mechanisms of reciprocity in primates: Testing for short-term contingency of grooming and food sharing in bonobos and chimpanzees. *Evolution and Human Behavior*, 34: 69–77.

Jaeggi A. V., Stevens J. M. G., and Van Schaik C. P. (2010) Tolerant food sharing and reciprocity is precluded by despotism among bonobos but not chimpanzees. *American Journal of Physical Anthropology*, 143: 41–51.

Jastorff J., Popivanov I. D., Vogels R., Vanduffel W., and Orban G. A. (2012) Integration of shape and motion cues in biological motion processing in the monkey STS. *Neuroimage*, 60: 911–921.

Kaas J. H. (2006) Evolution of the neocortex. *Current Biology,* 16: R910–R914.

Kaas J. H. (2013) The evolution of brains from early mammals to humans. *Interdisciplinary Reviews of Cognitive Science*, 4: 33–45.

Kaas J. H., Hackett T. A., and Tramo M. J. (1999) Auditory processing in primate cerebral cortex. *Current Opinion in Neurobiology,* 9: 164–170.

Kanwisher N., McDermott J., and Chun M. M. (1997) The fusiform face area: A module in human extrastriate cortex specialized for face perception. *Journal of Neuroscience*, 17: 4302–4311.

Kanwisher N., and Yovel G. (2006) The fusiform face area: A cortical region specialized for the perception of faces. *Proceedings of the Royal Society B Biological Sciences*, 361: 2109–2128.

Karg K., Schmelz M., Call J., and Tomasello M. (2016) Differing views: Can chimpanzees do Level 2 perspective-taking? *Animal Cognition*, 19: 555–564.

Kiefer M., and Pulvermüller F. (2012) Conceptual representations in mind and brain: Theoretical developments, current evidence and future directions. *Cortex*, 48(7): 805–825.

Klüver H., and Bucy P. C. (1937) "Psychic blindness" and other symptoms following bilateral temporal lobectomy in Rhesus monkeys. *American Journal of Physiology*, 119: 352–353.

Klüver H., and Bucy P. C. (1939) Preliminary analysis of functions of the temporal lobes in monkeys. *Archives of Neurology and Psychiatry*, 42: 979–1000.

Krachun C., Carpenter C. M., Call J., and Tomasello M. (2010) A new change-of-contents false belief test: Children and chimpanzees compared. *International Journal of Comparative Psychology*, 23: 145–165.

Kriegeskorte N., Formisano E., Sorger B., and Goebel R. (2007) Individual faces elicit distinct response patterns in human anterior temporal cortex. *Proceedings of the National Academy of Science USA*, 104: 20600–20605.

Lambon Ralph M. A., and Patterson K. (2008) Generalization and differentiation in semantic memory: Insights from semantic dementia. *Annals of the New York Academy of Science*, 1124: 61–76.

Lambon Ralph M. A., Sage K., Jones R. W., and Mayberry E. J. (2010) Coherent concepts are computed in the anterior temporal lobes. *Proceedings of the National Academy of Science USA*, 107: 2717–2722.

LeDoux J. (2007) The amygdala. *Current Biology*, 17: R868–R874.

Leslie A. M. (1987) Pretense and representation: The origins of "theory of mind." *Psychology Review*, 94: 412.

Levine B., Svoboda E., Turner G. R, Mandic M., and Mackey A. (2009) Behavioral and functional neuroanatomical correlates of anterograde autobiographical memory in isolated retrograde amnesic patient M.L. *Neuropsychologia*, 47: 2188–2196.

Lieberman M. D. (2007) Social cognitive neuroscience: A review of core processes. *Annual Review of Psychology*, 58: 259–289.

Livingstone M., and Hubel D. (1988) Segregation of form, color, movement, and depth: Anatomy, physiology, and perception. *Science*, 240: 740–749.

Livingstone M. S., Vincent J. L., Arcaro M. J., Srihasam K., Schade P. F., and Savage T. (2017) Development of the macaque face-patch system. *Nature Communications*, 8: 14897.

Lyras G. A. (2009) The evolution of the brain in Canidae (Mammalia: Carnivora). *Scripta Geologica*, 139: 1–93.

Machado C. J., Kazama A. M., and Bachevalier J. (2009) Impact of amygdala, orbital frontal, or hippocampal lesions on threat avoidance and emotional reactivity in nonhuman primates. *Emotion*, 9: 147–163.

Makris N., Papadimitriou G. M., Kaiser J. R., Sorg S., Kennedy D. N., and Pandya D. N. (2009) Delineation of the middle longitudinal fascicle in humans: A quantitative, in vivo, DT-MRI study. *Cerebral Cortex*, 19: 777–785.

Marchina S., Zhu L. L., Norton A., Zipse L., Wan C. Y., and Schlaug G. (2011) Impairment of speech production predicted by lesion load of the left arcuate fasciculus. *Stroke*, 42: 2251–2256.

Mars R. B., Neubert F.-X., Verhagen L., Sallet J., Miller K. L., Dunbar R. I. M., and Barton M. A. (2014) Primate comparative neuroscience using magnetic resonance imaging: Promises and challenges. *Frontiers in Neuroscience*, 8: 298.

Mason W. A., Capitanio J. P., Machado C. J., Mendoza S. P., and Amaral D. G. (2006) Amygdalectomy and responsiveness to novelty in rhesus monkeys (*Macaca mulatta*): Generality and individual consistency of effects. *Emotion*, 6: 73–81.

Menjot de Champfleur N., Lima Maldonado I., Moritz-Gasser S., Machi P., Le Bars E., Bonafé A., and Duffau H. (2013) Middle longitudinal fasciculus delineation within language pathways: A diffusion tensor imaging study in human. *European Journal of Radiology*, 82: 151–157.

Mishkin M., Ungerleider L. G., and Macko K. A. (1983) Object vision and spatial vision: Two cortical pathways. *Trends in Neuroscience*, 6: 414–417.

Moll J., Eslinger P. J., and Oliveira-Souza R. (2001) Frontopolar and anterior temporal cortex activation in a moral judgment task: Preliminary functional MRI results in normal subjects. *Arquivos de Neuro-Psiquiatria*, 59: 657–664.

Moll J., de Oliveira-Souza R., Bramati I. E., and Grafman J. (2002) Functional networks in emotional moral and nonmoral social judgments. *Neuroimage*, 16: 696–703.

Morris J. S., Frith C. D., Perrett D. I., Rowland D., Young A. W., Calder A. J., and Doland R. J. (1996) A differential neural response in the human amygdala to fearful and happy facial expressions. *Nature*, 383: 812–815.

Nasr S., Liu N., Devaney K. J., Yue X., Rajimehr R., Ungerleider L. G., and Tooteli R. B. H. (2011) Scene-selective cortical regions in human and nonhuman primates. *Journal of Neuroscience*, 31: 13771–13785.

Nucifora P. G. P., Verma R., Melhem E. R., Gur R. E., and Gur R. C. (2005) Leftward asymmetry in relative fiber density of the arcuate fasciculus. *Neuroreport*, 16: 791–794.

Olson I. R., McCoy D., Klobusicky E., and Ross L. A. (2013) Social cognition and the anterior temporal lobes: A review and theoretical framework. *Social Cognitive and Affective Neuroscience,* 8: 123–133.

Orban G. A., Van Essen D., and Vanduffel W. (2004) Comparative mapping of higher visual areas in monkeys and humans. *Trends in Cognitive Science*, 8: 315–324.

Pabba M. (2013) Evolutionary development of the amygdaloid complex. *Frontiers in Neuroanatomy*, 7: 27.

Parker G. J. M., Luzzi S., Alexander D. C., Wheeler-Kingshott C. A. M., Ciccarelli O., and Lambon Ralph M. A. (2005) Lateralization of ventral and dorsal auditory-language pathways in the human brain. *Neuroimage*, 24: 656–666.

Parr L. A., Hecht E., Barks S. .K, Preuss T. M., and Votaw J. R. (2009) Face processing in the chimpanzee brain. *Current Biology*, 19: 50–53.

Parr L. A., Siebert E., and Taubert J. (2011) Effect of familiarity and viewpoint on face recognition in chimpanzees. *Perception*, 40: 863–872.

Parr L. A., and Taubert J. (2011) The importance of surface-based cues for face discrimination in non-human primates. *Proceedings of the Royal Society B Biological Science*, 278: 1964–1972.

Parvizi J., Jacques C., Foster B. L., Witthoft N., Rangarajan V., Weiner K. S., and Grill-Spector K. (2012) Electrical stimulation of human fusiform face-selective regions distorts face perception. *Journal of Neuroscience*, 32: 14915–14920.

Passingham R. E., and Wise S. P. (2012) *The Neurobiology of the Prefrontal Cortex: Anatomy, Evolution, and the Origin of Insight*. Oxford: Oxford University Press.

Perrett D. I., Smith P. A., Potter D. D., Mistlin A. J., Head A. S., Milner A. D., and Jeeves M. A. (1984) Neurones responsive to faces in the temporal cortex: Studies of functional organization, sensitivity to identity and relation to perception. *Human Neurobiology*, 3: 197–208.

Pessoa L., McKenna M., Gutierrez E., and Ungerleider L. G. (2002) Neural processing of emotional faces requires attention. *Proceedings of the National Academy of Science USA*, 99: 11458–11463.

Petrides M., and Pandya D. N. (2007) Efferent association pathways from the rostral prefrontal cortex in the macaque monkey. *Journal of Neuroscience*, 27: 11573–11586.

Pitcher D., Duchaine B., Walsh V., Yovel G., and Kanwisher N. (2011) The role of lateral occipital face and object areas in the face inversion effect. *Neuropsychologia*, 49: 3448–3453.

Pobric G., Jefferies E., and Ralph M. A. L. (2007) Anterior temporal lobes mediate semantic representation: Mimicking semantic dementia by using rTMS in normal participants. *Proceedings of the National Academy of Science USA*, 104: 20137–20141.

Pobric G., Jefferies E., and Ralph M. A. L. (2010) Amodal semantic representations depend on both anterior temporal lobes: Evidence from repetitive transcranial magnetic stimulation. *Neuropsychologia*, 48: 1336–1342.

Povinelli D. J., and Eddy T. J. (1996) Chimpanzees: Joint visual attention. *Psychological Science,* 7: 129–135.

Povinelli D. J., Nelson K. E., and Boysen S. T. (1992a) Comprehension of role reversal in chimpanzees: Evidence of empathy? *Animal Behaviour*, 43(4): 633–640.

Povinelli D. J., Parks K. A., and Novak M. A. (1992b) Role reversal by rhesus monkeys, but no evidence of empathy. *Animal Behaviour*, 44: 269–281.

Powell H. W. R., Parker G. J. M., Alexander D. C. et al. (2006) Hemispheric asymmetries in language-related pathways: A combined functional MRI and tractography study. *Neuroimage*, 32: 388–399.

Premack D., and Woodruff G. (1978) Does the chimpanzee have a theory of mind? *Behavioral and Brain Sciences*, 1: 515–526.

Preuss T. M. (2011) The human brain: Rewired and running hot. *Annals of the New York Academy of Science*, 1225 (Suppl. 1): E182–E191.

Rausch R., Serafetinides E. A., and Crandall P. H. (1977) Olfactory memory in patients with anterior temporal lobectomy. *Cortex*, 13: 445–452.

Reiman E. M., Lane R. D., Ahern G. L. et al. (1997) Neuroanatomical correlates of externally and internally generated human emotion. *American Journal of Psychiatry*, 154: 918–925.

Rilling J. K. (2006) Human and nonhuman primate brains: Are they allometrically scaled versions of the same design? *Evolutionary Anthropology*, 15: 65–77.

Rilling J. K., Glasser M. F., Jbabdi S., Andersson J., and Preuss T. M. (2011) Continuity, divergence, and the evolution of brain language pathways. *Frontiers in Evolutionary Neuroscience*, 3: 11.

Rilling J. K., Glasser M. F., Preuss T. M., Ma X., Zhao T., Hu X., and Behrens T. E. (2008). The evolution of the arcuate fasciculus revealed with comparative DTI. *Nature Neuroscience*, 11: 426.

Rilling J. K., Scholz J., Preuss T. M., Glasser M. F., Errangi B. K., and Behrens T. E. (2012) Differences between chimpanzees and bonobos in neural systems supporting social cognition. *Social Cognitive and Affective Neuroscience,* 7: 369–379.

Rilling J. K., and Seligman R. A. (2002) A quantitative morphometric comparative analysis of the primate temporal lobe. *Journal of Human Evolution*, 42: 505–533.

Rivas E. (2005) Recent use of signs by chimpanzees (*Pan troglodytes*) in interactions with humans. *Journal of Comparative Psychology*, 119: 404–417.

Rogers Flattery C. N., Rosen R. F., Farberg A. S. et al. (2020). Quantification of neurons in the hippocampal formation of chimpanzees: Comparison to rhesus monkeys and humans. *Brain Structure and Function*, 1–11.

Rogers T. T., Lambon Ralph M. A., Garrard P., Bozeat S., McClelland J. L., Hodges J. R., and Patterson K. (2004) Structure and deterioration of semantic memory: A neuropsychological and computational investigation. *Psychological Review*, 111: 205–235.

Romanski L. M., Bates J. F., and Goldman-Rakic P. S. (1999) Auditory belt and parabelt projections to the prefrontal cortex in the rhesus monkey. *Journal of Comparative Neurology*, 403: 141–157.

Ross L. A., and Olson I. R. (2010) Social cognition and the anterior temporal lobes. *Neuroimage*, 49: 3452–3462.

Rossion B., Dricot L., Devolder A. et al. (2000). Hemispheric asymmetries for whole-based and part-based face processing in the human fusiform gyrus. *Journal of Cognitive Neuroscience*, 12(5): 793–802.

Rudrauf D., Mehta S. and Grabowski T. J. (2008) Disconnection's renaissance takes shape: Formal incorporation in group-level lesion studies. *Cortex*, 44: 1084–1096.

Rutishauser U., Mamelak A. N., and Adolphs R. (2015) The primate amygdala in social perception – Insights from electrophysiological recordings and stimulation. *Trends in Neuroscience*, 38: 295–306.

Rutishauser U., Tudusciuc O., Neumann D. et al. (2011) Single-unit responses selective for whole faces in the human amygdala. *Current Biology*, 21: 1654–1660.

Sallet J., Mars R. B., Noonan M. P. et al. (2011) Social network size affects neural circuits in macaques. *Science*, 334: 697–700.

Samson D., Apperly I. A., Chiavarino C., and Humphreys G. W. (2004) Left temporoparietal junction is necessary for representing someone else's belief. *Nature Neuroscience*, 7: 499–500.

Saur D., Kreher B. W., Schnell S. et al. (2008) Ventral and dorsal pathways for language. *Proceedings of the National Academy of Science USA*, 105: 18035–18040.

Savage-Rumbaugh S., McDonald K., Sevcik R. A., Hopkins W. D., and Rubert E. (1986) Spontaneous symbol acquisition and communicative use by pygmy chimpanzees (*Pan paniscus*). *Journal of Experimental Psychology: General*, 115: 211–235.

Saxe R., and Kanwisher N. (2003) People thinking about thinking people: The role of the temporo-parietal junction in "theory of mind." *Neuroimage*, 19: 1835–1842.

Saxe R., and Powell L. J. (2006) It's the thought that counts: Specific brain regions for one component of theory of mind. *Psychological Science*, 17: 692–699.

Schalk G., Kapeller C., Guger C. et al. (2017) Facephenes and rainbows: Causal evidence for functional and anatomical specificity of face and color processing in the human brain. *Proceedings of the National Academy of Science USA*, 114: 12285–12290.

Schmahmann J. D., Pandya D. N., Wang R., Dai G., D'Arceuil H. E., de Crespigny A. J., and Wedeen V. J. (2007) Association fibre pathways of the brain: Parallel observations from diffusion spectrum imaging and autoradiography. *Brain*, 130: 630–653.

Schmahmann J., and Pandya D. (2009) *Fiber Pathways of the Brain*. Oxford: Oxford University Press.

Schmolck H., and Squire L. R. (2001) Impaired perception of facial emotions following bilateral damage to the anterior temporal lobe. *Neuropsychology*, 15: 30–38.

Schoenemann P. T. (1997) An MRI study of the relationship between human neuroanatomy and behavioral ability. PhD Dissertation, University of California, Berkeley.

Schurz M., Radua J., Aichhorn M., Richlan F., and Perner J. (2014) Fractionating theory of mind: A meta-analysis of functional brain imaging studies. *Neuroscience and Biobehavioral Reviews*, 42: 9–34.

Seltzer B., and Pandya D. N. (1984) Further observations on parieto-temporal connections in the rhesus monkey. *Experimental Brain Research*, 55: 301–312.

Semendeferi K., and Damasio H. (2000) The brain and its main anatomical subdivisions in living hominoids using magnetic resonance imaging. *Journal of Human Evolution*, 38: 317–332.

Shapleske J., Rossell S. L., Woodruff P. W., and David A. S. (1999) The planum temporale: A systematic, quantitative review of its structural, functional and clinical significance. *Brain Research Brain Research Reviews*, 29: 26–49.

Simmons W. K., and Martin A. (2009) The anterior temporal lobes and the functional architecture of semantic memory. *Journal of the International Neuropsychology Society*, 15: 645–649.

Simmons W. K., Reddish M., Bellgowan P. S. F., and Martin A. (2010) The selectivity and functional connectivity of the anterior temporal lobes. *Cerebral Cortex*, 20: 813–825.

Skeide M. A., and Friederici A. D. (2016) The ontogeny of the cortical language network. *Nature Reviews Neuroscience*, 17: 323–332.

Small D. M., Jones-Gotman M., Zatorre R. J., Petrides M., and Evans A. C. (1997) A role for the right anterior temporal lobe in taste quality recognition. *Journal of Neuroscience*, 17: 5136–5142.

Sobolewski M. E., Brown J. L., and Mitani J. C. (2012) Territoriality, tolerance and testosterone in wild chimpanzees. *Animal Behaviour*, 84: 1469–1474.

Spocter M. A., Hopkins W.D., Barks S. K. et al. (2012) Neuropil distribution in the cerebral cortex differs between humans and chimpanzees. *Journal of Comparative Neurology*, 520: 2917–2929.

Spocter M. A., Hopkins W. D., Garrison A. R., Bauernfeind A. L., Stimpson C. D., Hof P. R., and Sherwood C. C. (2010) Wernicke's area homologue in chimpanzees (*Pan troglodytes*) and its relation to the appearance of modern human language. *Proceedings of the Royal Society of London B: Biological Sciences*, rspb20100011.

Stefanacci L., and Amaral D. G. (2002) Some observations on cortical inputs to the macaque monkey amygdala: An anterograde tracing study. *Journal of Comparative Neurology*, 451: 301–323.

Steiper M. E., and Seiffert E. R. (2012) Evidence for a convergent slowdown in primate molecular rates and its implications for the timing of early primate evolution. *Proceedings of the National Academy of Science USA*, 109: 6006–6011.

Stimpson C. D., Barger N., Taglialatela J. P., Gendron-Fitzpatrick A., Hof P. R., Hopkins W. D., and Sherwood C. C. (2016) Differential serotonergic innervation of the amygdala in bonobos and chimpanzees. *Social Cognitive and Affective Neuroscience*, 11: 413–422.

Sugiura M., Sassa Y., Watanabe J. et al. (2006) Cortical mechanisms of person representation: Recognition of famous and personally familiar names. *Neuroimage*, 31: 853–860.

Surbeck M., Girard-Buttoz C., Boesch C. et al. (2017) Sex-specific association patterns in bonobos and chimpanzees reflect species differences in cooperation. *Royal Society Open Science*, 4: 161081.

Tan J., Ariely D., and Hare B. (2017) Bonobos respond prosocially toward members of other groups. *Scientific Reports*, 7: 14733.

Tan J., and Hare B. (2013) Bonobos share with strangers. *PLoS ONE*, 8: e51922.

Taubert J., Wardle S., Flessert M., Leopold D., and Ungerleider L. (2017) Evidence for face pareidolia in rhesus monkeys. *Journal of Vision*, 17: 845–845.

Terrace H. S. (1979) *Nim*. New York: Alfred A. Knoff.

Thiebaut de Schotten M., Dell'Acqua F., Valabregue R., and Catani M. (2012) Monkey to human comparative anatomy of the frontal lobe association tracts. *Cortex*, 48: 82–96.

Thomas Schoenemann P. (1999) Syntax as an emergent characteristic of the evolution of semantic complexity. *Minds Mach*, 9: 309–346.

Thompson J. C., Clarke M., Stewart T., and Puce A. (2005) Configural processing of biological motion in human superior temporal sulcus. *Journal of Neuroscience*, 25: 9059–9066.

Tomonaga M., Tanaka M., Matsuzawa T. et al. (2004) Development of social cognition in infant chimpanzees (*Pan troglodytes*): Face recognition, smiling, gaze, and the lack of triadic interactions 1. *Japanese Psychological Research*, 46: 227–235.

Tsao D. Y., Moeller S., and Freiwald W. A. (2008) Comparing face patch systems in macaques and humans. *Proceedings of the National Academy of Science USA*, 105: 19514–19519.

Tsukiura T., Mano Y., Sekiguchi A. et al. (2010) Dissociable roles of the anterior temporal regions in successful encoding of memory for person identity information. *Journal of Cognitive Neuroscience*, 22: 2226–2237.

Turken A. U., and Dronkers N. F. (2011) The neural architecture of the language comprehension network: Converging evidence from lesion and connectivity analyses. *Frontiers in Systems Neuroscience*, 5: 1.

Ueno T., Saito S., Rogers T. T., and Lambon Ralph M. A (2011) Lichtheim 2: Synthesizing aphasia and the neural basis of language in a neurocomputational model of the dual dorsal-ventral language pathways. *Neuron*, 72: 385–396.

Ungerleider L. G., and Desimone R. (1986) Cortical connections of visual area MT in the macaque. *Journal of Comparative Neurology*, 248: 190–222.

Visser M., Jefferies E., Embleton K. V., and Ralph M. A. L. (2012) Both the middle temporal gyrus and the ventral anterior temporal area are crucial for multimodal semantic processing: Distortion-corrected fMRI evidence for a double gradient of information convergence in the temporal lobes. *Journal of Cognitive Neuroscience*, 24: 1766–1778.

Watson J. D., Myers R., Frackowiak R. S. et al. (1993) Area V5 of the human brain: Evidence from a combined study using positron emission tomography and magnetic resonance imaging. *Cerebral Cortex*, 3: 79–94.

Weiller C., Bormann T., Saur D., Musso M., and Rijntjes M. (2011) How the ventral pathway got lost – And what its recovery might mean. *Brain and Language*, 118: 29–39.

Whiten A. (1998) Imitation of the sequential structure of actions by chimpanzees (*Pan troglodytes*). *Journal of Comparative Psychology*, 112: 270–281.

Wong F. C. K., Chandrasekaran B., Garibaldi K., and Wong P. C. M. (2011) White matter anisotropy in the ventral language pathway predicts sound-to-word learning success. *Journal of Neuroscience*, 31: 8780–8785.

Yeatman J. D., Dougherty R. F., Rykhlevskaia E., Sherbondy A. J., Deutsch G. K., Wandell B. A., and Ben-Shacharet M. (2011) Anatomical properties of the arcuate fasciculus predict phonological and reading skills in children. *Journal of Cognitive Neuroscience*, 23: 3304–3317.

Yeo B. T. T., Krienen F. M., Sepulcre J. et al. (2011) The organization of the human cerebral cortex estimated by intrinsic functional connectivity. *Journal of Neurophysiology*, 106: 1125–1165.

Young L., Dodell-Feder D., and Saxe R. (2010) What gets the attention of the temporo-parietal junction? An fMRI investigation of attention and theory of mind. *Neuropsychologia*, 48: 2658–2664.

Zahn R., Moll J., Iyengar V., Huey E. D., Tierney M., Krueger F., and Grafman J. (2009) Social conceptual impairments in frontotemporal lobar degeneration with right anterior temporal hypometabolism. *Brain*, 132: 604–616.

Zahn R., Moll J., Krueger F., Huey E. D., Garrido G., and Grafman J. (2007) Social concepts are represented in the superior anterior temporal cortex. *Proceedings of the National Academy of Science USA*, 104: 6430–6435.

7 Role of Oxytocin and Vasopressin V1a Receptor Variation on Personality, Social Behavior, Social Cognition, and the Brain in Nonhuman Primates, with a Specific Emphasis on Chimpanzees

William D. Hopkins and Robert D. Latzman

Introduction

Primates engage in a variety of complex social behaviors. Broadly speaking, these social behaviors can range from agonistic to affiliative depending on the context of a given interaction and a variety of other factors such as the sex, age, familiarity, and rank of individuals. Social interactions of any kind – whether cooperative or "prosocial," as they is often termed, or conflict- and aggression-based, often termed "antisocial" – are based on the individual's personality and cognitive traits and are manifest in their communication and behaviors directed toward others. (Chapter 5 discusses the problems associated with this terminology.) In other words, similar to humans, within different primate groups there are individual differences in the frequency of behaviors that reflect the range of social behaviors that are expressed during social interactions. Understanding how or why this cluster of traits varies among individuals is therefore important for understanding social interactions. It is now clear that one source of individual variation in both competitive and cooperative behavior is genes. Two of the most widely studied are genes that regulate the receptor distribution of oxytocin (*OXTR*) and vasopressin (*AVPRA*, *AVPR1B* and *AVPR2*). (See Box 7.1 for an overview of terminology and concepts associated with genetic variation.)

Oxytocin (OXY) and vasopressin (AVP) are neuropeptides implicated in the development and maintenance of pair bonding and social relationships, including attachment, social cognition, and social communication in mammals (Goodson and Bass, 2001; Hammock and Young, 2006; Donaldson and Young, 2008; Meyer-Lindenberg et al., 2011; Caldwell, 2017; see also Chapter 5) and other vertebrates

Portions of this research were supported by NIH grants MH-92923 and NS-42867 to WDH. We thank all the research assistants, students, and veterinarians for their assistance in data collection. We also thank Dr. Larry Young and his staff for genotyping of *AVPR1A* in some of the chimpanzee findings discussed in this manuscript.

Box 7.1 Genetic and Genomic Concepts

Studying gene sequence (genetics) variation or variation in the expression of those genes over the life span (genomics) has become an important aspect of investigating the mechanism and evolution of social behavior. Here and in other studies, the terminology used in presenting this information may not be familiar, nor might some of the overarching ideas about such variation that have emerged during the study of the genomics of social behavior. Here we provide a simplified overview of both.

Although it is tempting to assume that there is a "gene" for a trait – for example a specific gene for monogamy present in species that express that social behavior and absent in those that do not – many years of research have shown this is not how the molecular underpinnings of complex behaviors (and likely other traits) are regulated. Rather, it is variation in the *expression* of widely shared genes that maps best onto variation in behavior. *Expression* refers to the process by which the gene (DNA) is translated into its product (such as a protein). Variation in the amount, spatial distribution, or timing of gene expression most often relates to variation in social traits within and across species. Relevant to this chapter, the expression of the gene encoding a receptor for vasopressin, the V1a receptor, differs in monogamous and promiscuous species of voles, leading to differences in the amount of the receptor and its distribution across brain areas. That is, there is no "monogamy gene." Instead, genetic-based variation in the amount and distribution of an important peptide receptor in the brain shared by all vertebrates leads to fundamental differences in social behavior.

Expression variation occurs because there exist *gene polymorphisms*, that is, variation in the DNA sequence making up the coding region of a gene (its *exon*; note that not all of the DNA sequence is translated into a product) or areas that regulate its expression (*promotor* regions). Such polymorphisms can result from a point mutation changing a single DNA nucleotide, the deletion of part of a gene sequence or all of the gene, or the insertion of additional nucleotides into the gene. Many polymorphisms have no functional consequences, but variation in the critical regions of the exon may affect the function of the protein that the gene codes. Variation in, that is, polymorphisms of, promotor regions can also ultimately affect how a gene is acting by resulting in differences in the amount of that gene's expression in different individuals or in different species rather than affecting the function of the gene product itself. A particular type of variation occurs in *microsatellite* regions, that is, areas of repetitive DNA of varying lengths. Microsatellite variation can be manifest as either deletions or insertions of the repetitive DNA and hence this type of variation if often termed an *indel* mutation (for "insertion or deletion"). Variation in microsatellite regions associated with promotor regions of key brain receptors appear to play a large role in variation in social behavior, as explained in this chapter with regard to the vasopressin V1a receptor.

Finally, it is worth a reminder that vertebrates and most invertebrates have their chromosomes in pairs, and therefore have genes matched up in pairs. The same

Box 7.1 (*cont.*)

gene polymorphism, or mutations resulting in deletions/insertions in the genes, may be found in both genes of the pair, in neither, or in only one in any individual. These are designated by a shorthand using + and −. In this chapter, for example, DupB−/− means that a particular microsatellite deletion is found on both chromosomes (the individual is *homozygous* for the mutation resulting in the deletion), DupB+/− means that only one of the paired genes has the microsatellite deletion (the individual is *heterozygous* for that deletion), and DupB+/+ indicates the individual lacks the deletion on both chromosomal pairs. Generally speaking, the homozygous condition manifests the greatest effect of the genetic mutation, whereas the heterozygous condition may have a phenotype intermediate from the −/− and +/+ conditions, but is often not significantly different from the +/+ condition, presumably because of some compensatory effects.

(Goodson and Bass, 2001; Wilczynski et al., 2017). With specific reference to human and nonhuman primates, numerous studies have implicated endogenous variation in AVP and particularly OXY assayed from blood, salvia, or cerebrospinal fluid with variation in typical and atypical social behavior and cognition (Ebstein et al., 2012; Uzefovsky et al., 2012; Taylor and French, 2015; Ebert and Brune, 2017; Parker et al., 2018). Additional studies, albeit typically with small sample sizes, have demonstrated the direct effect of intranasal administration of vasopressin and oxytocin on a variety of dimensions of social behavior and cognition in both human (Guastella et al., 2010a, 2010b; Evans et al., 2014; Leng and Ludwig, 2016) and nonhuman primates (Parr et al., 2013, 2018; Brosnan et al., 2015; Madrid et al., 2017; Samuni et al., 2017; Ziegler and Crockford, 2017; Bauman and Schumann, 2018; Bauman et al., 2018; Cavanaugh et al., 2018). For example, intranasal OXY has been shown to increase social abilities among children with autism spectrum disorder ($N = 32$), with this effect strongest for individuals lowest on blood OXY concentration (Parker et al., 2018). Among nonhuman primates, intranasally administered OXY has similarly been found to increase sociability and reduce dominance among macaque monkeys ($N = 12$; Jiang and Platt, 2018). It is important to acknowledge that some of these studies have significant limitations, not the least of which is sample size and statistical power. Indeed, in humans, more well-powered studies on the effects of intranasal oxytocin and vasopressin administration on social behavior and cognition have failed to find associations (Tabak et al., 2019).

From a comparative perspective, it has often been claimed that eutherian mammals have one receptor for oxytocin (*OXTR*) and three receptors for vasopressin (*AVPR1A*, *AVPR1B*, and *AVPR2*), and that these systems are highly conserved across species. However, some of these assumptions have recently been challenged by a series of findings on phylogenetic variation in coding regions for vasopressin and, specifically, oxytocin. (For review see French et al., 2016.) Notably, within New World primates, the genetic structure of oxytocin is highly variable and appears to be associated with taxon or species variation in social monogamy. Thus, rather than focus on phylogenetic variation, in this review, we focus on the role of polymorphisms in *OXTR* and *AVPR1A* in intraspecies variation in social behavior and cognition. In humans, single nucleotide polymorphisms (SNPs) in *OXTR* are associated with a variety of social behaviors, social cognition, and neural functions (Donaldson and Young, 2008; Meyer-Lindenberg et al., 2011; Skuse et al., 2014). Though less studied, there is also evidence in humans that polymorphisms in the V1 receptor gene (*AVPR1A*) are associated with individual variation in personality, social behavior, and cognition. For instance, *AVPR1A* polymorphic variation in the microsatellite element RS3 has been linked to variation in social behavior, including empathy, altruistic behavior, and pair bonding in males (Walum et al., 2008; Avinum et al., 2011; Ebstein et al., 2012; Uzefovsky et al., 2015). Other studies in humans have reported associations between *AVPR1A* and age at first intercourse (Yirmiya et al., 2006), increased Novelty Seeking, decreased Harm Avoidance (Walum et al., 2008), and increased Reward Dependence (Bachner-Melman et al., 2005). It has also been suggested that *OXTR* and *AVPR1A* may be candidate susceptibility genes for autism, a neurodevelopmental disorder that is marked by problems in the development of normal social relationships, socio-communicative and socio-cognitive abilities (Kim et al., 2002; Hammock and Young, 2006; LoParo and Waldman, 2015; Francis et al., 2016; Zhang et al., 2017).

Studies in nonhuman primates (largely chimpanzees) have examined the association between polymorphisms in *OXTR* and *AVPR1A* with personality, social behavior, and brain variations. In particular, vasopressin receptor variation is highly conserved in nonhuman primates; however, one of the remarkable aspects of the *AVPR1A microsatellite* in nonhuman primates is that the polymorphic variation in chimpanzees is, in essence, dichotomous (Figure 7.1). Chimpanzees are polymorphic at the same *AVPR1A* locus as are many primates (including humans) but chimpanzee alleles either have both the STR1 and RS3 microsatellite duplication, like humans, or completely lack the behaviorally linked RS3 (Hammock and Young, 2005; Donaldson et al., 2008; Rosso et al., 2008). In short, this polymorphism creates essentially two populations of chimpanzees that are either (a) homozygous for alleles lacking the RS3 (DupB−/−) or (b) heterozygous with one allele having both the RS3 and STR1 repeat (DupB+/−), as in humans and other great apes.

Polymorphisms in the *AVPR1A* Gene Are Associated with Different Dimensions of Personality

Measuring Personality in Nonhuman Primates

Several studies have tested whether polymorphisms in the V1a receptor gene, *AVPR1A*, are associated with dimensions of personality in chimpanzees, bonobos, and marmosets. Personality in nonhuman primates is primarily measured by having care or research staff rate individual monkeys or apes on a set of item-level adjectives that include a behavioral description characterizing different dispositional traits

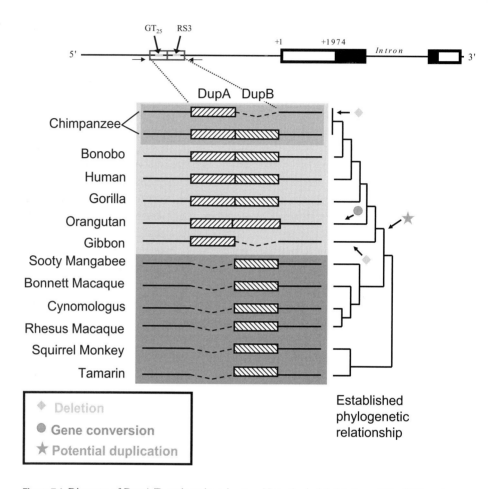

Figure 7.1 Diagram of DupA/B regions in primates. Note the indel deletion of the RS3 region in some chimpanzees which is absent in other great ape species including humans. (From Donaldson et al., 2008.) (A black and white version of this figure will appear in some formats. For the color version, please refer to the plate section.)

(Freeman and Gosling, 2010; Weiss et al., 2011). For instance, the Hominoid Personality Assessment instrument originally developed by King and Figueredo (1977) and expanded upon by Weiss et al. (2006) includes the adjective "manipulative" along with the description "Subject is adept at forming social relationships for its own advantage, especially using alliances and friendships to increase its social standing. Subjects seems able and willing to use others." Based on this adjective and description, raters assign a value from 1 (displays either total absence or negligible amounts of the trait) to 7 (displays extreme amount of the trait) that reflect the degree to which a given primate expresses a given behavioral trait. Typically, the rating scores are then subjected to factor analysis to explicate different dimensions of personality within and between species.

For instance, Weiss et al. (2007) used this instrument to evaluate personality in a sample of 202 zoo-living chimpanzees and 175 chimpanzees housed at the Yerkes National Primate Research Center (see Table 7.1). The ratings for each chimpanzee and item within each cohort as well as for the entire sample were subjected to factor analysis and the results revealed four personality dimensions: dominance, conscientiousness, extraversion, and agreeableness. The items positively and negatively loading on each personality dimension are also shown in Table 7.1. As can be seen, for the personality dimension dominance (D_{CH}), the following items loaded either positively or negatively (dominant+, submissive−, dependent−, timid−, persistent+, bullying+, independent+, decisive+, aggressive+, fearful−, cautious−, defiant+, stingy+, gentle−, manipulative+). In contrast, different questionnaire items load positively or negatively on the personality dimensions of extraversion (E_{CH}), conscientiousness (C_{CH}), and agreeableness (A_{CH}).

Several subsequent studies have largely revealed consistent results in additional captive and wild chimpanzee populations (Freeman and Gosling, 2010; Freeman et al., 2013). Moreover, similar if not identical personality questionnaires have been used with other great ape species including bonobos, orangutans, and gorillas (Weiss et al., 2006, 2015; Schaefer and Steklis, 2014; Eckardt et al., 2015). An obvious advantage of personality rating scales is that they can be obtained in relatively large samples of individuals in a short time period and, assuming that they have some construct validity in the context of how nonhuman primates behave in their day-to-day interactions, have potentially widespread application (Murray, 2011). Indeed, similar kinds of rating systems have been developed for characterizing sociality and related behavioral constructs, such as temperament, in chimpanzees and macaque monkeys (Marrus et al., 2011; Feczko et al., 2016).

The Relationship between Personality and Oxytocin and Vasopressin Receptor Gene Polymorphisms

Regarding oxytocin and vasopressin, five different studies in chimpanzees, one study in bonobos, and one in marmosets have examined whether polymorphisms in the *OXTR* and *AVPR1A* genes are associated with personality. Using the personality data from the study by Weiss et al. (2007), Hopkins et al. (2012) genotyped 83 of these chimpanzees for the *AVPR1A* gene and found that DupB+/− males had significantly higher dominance and lower conscientiousness scores than females with the same genotype. When considering the individual item-level adjectives, Dup+/− males had higher ranked scores than Dup+/− females on the following items: dominant, personable, reckless, playful, bullying, aggressive, manipulative, excitable, impulsive, defiant, and erratic. In contrast, DupB+/− females had significantly higher ranked scores than DupB+/− males on the following items: fearful, cautious, timid, sympathetic, submissive, and dependent. All told, both scale-level trait scores, as well as rank scores on individual items, suggest an important sex-specific role for AVP receptor variation in the explanation of dominance- and conscientiousness-related behaviors. (See Chapter 5 for further discussion of sex-specific actions of AVP and oxytocin.)

Table 7.1 Structures of zoo, Yerkes, and pooled samples derived via principal components analysis

Item	Zoo[a]				Yerkes[b]				Pooled[a]			
	D_{CH}	E_{CH}	C_{CH}[c]	A_{CH}	D_{CH}	E_{CH}	C_{CH}[c]	A_{CH}	D_{CH}[c]	E_{CH}	C_{CH}[c]	A_{CH}
Dominant	**0.83**	−0.24	0.03	0.13	**0.83**	−0.04	−0.15	0.02	**0.81**	−0.10	−0.25	0.08
Submissive	**−0.81**	0.09	−0.21	0.02	**−0.84**	−0.04	−0.13	0.08	**−0.85**	−0.05	0.05	0.05
Dependent	**−0.79**	0.31	−0.21	−0.06	**−0.77**	0.13	−0.24	0.11	**−0.83**	0.18	−0.03	0.03
Timid	**−0.78**	−0.16	−0.27	−0.06	**−0.70**	−0.33	−0.36	0.08	**−0.78**	−0.33	−0.12	0.00
Persistent	**0.75**	0.24	−0.13	0.11	**0.62**	0.26	−0.28	0.02	**0.58**	0.24	**−0.40**	0.09
Bullying	**0.74**	−0.08	−0.33	−0.27	**0.74**	0.05	−0.33	−0.24	**0.63**	−0.01	**−0.52**	−0.23
Independent	**0.73**	−0.20	0.21	0.13	**0.73**	−0.16	0.14	0.04	**0.77**	−0.14	0.02	0.08
Decisive	**0.72**	−0.01	0.22	0.26	**0.66**	0.00	0.11	0.16	**0.69**	0.06	−0.02	0.23
Aggressive	**0.72**	−0.09	−0.36	−0.34	**0.67**	0.04	**−0.46**	−0.30	**0.57**	0.00	**−0.60**	−0.27
Fearful	**−0.71**	−0.06	**0.47**	0.04	**−0.59**	0.11	−0.36	0.10	**−0.72**	0.02	−0.20	0.11
Cautious	**−0.70**	−0.14	−0.13	0.25	**−0.65**	−0.06	0.02	**0.41**	**−0.66**	−0.13	0.16	0.33
Defiant	**0.59**	0.17	**−0.43**	−0.25	**0.60**	0.05	**−0.49**	−0.09	**0.44**	0.09	**−0.63**	−0.12
Stingy	**0.59**	−0.09	**−0.40**	−0.04	**0.62**	−0.10	−0.38	−0.14	**0.49**	−0.10	**−0.54**	−0.09
Gentle	**−0.57**	0.01	**0.43**	**0.51**	**−0.50**	0.06	0.34	**0.58**	**−0.42**	0.05	**0.53**	**0.52**
Manipulative	**0.53**	0.13	−0.32	0.17	**0.55**	0.06	−0.34	0.29	**0.42**	0.04	**−0.45**	0.23
Active	−0.05	**0.90**	−0.10	−0.17	0.11	**0.80**	−0.20	−0.22	−0.08	**0.85**	−0.21	−0.12
Playful	−0.14	**0.88**	−0.10	−0.09	0.05	**0.86**	0.03	0.03	−0.13	**0.88**	−0.05	0.04
Lazy	0.09	**−0.85**	−0.03	0.15	−0.12	**−0.82**	0.05	0.18	0.04	**−0.86**	0.07	0.08
Solitary	−0.06	**−0.78**	−0.12	−0.01	−0.10	**−0.78**	−0.22	−0.04	−0.06	**−0.79**	−0.12	−0.06
Sociable	0.01	**0.76**	0.18	0.32	0.05	**0.61**	0.27	**0.47**	0.02	**0.70**	0.22	**0.42**
Inquisitive	0.21	**0.76**	−0.07	0.22	0.17	**0.77**	0.06	**0.16**	0.12	**0.76**	−0.07	**0.26**

Imitative	−0.16	**0.74**	−0.16	−0.06	−0.25	**0.33**	−0.34	**0.32**	−0.33	**0.51**	−0.21	**0.19**
Friendly	−0.26	**0.66**	0.26	0.47	−0.25	**0.42**	0.34	**0.53**	−0.21	**0.54**	0.36	**0.50**
Depressed	−0.09	**−0.67**	−0.37	−0.12	−0.24	**−0.69**	−0.41	**−0.05**	−0.21	**−0.74**	−0.31	**−0.12**
Inventive	0.31	**0.60**	−0.01	0.30	0.35	**0.57**	−0.07	**0.27**	0.24	**0.59**	−0.14	**0.35**
Affectionate	−0.10	**0.58**	0.27	0.50	−0.14	**0.50**	0.36	**0.54**	−0.09	**0.56**	0.35	**0.53**
Unemotional	0.08	**−0.53**	0.35	−0.01	−0.32	**−0.51**	0.12	**0.12**	−0.05	**−0.56**	0.27	**−0.01**
Erratic	0.06	**−0.06**	−0.73	−0.27	0.18	**−0.18**	−0.76	**−0.26**	−0.05	**−0.20**	−0.76	**−0.25**
Impulsive	0.23	**0.27**	−0.71	−0.20	0.34	**0.17**	−0.65	**−0.25**	0.09	**0.23**	−0.75	**−0.14**
Excitable	−0.06	**−0.05**	−0.70	−0.16	0.21	**0.19**	−0.74	**−0.15**	−0.09	**0.10**	−0.73	**−0.06**
Stable	0.26	**0.02**	0.62	0.29	−0.02	**−0.17**	0.60	**0.51**	0.24	**−0.06**	0.60	**0.38**
Jealous	0.46	**0.11**	−0.58	−0.17	0.51	**0.18**	−0.43	**−0.11**	0.33	**0.10**	−0.62	**−0.12**
Disorganized	−0.20	**−0.02**	−0.57	−0.19	−0.26	**−0.31**	−0.56	**−0.21**	−0.32	**−0.33**	−0.47	**−0.25**
Predictable	−0.13	**−0.31**	0.56	0.27	0.00	**−0.13**	0.40	**0.40**	0.09	**−0.21**	0.48	**0.27**
Autistic	−0.14	**−0.11**	−0.53	0.16	−0.10	**−0.23**	−0.57	**0.11**	−0.24	**−0.27**	−0.50	**0.14**
Irritable	0.48	**−0.38**	−0.52	−0.19	0.38	**−0.26**	−0.64	**−0.30**	0.28	**−0.35**	−0.67	**−0.23**
Reckless	0.41	**0.31**	−0.48	−0.44	0.56	**0.18**	−0.46	**−0.37**	0.34	**0.26**	−0.61	**−0.37**
Clumsy	−0.17	**−0.13**	−0.46	0.05	−0.27	**−0.36**	−0.55	**0.08**	−0.30	**−0.40**	−0.41	**−0.01**
Sympathetic	−0.14	**0.03**	0.14	0.86	−0.30	**0.15**	0.25	**0.71**	−0.17	**0.09**	0.30	**0.78**
Sensitive	0.03	**0.02**	0.08	0.79	−0.25	**0.29**	0.19	**0.55**	−0.08	**0.12**	0.21	**0.69**
Helpful	−0.12	**0.31**	0.17	0.75	−0.06	**0.20**	0.03	**0.72**	−0.09	**0.25**	0.14	**0.74**
Protective	0.37	**−0.06**	0.14	0.70	0.10	**0.07**	−0.09	**0.78**	0.22	**0.01**	0.02	**0.77**
Intelligent	0.42	**0.11**	0.32	0.53	0.39	**0.33**	0.21	**0.41**	0.41	**0.30**	0.16	**0.51**

Salient loadings are in boldface. Taken directly from Weiss et al. (2007).

More recently, Latzman et al. (2014) extended this work to further evaluate the influence of polymorphisms in the *AVPR1A* gene on personality in a sample of 174 chimpanzees. In this study, the DupB+/− and DupB−/− chimpanzees were compared across the personality hierarchy from more general, broad dimensions to more narrow and specific traits. *AVPR1A* was associated with alpha, a dimension reflecting a tendency to behave in an uncontrolled, agonistic manner, at the most basic two-factor level of the hierarchy and with disinhibition (low conscientiousness) and dominance, dimensions that emerge from alpha, at the three-factor level. Similar to findings from Hopkins et al. (2012), these associations were found to vary by sex. Specifically, whereas chimpanzees homozygous for the short allele DupB−/− did not differ by sex, males with one copy of the DupB+/− allele evidenced lower levels of alpha and higher levels of disinhibition than DupB+/− females. Additionally, DupB−/− males displayed lower levels of dominance than DupB−/− females. These results are consistent with previous findings with regard to dominance and conscientiousness (Hopkins et al., 2012), traits that emerge from dominance and disinhibition at a lower level of the hierarchy. A more well-explicated model of the hierarchical structure of chimpanzee personality is shown in Figure 7.2.

Wilson et al. (2018) measured *AVPR1A* genotypes and personality in 129 chimpanzees using a slightly expanded personality measurement instrument compared to the one used in the Hopkins et al. (2012) and Latzman et al. (2014) studies. Based on the individual items, these authors derived the same six personality dimensions in the chimpanzees described in King and Figueredo (1997) and also three hierarchical dimensions taken from Latzman et al. (2014) including (low) alpha, disinhibition, and negative emotionality / low dominance. For the full model analysis, a significant effect of *AVPR1A* genotype was found for the personality dimension of conscientiousness with DupB+/− apes (which included DupB+/+ individuals) having higher scores than DupB−/− apes, particularly in females.

Staes et al. (2015) tested for associations between and both single nucleotide polymorphisms in the *OXTR* gene and the indel deletion polymorphism in the *AVPR1A* gene in a sample of 62 chimpanzees residing in European zoos. Rather than use subjective ratings to determine personality, these authors used observational methods to quantify individual variation in social behaviors of the apes. From these observational data, they derived four basic dimensions of personality using factor analysis, including (1) sociability, (2) grooming equitability, (3) positive affect, and (4) anxiety. Results from this study showed that *AVPR1A* genotypes in both males and females accounted for a significant amount of variation in the sociability personality dimension. For both males and females, Dup+/+ individuals had significantly higher scores than DupB+/− and DupB−/− chimpanzees. Note that the primary behavioral measure that comprises the sociability personality dimension was the frequency in giving and receiving of grooming to all other social partners. No significant associations were found between the *OXTR* single nucleotide polymorphism and any of the personality traits.

Finally, Anestis et al. (2014) examined the association between *AVPR1A* polymorphisms and what they termed "behavioral styles." Like Staes et al. (2015), rather

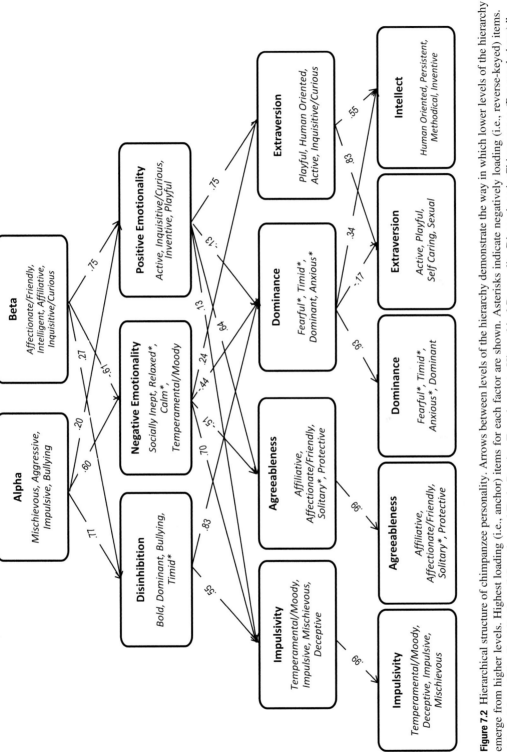

Figure 7.2 Hierarchical structure of chimpanzee personality. Arrows between levels of the hierarchy demonstrate the way in which lower levels of the hierarchy emerge from higher levels. Highest loading (i.e., anchor) items for each factor are shown. Asterisks indicate negatively loading (i.e., reverse-keyed) items. Adapted from "The Contribution of Genetics and Early Rearing Experiences to Hierarchical Personality Dimensions in Chimpanzees (Pan troglodytes)," by R. D. Latzman, H. D. Freeman, S. J. Schapiro, and W. D. Hopkins (2015) *Journal of Personality and Social Psychology*, 109: 894. Copyright 2015 by the American Psychological Association.

than use subjective ratings of personality items, these authors quantified the frequency and percentage of occurrence of 15 different behavioral styles in a sample of 64 captive chimpanzees. From the 15 behavioral style measures, they subjected the data to factor analysis and derived six basic dimensions of personality labeled as (1) affiliative, (2) aggressive, (3) friendly, (4) mellow, (5) playful, and (6) smart. Anestis et al. (2014) found a significant main effect of *AVPR1A* genotype on the personality style smart, which included the specific behavioral characteristics of using coalitions, getting groomed frequently, and having their play offers accepted by other group members. For the smart dimension, DupB+/− apes had higher scores than DupB−/− apes. Anestis et al. (2014) also found a significant interaction between sex and *AVPR1A* genotype on the behavioral style friendly. For this dimension, DupB+/− males had higher scores than DupB−/− males, while no difference was found between females with these two genotypes. These authors also found that DupB+/− chimpanzees that were high ranking had particularly high friendly scores compared to medium- and low-ranking DupB+/− and DupB−/− individuals. The behavioral styles comprising the friendly personality dimension included the directing of affiliative behaviors to all group members.

In the lone study on *AVPR1A* genotypes and personality in bonobos, Staes et al. (2016) genotyped 113 individuals for which personality data were available. Personality was assessed using a 54-item subjective questionnaire identical to the one used by Wilson et al. (2018) in chimpanzees. In the bonobos, factor analysis of the 54 items on the questionnaire revealed six personality dimensions including (1) assertiveness, (2) conscientiousness, (3) openness, (4) attentiveness, (5) agreeableness, and (6) extraversion. As a reminder, unlike chimpanzees, bonobos do not exhibit an indel deletion of the RS3 portion of the V1a receptor (*AVPR1A*). Thus, to characterize *AVPR1A*, Staes et al. (2016) divided their bonobo sample into long (LL) or short (SS) groups based on whether they had greater or fewer than 476 base pair alleles. In the bonobos, *AVPR1A* genotype was associated with openness and attentiveness. LL bonobos scored lower on openness (e.g., active, playful, inquisitive) and higher on attentiveness (e.g., disorganized (−), intelligent, clumsy (−)) than SS bonobos.

Finally, Inoue-Murayama et al. (2018) assessed the personality of a sample of 77 common marmosets (*Callitrix jacchus*) (largely males, $n = 66$) and tested for its association with *AVPR1A* polymorphisms (and two other genes). In this study, personality was assessed using a 54-item Hominoid personality questionnaire and subsequently subjected to factor analysis to derive personality constructs. This analysis revealed three personality factors labeled dominance (e.g., defiant, stingy/greedy, jealous, aggressive), sociability (e.g., helpful, solitary (−), imitative, dependent), and neuroticism (e.g., timid, stable (−), autistic, fearful). For *AVPR1A*, these authors found 10 base pair length alleles and, to increase statistical power, they classified alleles as long (212–223 base pairs) or short (202–210 base pairs). From the classification data, subjects were then characterized as homozygotic short (SS), homozygotic long (LL), or heterogenous (SL) for the *AVPR1A* gene. The association analyses revealed significant early rearing X genotype interactions on the personality dimensions of dominance and sociability. Specifically, for dominance, SS and SL mothered-reared (deemed normal or typical) males scored significantly lower than the LL males. In contrast, SS and SL males that were atypically raised had higher scores on dominance than LL males.

Summary of Findings

All told, the extant literature clearly suggests that polymorphisms in the *AVPR1A* gene influence different dimensions of personality, notably dominance and conscientiousness. The variability in methods across studies makes it difficult to conclude anything more specific regarding which social behaviors, more or less, are associated with variation in the *AVPR1A* gene. The most consistent behavioral attributes that make up the personality traits of dominance and conscientiousness across studies and methods include grooming (both receiving and giving), play, social sophistication or manipulation, and disinhibition. It will thus be critical for future studies to begin to harmonize methodological approaches to allow for better integration across studies and species so that more nuanced conclusions can be made. Sex also appears to be an important factor as it relates to the influence of *AVPR1A* genotypes on personality. Finally, there is only a single study on the association between personality and polymorphisms in the *OXTR* gene and no effects were reported in this report. Thus, further studies are needed on the effects of *OXTR* polymorphic variation on personality.

Influence of Polymorphisms in the *AVPR1A* Gene on Psychopathic Personality (Psychopathy) Traits

Personality pathology can be conceptualized as a configuration of (personality) traits that differ from normality in degree rather than kind and result in significant impairment to oneself or others (Lilienfeld and Latzman, 2018). Thus, through an understanding the neurobiological and evolutionary basis of personality, the pathophysiology underlying personality pathology can be better elucidated (Latzman et al., 2017). One example of the utility of chimpanzee models for elucidating the neurobiological foundations of personality pathology is the recent development of a chimpanzee model of psychopathic personality (psychopathy) dimensions. Psychopathic personality (psychopathy) is a condition characterized by distinct affective and interpersonal features accompanied by persistent behavioral deviancy. Psychopathy is a multifaceted construct (Patrick et al., 2009; Lilienfeld et al., 2015). Recently, though, scientists have begun to coalesce around an understanding that psychopathic tendencies are grounded in basic biobehavioral dispositions that vary continuously within the human population and, more recently, in other species (i.e., chimpanzees) as well. Consistent with this conceptualization, recent theoretical and empirical work has sought to more accurately capture the dimensions of the construct, through the elucidation of its component traits. Developed for this purpose, the triarchic model (Patrick et al., 2009) characterizes psychopathy as a configuration of three-dimensional trait constructs with distinct biological referents: *boldness*, *meanness*, and *disinhibition.* Disinhibition and meanness (callous-aggression) are anchor dimensions of the externalizing spectrum of psychopathology (Krueger et al., 2007), whereas boldness reflects more adaptive aspects of psychopathy (e.g., social efficacy, stress immunity, venturesomeness) that can be viewed in turn as facets of fear/fearlessness (Kramer et al., 2012).

Within this framework, investigation of these dispositional dimensions of psychopathy has been extended to chimpanzees (Latzman et al., 2016), providing a basis for comparative research on their behavioral and neurobiological foundations. Although findings suggest a strong genetic contribution to individual variation in psychopathic personality dimensions in both chimpanzees and humans (Latzman et al., 2017), little is known concerning specific genes that might explain this heritability. Latzman et al. (2017) thus investigated associations between CHMP-Tri scales and polymorphisms in the *AVPR1A* gene in a sample of 164 chimpanzees including 119 mother-reared and 59 nursery-reared apes. Mother-reared apes were those that were born and raised by their conspecific mother in a typical nuclear family setting. In contrast, nursery-reared apes were individuals that were raised in a human nursery setting with same-age peers for the first three years of their life, whereupon they were integrated into conspecific groups. Because Latzman et al. (2017) had previously found that the CHMP-Tri scales were heritable only in mother-reared and not in nursery-reared chimpanzees, the influence of *AVPR1A* on these outcome measures were analyzed separately between these cohorts. In the mother-reared apes, *AVPR1A* variation was found to uniquely explain variability in disinhibition, with DupB−/− chimpanzees having higher scores than DupB+/− individuals. Additionally, sex-related *AVPR1A* genotype interactions were found for boldness and a total psychopathy score. For boldness, DupB+/− males had higher scores than DupB−/− males, and both DupB−/− and DupB+/− females. For the total psychopathy score, DupB−/− males had significantly higher scores than DupB+/− males as well as DupB−/− and DupB+/− females. No significant associations were found between *AVPR1A* and any of the psychopathy dimensions in nursery-raised chimpanzees. When considered in its entirety, these results suggest an important contributory influence of vasopressin genotype variation in the explanation of the development of psychopathy under some, but not all, early rearing conditions.

Influence of Polymorphisms in the *AVPR1A* Gene on Social Behavior

There are three studies that have examined the association between polymorphisms in the *AVPR1A* gene and social behavior in nonhuman primates, two in chimpanzees and one in rhesus monkeys. In one study, Mahovetz et al. (2016) tested whether DupB+/− and DupB−/− chimpanzees differed in their behavioral responses when confronted with a mirror. Mirror self-recognition (MSR) abilities are a sophisticated form of social cognition observed in great apes but not in more distantly related Old and New World monkeys nor in prosimians (see Anderson and Gallup, 2015, for review). Chimpanzees and other great apes show considerable individual variation in response

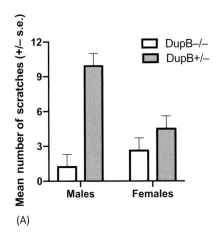
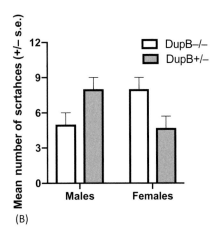

(A)

(B)

Figure 7.3 Mean frequency of gentle and rough scratching (+/− s.e.) in male and female chimpanzees with the DupB−/− and DupB+/− *AVPR1A* genotypes in response to (A) their mirror image and (B) videos of unfamiliar chimpanzees. From data reported in Mahovetz et al. (2016) and Latzman et al. (2016) respectively.

to their mirror reflection; thus, Mahovetz et al. (2016), in addition to assessing their MSR abilities, also measured the social behaviors that the chimpanzees produced in front of the mirror during two 10-minute tests. These authors found significant sex by *AVPR1A* genotype effects on two behaviors, including scratching and agonistic responses. For both behaviors, DupB+/− males showed significantly higher frequencies than DupB−/− males as well as females with both genotypes (Figure 7.3A). It should be noted here that scratching, and particularly gentle and rough scratching, is considered a measure of anxiety in chimpanzees and other nonhuman primates (Pavani et al., 1991; Schino et al., 1991, 1996; Troisi et al., 1991; Baker and Aureli, 1997; Leavens et al., 1997); thus, DupB+/− male chimpanzees engaged in significantly more anxious and agonistic behaviors compared to all other sex and genotype cohorts.

Latzman et al. (2016) compared a sample of 76 Dup+/− and DupB−/− male and female chimpanzees on their rate of scratching in response to social stimuli. Specifically, in a previous behavioral study in captive chimpanzees, Hopkins et al. (2006) measured the frequency in rubs, and gentle and rough scratches in response to videos of unfamiliar chimpanzees fighting or otherwise engaging in agonistic behaviors compared to control conditions in which the television and experimenter were present but no video was played. After accounting for the number of scratches in the baseline compared to experimental condition, Latzman et al. (2016) found that DupB +/− males showed significantly higher rates of scratching compared to DupB−/− males. In contrast, among females, DupB−/− females showed significantly higher rates of scratching compared to DupB+/− individuals (Figure 7.3B).

Madlon-Kay et al. (2018) genotyped 201 rhesus monkeys from the island of Cayo Santiago for polymorphisms in *AVPR1A*, *AVPR1B*, and *OXTR* genes and tested for their associations with five social behaviors, including contact aggression, non-contact aggression, approach, grooming, and passive contact. Within each category, save passive contact, they also characterized whether the focal subjects received or initiated each behavior. Ten-minute focal observations were made on each behavior and the range of focal observations per monkey was between 11 and 174 across the time period of data collection. After performing tests of disequilibrium between the different polymorphisms, these authors identified 12 genotypes that were used in subsequent analyses. No significant associations were found between any of the behaviors and the genotypes for either *OXTR*, *AVPR1A*, or *AVPR1B*. Though not significant, the largest differences reported between individuals with the major and minor alleles for each gene were for *AVPR1A* and approach behavior (both receiving and initiating). Nearly 6% of the variance in the approach behavior was accounted for by these polymorphisms in the two *AVPR1A* single nucleotide polymorphisms.

The Role of Polymorphisms in the *AVPR1A* Gene and Social Cognition

In children, the initiation of joint attention (IJA) and response to joint attention cues (RJA) are early developing socio-communicative skills that reflect children's motivation and ability to respond to and/or follow communicative behaviors from adults, including gaze-following and pointing. Some have suggested that the development of joint attention is uniquely human and is fundamentally tied to increasing selection for cooperative behavior and language (Warneken et al., 2006; Tomasello and Carpenter, 2007; Warneken and Tomasello, 2009). Indeed, a number of studies have shown that IJA and RJA skills predict the subsequent age of onset of spoken language and vocabulary development (Baldwin, 1995; Carpenter et al., 1998; Charman et al., 2000; Morales et al., 2000; Mundy et al., 2007); thus, IJA and RJA play a role in facilitating typical language development. There are also a number of studies that have shown that children at risk or with an autism spectrum disorder (ASD) diagnosis perform more poorly on IJA and/or RJA compared to neurotypical controls (Dawson et al., 2002, 2004; Presmanes et al., 2007; Bottema-Beutel, 2016; Mundy, 2018). In light of the fact that some have suggested that *AVPR1A* may be a risk factor gene for ASD, Hopkins et al. (2014) tested for differences between DupB+/− and DupB−/− chimpanzees' performance on a measure of RJA in 232 captive chimpanzees (herein the Dawson task). To assess RJA, these authors adopted a measure that had been previously shown to discriminate children with ASD from neurotypical controls (Dawson et al., 2002). Briefly, the RJA tasked evaluated the number of social cues needed to elicit an orienting response starting with the use of gaze cues alone (score = 3), followed by gaze cues plus pointing (score = 2), followed by gaze cues, pointing, and the use of an auditory cue (score = 1), and finally no response (score = 0). Each chimpanzee was tested twice, separated by at least one day; thus, scores could range from 0 to 6, with higher scores reflecting better performance. Hopkins et al. (2014) found that Dup+/− males performed significantly better (i.e., needed fewer cues to respond) than DupB−/− males, whereas no difference in performance was found between females of both genotypes (Figure 7.4A).

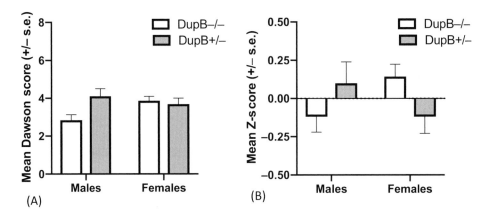

Figure 7.4 (A) Mean performance (+/− s.e.) on the RJA task (Dawson). (B) Mean standardized z-scores (+/− s.e.) for the tasks measuring joint attention.

In a follow-up unpublished study, we have recently tested 187 chimpanzees on an alternative measure of RJA developed by Mundy and colleagues for use in typically developing and ASD children (Morales et al., 2000; Mundy et al., 2000, 2003, 2007; herein the Mundy task). For this task, two objects were placed equidistant from a centrally positioned chimpanzee. The human experimenter sat directly in front of the chimpanzee, facing it. When the human and chimpanzee were engaged in mutual eye gaze, the experimenter would point to one of the two laterally displaced objects and the experimenter recorded whether the chimpanzee looked to or touched the correct object within a 30-sec response window. Subjects received four 6-trial blocks of testing that were separated by at least one day, for a total of 24 test trials.

We have also assessed IJA, or as defined by some, the initiation of a behavioral request, in nearly 200 chimpanzees using several similar paradigms (Leavens et al., 1996, 2004, 2015; Leavens and Hopkins, 1998). Basically, in these socio-communicative tests, food or some required object (such as a tool) is placed outside the apes' home enclosure. Thereafter, the chimpanzees' type of communication and the degree to which they point or gesture to the food/object while alternating their gaze between the food item and the human experimenter – who can be either present or absent, be oriented toward or away from the ape, or have their eyes closed or open to the focal subject – are assessed (Hostetter et al., 2001, 2007). Successful IJA is scored if the chimpanzees point or gesture toward the object while simultaneously alternating their gaze between the human experimenter and the referent.

When combining the data from our chimpanzee sample, we have two measures of RJA and have one measure of IJA as well as *AVPR1A* genotype data in 159 chimpanzees. Here we compared performance on these three measures between *AVPR1A* genotypes, groups, and sexes to determine whether the reported effects on RJA for the Dawson task used by Hopkins et al. (2014) were specific to this measure. To compare measures, we converted the raw performance data to standardized z-scores within each task and then used these scores as repeated measures in a mixed model analysis of variance while sex and *AVPR1A* genotype were between group factors. This analysis revealed a significant two-way interaction between *AVPR1A* genotype and sex, $F(1, 155) = 4.451$, $p = .035$. The mean standardized z-scores in male and female DupB+/− and DupB−/− individuals are shown in Figure 7.4B. DupB+/− males performed significantly better than DupB−/− males whereas the opposite effect was found for females.

AVPR1A Genetic Variation Is Associated with Brain Variation

Despite the evidence implicating polymorphisms in the *AVPR1A* gene to different dimensions of personality, social behavior, and cognition in chimpanzees, there is little data on their neurobiological, neuroanatomical, or neurofunctional correlates in primates. In the only study to date, Latzman et al. (2016) compared differences in whole brain gray matter between DupB+/− and DupB−/− chimpanzees using voxel-based morphometry. These found significant differences in gray matter between DupB +/− and DupB−/− chimpanzees across 12 clusters largely in the frontal cortex including, most notably, the right superior precentral sulcus/gyrus and the dorsal prefrontal cortex bilaterally. Several studies have examined the distribution of receptor densities for vasopressin and oxytocin in both Old and New World monkeys (Freeman et al., 2014a, 2014b; French et al., 2016). These studies generally show that the distribution of oxytocin and vasopressin is relatively sparse in the neocortex but more prevalent in subcortical, limbic, and brainstem structures. In a recent paper, Rogers et al. (2018) reported species similarities and differences in oxytocin and vasopressin fibers in regions of the cortex in humans, chimpanzees and rhesus monkeys. With respect to vasopressin, fibers were found in all three species in the anterior olfactory nucleus, primary olfactory cortex, and subgenual region of the anterior cingulate cortex. Rogers et al. (2018) further reported that vasopressin fibers were present in humans and chimpanzees but absent in rhesus monkeys in the agranular insula cortex and were only found in humans within the frontal operculum.

Comparative Data between Chimpanzees and Bonobos

Bonobos and chimpanzees are closely related species within the Genus *Pan*. Despite their close genetic and phylogenetic proximity, they exhibit a number of behavioral, cognitive, and neurological differences that some have hypothesized are associated with the neuropeptides vasopressin and oxytocin (Staes et al., 2014). As noted above, the potential role that variation in vasopressin and particularly oxytocin play in phylogenetic variation in cooperative behavior and social monogamy among New World primates is quite compelling; thus, the working phylogenetic hypothesis is that genetic, ligand, or receptor variation in either oxytocin or vasopressin between chimpanzees and bonobos might similarly explain different dimensions of behavior, cognition, and brain variation. We know that (1) chimpanzees have a deletion of the RS3 region of the *AVPR1A* that is absent in bonobos and (2) that a higher percentage of bonobos show longer base pair allele lengths for the RS3 compared to chimpanzees (Staes et al., 2018). However, the extent to which this variation maps onto or explains interspecies variation in brain, cognition, and behavior remains largely unknown.

For instance, though subject to some debate, it has been suggested that bonobos are more prosocial and perform better on measures of social cognition and cooperation compared to chimpanzees (Hare and Yamamoto, 2017). In contrast, chimpanzees have been reported to engage in more tool manufacture and use, to be more territorial, and to engage in more lethal and severe aggression during interindividual and inter-group encounters. Bonobos also live in more female-dominated societies compared to chimpanzees. There is also evidence that chimpanzees and bonobos differ in gray matter variation, cortical connectivity, and neuropil space within limbic and cortical areas that are implicated in socio-emotional, communicative, and social cognition functions such as the amygdala, hippocampus, and anterior cingulate, orbital frontal, and insular cortices (Rilling et al., 2012; Bauernfeind et al., 2013; Hopkins et al., 2017; Issa et al., 2018). Unfortunately, for each of these brain and behavioral phenotypes, there are no available data that directly implicate *AVPR1A* or *OXTR* polymorphisms in their differential function. There is one report that shows that bonobos have an increased density of serotonergic axons within the basal and central nuclei of the amygdala compared to chimpanzees (Stimpson et al., 2016). In the only study to date on OXT and AVP in chimpanzees and bonobos, Hopkins et al. (2017) used immunohistochemistry to label AVP- and OXT-containing neurons in the para-ventricular nucleus (PVN) and supraoptic nucleus (SON) of the hypothalamus in five bonobos and seven chimpanzees. The number of neurons that expressed AVP and OXT did not significantly differ between species, though there was a trend toward bonobos having higher values than chimpanzees.

Conclusions

The majority of nonhuman primate studies that have examined associations between *AVPR1A* genotypes and (1) personality, (2) social behavior, and (3) social cognition have been carried out in chimpanzees, likely due to the fact that this species shows an unusual deletion of the RS3 portion of the gene and because of their greater numbers in captive settings. Importantly, a review of the existing literature underscores the fact that variation in the AVP receptor influences not just one characteristic but individual trait variation across multiple dimensions of trait personality, in vivo social behavior, and experimentally studied social cognition. Indeed, it is clear that AVP variation affects sociality across domains, and sometimes in opposing ways. For example, in opposing directions, AVP influences not only traits associated with increased levels of dependency and submissiveness (i.e., dominance) but also traits associated with aggression, low levels of sympathy, and increased levels of bullying (i.e., disinhibition; Latzman et al., 2014).

With respect to personality more specifically, all of the studies have reported significant associations between *AVPR1A* polymorphisms and at least one dimension of personality. In particular, personality items that load high on the constructs dominance and conscientiousness are somewhat but not entirely consistent across studies. It is worth noting that the *AVPR1A*-personality studies in chimpanzees have been done in four genetically distinct populations or laboratories of individuals; thus, significant associations between personality and *AVPR1A* polymorphism appear to be repeatable across populations. If possible, in some future study, it would be interesting to determine the influence of *AVPR1A* polymorphisms on personality for the four chimpanzee populations combined into a single sample, which should be feasible because basically the same personality instrument was utilized between the extant groups. This type of analysis would be of interest for at least two reasons. First, there would likely be increased stability in personality dimensions derived from the questionnaire in this much larger, combined sample. Second, though not often discussed, there is a third genotype group for *AVPR1A* in chimpanzees, sometimes labeled as DupB+/+ but it has an extremely low frequency (<5% occurrence; Donaldson et al., 2008). In most studies, these individuals are either dropped from the study or combined with the DupB+/− group. Combining the *AVPR1A* and personality data across the different populations would likely yield a large enough sample of DupB+/+ to include this third genotype as a distinct group.

Many fewer studies have examined associations between *AVPR1A* and social behavior and this is unfortunate. Personality assessment is a good proxy to quantify how people (human raters) perceive animals. It is important, though, to supplement these data with quantitative measures of observed behavior. Of the studies that do exist, significant associations were found between *AVPR1A* and both agonistic and anxious behaviors in chimpanzees, and in particular for rough and gentle scratching. In the lone study in rhesus monkeys, no associations were found between SNPs (single nucleotide polymorphisms) in *OXTR*, *AVPR1A*, or *AVPR1B* with five social behavior measures. Though the study by Madlon-Kay et al. (2018) had sufficient power to detect significant associations, none were found, which was somewhat surprising and unexpected. It may be the case that the influence of *AVPR1A* polymorphisms on social behavior differs between chimpanzees and rhesus monkeys, but this conclusion would be premature in light of the fact that the method and behaviors of interest were not the same between these studies. Notably, in chimpanzees, polymorphisms in the *AVPR1A* gene were associated with agonistic and anxiety behaviors (i.e., scratching), whereas none of these types of measures were included in the rhesus monkey ethogram, with the exception of aggression. Further, in the Madlon-Kay et al. (2018) study, the collection of the behavior was opportunistic and extended over a long period of time (as many as five years). Changes in social dynamics and social hierarchies within the group may have occurred during the observation period, which could have differentially influenced the outcome. Finally, there was also considerable individual variation in the number of focal observations that were made on each monkey and it did not appear that this factor was taken into consideration in the statistical models reported by these authors. We emphasize again here, though, that the largest effects reported in the Madlon-Kay et al. (2018) paper were for *AVPR1A* variation and the proportion of variance in behavior that was accounted for by polymorphisms in this gene were ~6%, which is not all that different than findings in the chimpanzee studies, which all reported small to moderate effect sizes.

There are only three studies on *AVPR1A* and social cognition in nonhuman primates and these findings are all from a single population of chimpanzees studied by our group. Thus, some caution should be exercised when making strong inferences about these associations until they are further studied and hopefully replicated in a separate population. The three sets of data that were presented in this review did come from two genetically isolated colonies of chimpanzees including those housed at the Yerkes National Primate Research Center (YNPRC) and the National Center for Chimpanzee Care (NCCC) at the University of Texas MD Anderson Cancer Center. None of the effects involving the *AVPR1A* gene significantly interacted with the colony variable; thus, as we have found in other measures of behavior between the NCCC and YNPRC, such as tool-use skill (Hopkins et al., 2015, 2018), there can be significant differences in the outcome measures of interest between chimpanzee colonies, but these differences appear to exist independent of the genetic effects.

Finally, in a number of studies, the influence of *AVPR1A* genotypes on personality, social behavior, and cognition were mediated by sex, and this is not all that surprising. Notably, the influence of *AVPR1A* seemed to mediate the behavior of males much more than females. Consistent with these findings, vasopressin systems in the brain have been found to be sexually dimorphic and thought to regulate social behaviors in sex-specific ways (de Vries, 2008; Terranova et al., 2016). It will thus be critical for future research to consider the sex-specific nature of these associations as well as the way in which neural processes may mediate these associations.

All told, by comparison to studies in other mammalian model species, studies on the role of polymorphisms in the *OXTR* and, particularly the *AVPR1A* genes on social behavior and cognition in primates (including humans) are in their infancy. Notwithstanding, we believe that continued research with nonhuman primates will provide a uniquely powerful model for investigating associations between these genes and personality, social behavior, and social cognition. Moreover, because more sophisticated behavioral and cognitive phenotypes can be quantified in nonhuman primates compared to more distantly related species, this arguably enhances the translational value of this research to our understanding of human behavior and mental health. Research to date, however, appears to suggest an inseparability between affiliative and conflict behaviors (an idea discussed in multiple chapters in this volume). Indeed, it appears relatively clear that affiliation and antagonism fall at opposing ends of a continuous bipolar dimension (Palumbo and Latzman, 2019). There is still much work to be done to more fully recognize the role that oxytocin and vasopressin systems play in individual sociality and we hope that the preceding review provides a springboard for this promising work.

References

Anderson J. R., and Gallup G. G., Jr. (2015) Mirror self-recognition: A review and critique of attempts to promote and engineer self-recognition in primates. *Primates*, 56: 317–326.

Anestis S. F., Webster T. H., Kamilar J. M., Fontenot M. B., Watts D. P., and Bradley B. J. (2014) AVPR1A variation in chimpanzees (*Pan troglodytes*): Population differences and association with behavioral style. *International Journal of Primatology*, 35(1): 305–324.

Avinum R., Israel S., Shalev I., Gritsenko I., Bornstein G., Ebstein R. P., and Knafo A. (2011) AVPR1A variant associated with preschoolers' lower altruistic behavior. *PLoS ONE*, 6(9): E25274.

Bachner-Melman R., Dina C., Zohar A. H. et al. (2005) AVPR1a and SLC6A4 gene polymorphisms are associated with creative dance performance. *PLoS Genetics*, 1(3): e42.

Baker K. C., and Aureli F. (1997) Behavioural indicators of anxiety: An empirical test in chimpanzees. *Behaviour*, 134: 1031–1050.

Baldwin D. A. (1995) Understanding the link between joint attention and language. In Moore C., and Dunham P. J., eds., *Joint Attention: Its Origins and Role in Development*. Hillsdale, NJ: Erlbaum, pp. 131–158.

Bauernfeind A. A., Sousa A. M. M., Avashti T. et al. (2013) A volumetric comparison of the insular cortex and its subregions in primates. *Journal of Human Evolution*, 64: 263–279.

Bauman M. D., Murai T., Hogrefe C. E., and Platt M. L. (2018) Opportunities and challenges for intranasal oxytocin treatment studies in nonhuman primates. *American Journal of Primatology*, 80(10): e22913.

Bauman M. D., and Schumann C. M. (2018) Advances in nonhuman primate models of autism: Integrating neuroscience and behavior. *Experimental Neurology*, 299(Pt A): 252–265.

Bottema-Beutel K. (2016) Associations between joint attention and language in autism spectrum disorder and typical development: A systematic review and meta-regression analysis. *Autism Research*, 10: 1021–1035.

Brosnan S. F., Talbot C. F., Essler J. L., Leverett K., Felemming T. P. G., Heyler C., and Zak P. J. (2015) Oxytocin reduces food sharing in capuchin monkeys by modulating social distance. *Behaviour* (152): 941–961.

Caldwell H. K. (2017) Oxytocin and vasopressin: Powerful regulators of social behavior. *Neuroscientist,* 23(5): 517–528.

Carpenter M., Nagell K., Tomasello M., Butterworth G., and Moore C. (1998) Social cognition, joint attention, and communicative competence from 9 to 15 months of age. *Monographs of the Society for Research in Child Development*, 63(4).

Cavanaugh J., Mustoe A., Womack S. L., and French J. A. (2018) Oxytocin modulates mate-guarding behavior in marmoset monkeys. *Hormones and Behavior,* 106: 150–161.

Charman T., Baron-Cohen S., Swettenham J., Baird G., Cox A., and Drew A. (2000) Testing joint attention, imitation and play as infancy precursors to language and theory of mind. *Cognitive Development,* 15: 481–498.

Dawson G., Munson J., Estes A. et al. (2002) Neurocognitive function and joint attention ability in young children with autism spectrum disorder versus developmental delay. *Child Development*, 73(2): 345–358.

Dawson G., Toth K., Abbott R., Osterling J., Munson J., Estes A., and Liaw J. (2004) Early social attention impairments in autism: Social orienting, joint attention and attention to distress. *Developmental Psychology*, 40(2): 271–283.

de Vries G. J. (2008) Sex differences in vasopressin and oxytocin innervation of the brain. *Progress in Brain Research,* 170: 17–27.

Donaldson Z. R., Bai Y., Kondrashov F. A., Stoinski T. L., Hammock E. A. D., and Young L. J. (2008) Evolution of a behavior-linked microsatellite-containing element of the 5′ flanking region of the primate *AVPR1A* gene. *BMC Evolutionary Biology*, 8: 180–188.

Donaldson Z. R., and Young L. J. (2008) Oxytocin, vasopressin and the neurogenetics of sociality. *Science*, 322: 900–904.

Ebert A., and Brune M. (2017) Oxytocin and social cognition. In Hurlemann R., and Grinevich V., eds., *Behavioral Pharmacology of Neuropeptides: Oxytocin*. Switzerland: Springer, pp. 375–388.

Ebstein R. P., Knafo A., Mankuta D., Chew S. H., and Lai P. S. (2012) The contributions of oxytocin and vasopressin pathway genes to human behavior. *Hormones and Behavior*, 61(3): 359–379.

Eckardt W., Steklis H. D. Steklis N. G., Fletcher A. W., Stoinski T. S., and Weiss A. (2015) Personality dimensions and their behavioral correlates in wild Virunga mountain gorillas (Gorilla beringei beringei). *Journal of Comparative Psychology*, 129(1): 26–41.

Evans S. L., Dal Monte O., Noble P., and Averbeck B. B. (2014) Intranasal oxytocin effects on social cognition: A critique. *Brain Research*, 1580: 69–77.

Feczko E. J., Bliss-Moreau E., Walum H., Pruett J. R. Jr., and Parr L. A. (2016) The Macaque Social Responsiveness Scale (mSRS): A rapid screening tool for assessing variability in the social responsiveness of Rhesus monkeys (*Macaca mulatta*). *PLoS ONE*, 11(1): e0145956.

Francis S. M., Kim S. J., Kistner-Griffin E., Guter S., Cook E. H., and Jacob S. (2016) ASD and genetic associations with receptors for oxytocin and vasopressin-AVPR1A, AVPR1B, and OXTR. *Frontiers in Neuroscience*, 10: 516.

Freeman H. D., Brosnan S. F., Hopper L. M., Lambeth S. P., Schapiro S. J., and Gosling S. D. (2013) Developing a comprehensive and comparative questionnaire for measuring personality in chimpanzees using a simultaneous top-down/bottom-up design. *American Journal of Primatology*, 75: 1042–1053.

Freeman H. D., and Gosling S. D. (2010) Personality in nonhuman primates: a review and evaluation of past research. *American Journal of Primatology*, 72(8): 653–671.

Freeman S. M., Inoue K., Smith A. L., Goodman M. M., and Young L. J. (2014a) The neuroanatomical distribution of oxytocin receptor binding and mRNA in the male rhesus macaque (*Macaca mulatta*). *Psychoneuroendocrinology*, 45: 128–141.

Freeman S. M., Walum H., Inoue K., Smith A. L., Goodman M. M., Bales K. L., and Young L. J. (2014b) Neuroanatomical distribution of oxytocin and vasopressin 1a receptors in the socially monogamous coppery titi monkey (*Callicebus cupreus*). *Neuroscience*, 273: 12–23.

French J. A., Taylor J. H., Mustoe A. C., and Cavanaugh J. (2016) Neuropeptide diversity and the regulation of social behavior in New World primates. *Frontiers in Neuroendocrinology*, 42: 18–39.

Goodson J. L., and Bass A. H. (2001) Social behavior functions and related anatomical characteristics of vasotocin/vasopressin systems in vertebrates. *Brain Research Reviews*, 35: 246–265.

Guastella A. J., Einfeld S. L., Gray K. M., Rinehart N. J., Tonge B. J., Lambert T. J., and Hickie I. B. (2010a) Intranasal oxytocin improves emotion recognition for youth with autism spectrum disorders. *Biological Psychiatry*, 67(7): 692–694.

Guastella A. J., Kenyon A. R., Alvares G. A., Carson D. S., and Hickie I. B. (2010b) Intranasal arginine vasopressin enhances the encoding of happy and angry faces in humans. *Biological Psychiatry*, 67(12): 1220–1222.

Hammock E. A., and Young L. J. (2005) Microsatellite instability generates diversity in brain and sociobehavioral traits. *Science*, 308: 1630–1634.

Hammock E. A., and Young L. J. (2006) Oxytocin, vasopressin and pair bonding: Implications for autism. *Philosophical Transactions of the Royal Society of London Series B Biological Sciences*, 361(1476): 2187–2198.

Hare B., and Yamamoto S. (2017) *Bonobos: Unique in Mind, Brain and Behavior*. Oxford: Oxford University Press.

Hopkins W. D., Donaldson Z. R., and Young L. Y. (2012) A polymorphic indel containing the RS3 microsatellitein the 5′ flanking region of the vasopressin V1a receptor gene is associated with chimpanzee (*Pan troglodytes*) personality. *Genes, Brain and Behavior,* 11: 552–558.

Hopkins W. D., Keebaugh A. C., Reamer L. A., Schaeffer J., Schapiro S. J., and Young L. J. (2014) Genetic influences on receptive joint attention in chimpanzees (*Pan troglodytes*). *Scientific Reports* 4(3774): 1–7.

Hopkins W. D., Latzman R. D., Mareno M. C., Schapiro S. J., Gomez-Robles A., and Sherwood C. C. (2018) Heritability of gray matter structural covariation and tool use skills in Chimpanzees (*Pan troglodytes*): A source-based morphometry and quantitative genetic analysis. *Cerebral Cortex*, 29: 3702–3711.

Hopkins W. D., Reamer L., Mareno M. C., and Schapiro S. J. (2015) Genetic basis for motor skill and hand preference for tool use in chimpanzees (*Pan troglodytes*). *Proceedings of the Royal Society London B Biological Sciences*, 282: 1800.

Hopkins W. D., Russell J. L., Freeman H., Reynolds E. A. M., Griffis C., and Leavens D. A. (2006) Lateralized scratching in chimpanzees (*Pan troglodytes*): Evidence of a functional asymmetry in arousal. *Emotion*, 6(4): 553–559.

Hopkins W. D., Stimpson C. D., and Sherwood C. C. (2017) Social cognition and brain organization in chimpanzees (*Pan troglodytes*) and bonobos (*Pan paniscus*). In Hare B., and Yamamoto S. eds., *Bonobos: Unique Mind, Brain and Behavior*. Oxford: Oxford University Press, pp. 199–213.

Hostetter A. B., Cantero M., and Hopkins W. D. (2001) Differential use of vocal and gestural communication by chimpanzees (*Pan troglodytes*) in response to the attentional status of a human (*Homo sapiens*). *Journal of Comparative Psychology*, 115(4): 337–343.

Hostetter A. B., Russell J. L., Freeman H., and Hopkins W. D. (2007) Now you see me, now you don't: Evidence that chimpanzees understand the role of the eyes in attention. *Animal Cognition*, 10: 55–62.

Inoue-Murayama M., Yokoyama C., Yamanashi Y., and Weiss A. (2018) Common marmoset (*Callithrix jacchus*) personality, subjective well-being, hair cortisol level and AVPR1a, OPRM1, and DAT genotypes. *Science Reports*, 8(1): 10255.

Issa H. A., Staes N., Diggs-Galligan S. et al. (2018) Comparison of bonobo and chimpanzee brain microstructure reveals differences in socio-emotional circuits. *Brain Structure and Function*, 224: 239–251.

Jiang Y., and Platt M. L. (2018) Oxytocin and vasopressin flatten dominance hierarchy and enhance behavioral synchrony in part via anterior cingulate cortex. *Science Reports*, 8(1): 8201.

Kim H. S., Young L. J., Gonen D. et al. (2002) Transmission disequilibrium testing of arginine vasopressin receptor 1A (AVPR1A) polymorphisms in autism. *Molecular Psychiatry*, 7: 503–507.

King J. E., and Figueredo A. J. (1997) The five-factor model plus dominance in chimpanzee personality. *Journal of Research in Personality*, 31(2), 257–271, DOI: https://doi.org/10.1006/jrpe.1997.2179.

Kramer M. D., Patrick C. J., Krueger R. F., and Gasperi M. (2012) Delineating physiologic defensive reactivity in the domain of self-report: Phenotypic and etiologic structure of dispositional fear. *Psychological Medicine*, 42(6): 1305–1320.

Krueger R. F., Markon K. E., Patrick C. J., Benning S. D., and Kramer M. D. (2007) Linking antisocial behavior, substance use, and personality: An integrative quantitative model of the adult externalizing spectrum. *Journal of Abnormal Psychology*, 116(4): 645–666.

Latzman R. D., Drislane L. E., Hecht L. K. et al. (2016) A chimpanzee model of triarchic psychopathy constructs: development and initial validation. *Clinical Psychological Science*, 4(1): 50–66.

Latzman R. D., Green L. M., and Fernandes M. A. (2017) The importance of chimpanzee personality research to understanding processes associated with human mental health. *International Journal of Comparative Psychology*, 30: 34268.

Latzman R. D., Hopkins W. D., Keebaugh A. C., and Young L. J. (2014) Personality in chimpanzees (*Pan troglodytes*): Exploring the hierarchical structure and associations with the vasopressin V1A receptor gene. *PLoS ONE*, 9(4): e95741.

Latzman R. D., Patrick C. J., Freeman H. D., Schapiro S. J., and Hopkins W. D. (2017) Etiology of triarchic psychopathy dimensions in Chimpanzees (*Pan troglodytes*). *Clinical Psychological Science*, 5(2): 341–354.

Latzman R. D., Schapiro S. J., and Hopkins W. D. (2017) Triarchic psychopathy dimensions in Chimpanzees (*Pan troglodytes*): Investigating associations with genetic variation in the vasopressin receptor 1A gene. *Frontiers in Neuroscience*, 11: 407.

Latzman R. D., Young L. J., and Hopkins W. D. (2016) Displacement behaviors in chimpanzees (*Pan troglodytes*): A neurogenomics investigation of the RDoC Negative Valence Systems domain. *Psychophysiology*, 53: 355–363.

Leavens D. A., Aureli F., and Hopkins W. D. (1997) Scratching and cognitive stress: Performance and reinforcement effects on hand use, scratch type, and afferent cutaneous pathways during computer cognitive testing by a chimpanzee (*Pan troglodytes*). *American Journal of Primatology*, 42: 126–127.

Leavens D. A., and Hopkins W. D. (1998) Intentional communication by chimpanzee (*Pan troglodytes*): A cross-sectional study of the use of referential gestures. *Developmental Psychology*, 34: 813–822.

Leavens D. A., Hopkins W. D., and Bard K. A. (1996) Indexical and referential pointing in chimpanzees (*Pan troglodytes*). *Journal of Comparative Psychology*, 110(4): 346–353.

Leavens D. A., Hopkins W. D., and Thomas R. (2004) Referential communication by chimpanzees (*Pan troglodytes*). *Journal of Comparative Psychology*, 118: 48–57.

Leavens D. A., Reamer L. A., Mareno M. C., Russell J. L., Wilson D. C., Schapiro S. J., and Hopkins W. D. (2015) Distal communication by chimpanzees (*Pan troglodytes*): Evidence for common ground? *Child Development*, 86(5): 1623–1638.

Leng G., and Ludwig M. (2016) Intranasal oxytocin: Myths and delusions. *Biological Psychiatry*, 79(3): 243–250.

Lilienfeld S. O., and Latzman R. D. (2018) Personality disorders: Current scientific status and ongoing controversies. In J. N. Butcher, ed., *APA Handbook of Psychopathology: Psychopathology: Understanding, Assessing, and Treating Adult Mental Disorders*. Washington, DC: American Psychological Association, pp. 557–606.

Lilienfeld S. O., Watts A. L., Francis Smith S., Berg J. M., and Latzman R. D. (2015) Psychopathy deconstructed and reconstructed: Identifying and assembling the personality building blocks of Cleckley's Chimera. *Journal of Personality*, 83(6): 593–610.

LoParo D., and Waldman I. D. (2015) The oxytocin receptor gene (OXTR) is associated with autism spectrum disorder: A meta-analysis. *Molecular Psychiatry*, 20: 640–646.

Madlon-Kay S., Montague M. J., Brent L. J. N. et al. (2018) Weak effects of common genetic variation in oxytocin and vasopressin receptor genes on rhesus macaque social behavior. *American Journal of Primatology*, 80(10): e22873.

Madrid J. E., Oztan O., Sclafani V. et al. (2017) Preference for novel faces in male infant monkeys predicts cerebrospinal fluid oxytocin concentrations later in life. *Science Reports*, 7(1): 12935.

Mahovetz L. M., Young L. J., and Hopkins W. D. (2016) The influence of AVPR1A genotype on individual differences in behaviors during a mirror self-recognition task in chimpanzees (*Pan troglodytes*). *Genes Brain and Behavior*, 15(5): 445–452.

Marrus N., Faughn C., Shuman J., Petersen S. E., Constantino J. N., Povinelli D. J., and Pruett J. R., Jr. (2011) Initial description of a quantitative, cross-species (chimpanzee-human) social responsiveness measure. *Journal of the American Academy of Child and Adolescent Psychiatry*, 50(5): 508–518.

Meyer-Lindenberg A., Domes G., Kirsch P., and Heinrichs M. (2011) Oxytocin and vasopressin in the human brain: Social neuropeptides for translational medicine. *Nature Neuroscience Reviews*, 12: 524–538.

Morales M., Mundy P., Delgado C. E. F., Yale M., Messinger D., Neal R., and Schwartz H. K. (2000) Responding to joint attention across the 6- through 24-month age period and early language acquisition. *Journal of Applied Developmental Psychology*, 21(3): 283–298.

Mundy P. (2018) A review of joint attention and social-cognitive brain systems in typical development and autism spectrum disorder. *European Journal of Neuroscience*, 47(6): 497–514.

Mundy P., Block J., Delgado C., Pomares Y., Van Hecke A. V., and Parlade M. V. (2007) Individual differences and the development of joint attention in infancy. *Child Development*, 78(3): 938–954.

Mundy P., Card J., and Fox N. (2000) EEG correlates of the development of infant joint attention skills. *Developmental Psychobiology*, 36: 325–338.

Mundy P., Delgado P., Block J., Venezia M., Hogan A., and Siebert J. (2003) *A Manual for the Abridged Early Social Communication Scales (ESCS)*. Coral Gables, FL: University of Miami Press.

Murray L. (2011) Predicting primate behavior from personality ratings. In Weiss A., King J. E., and Murray L., eds., *Personality and Temperament in Nonhuman Primates*. New York: Springer, pp. 129–166.

Palumbo I. M., and Latzman R. D. (2019) Translational value of nonhuman primate models of antagonism. In Lyman D. R., and Miller J. D., eds., *The Handbook of Antagonism*. San Diego, CA: Elsevier, pp. 113–126.

Parker K. J., Garner J. P., Oztan O. et al. (2018) Arginine vasopressin in cerebrospinal fluid is a marker of sociality in nonhuman primates. *Science Translational Medicine*, 10: 439.

Parr L. A., Mitchell T., and Hecht E. (2018) Intranasal oxytocin in rhesus monkeys alters brain networks that detect social salience and reward. *American Journal of Primatology*, 80(10): e22915.

Parr L. A., Modi M., Siebert E., and Young L. J. (2013) Intranasal oxytocin selectively attenuates rhesus monkeys' attention to negative facial expressions. *Psychoneuroendocrinology*, 38(9): 1748–1756.

Patrick C. J., Fowles D. C., and Krueger R. F. (2009) Triarchic conceptualization of psychopathy: Developmental origins of disinhibition, boldness, and meanness. *Developmental Psychopathology*, 21(3): 913–938.

Pavani S., Maestripieri D., Schino G., Giovanni Turillazzi P., and Scucchi S. (1991) Factors influencing scratching behaviour in long-tailed macaques (*Macaca fascicularis*). *Folia Primatologica*, 57: 34–38.

Presmanes A. G., Walden T. A., Stone W. L., and Yoder P. J. (2007) Effects of different attentional cues on responding to joint attention in younger siblings of children with autism spectrum disorders. *Journal of Autism and Developmental Disorders*, 37: 133–144.

Rilling J. K., Scholz J., Preuss T. M., Glasser M. F., Errangi B. K., and Behrens T. E. (2012) Differences between chimpanzees and bonobos in neural systems supporting social cognition. *Social, Cognitive and Affective Neuroscience,* 7(4): 369–379.

Rogers C. N., Ross A. P., Sahu S. P. et al. (2018) Oxytocin- and arginine vasopressin-containing fibers in the cortex of humans, chimpanzees, and rhesus macaques. *American Journal of Primatology*, 80: e22875.

Rosso L., Keller L., Kaessmann H., and Hammond R. L. (2008) Mating systems and *avpr1a* promoter variation in primates. *Biology Letters*, 4: 375–378.

Samuni L., Preis A., Mundry R., Deschner T., Crockford C., and Wittig R. M. (2017) Oxytocin reactivity during intergroup conflict in wild chimpanzees. *Proceedings of the National Academy of Science USA*, 114(2): 268–273.

Schaefer S. A., and Steklis H. D. (2014) Personality and subjective well-being in captive male western lowland gorillas living in bachelor groups. *American Journal of Primatology*, 76(9): 879–889.

Schino G., Peretta G., Taglioni A. M., Monaco V., and Troisi A. (1996) Primate displacement activities as an ethopharmacological model of anxiety. *Anxiety*, 2: 186–191.

Schino G., Troisi A., Perretta G., and Monaco V. (1991) Measuring anxiety in nonhuman primates: Effect of lorazepam on macaque scratching. *Pharmacology, Biochemistry and Behavior,* 38: 889–891.

Skuse D. H., Lori A., Cubelis J. F. et al. (2014) Common polymorphism in the oxytocin receptor gene (OXTR) is associated with human social recognition skills. *Proceedings of the National Academy of Science USA*, 111(5): 1987–1992.

Staes N., Bradley B. J., Hopkins W. D., and Sherwood C. C. (2018) Genetic signatures of socio-communicative abilities in primates. *Current Opinion in Behavioral Sciences,* 21: 33–38.

Staes N., Koski S. E., Helsen P., Fransen E., Eens M., and Stevens J. M. (2015) Chimpanzee sociability is associated with vasopressin (Avpr1a) but not oxytocin receptor gene (*OXTR*) variation. *Hormones and Behavior*, 75: 84–90.

Staes N., Stevens J. M. G., Helsen P., Hillyer M., Korody M., and Eens M. (2014) Oxytocin and vasopressin receptor gene variation as a proximate base for Inter- and intraspecific behavioral differences in bonobos and chimpanzees. *PLoS ONE*, 9(11): e113364.

Staes N., Weiss A., Helsen P., Korody M., Eens M., and Stevens J. M. (2016) Bonobo personality traits are heritable and associated with vasopressin receptor gene 1a variation. *Science Reports*, 6: 38193.

Stimpson C. D., Hopkins W. D., Taglialatela J., Barger N., Hof P. R., and Sherwood C. C. (2016) Differential serotonergic innervation of the amygdala in bonobos and chimpanzees. *Social, Cognitive and Affective Neuroscience,* 11(3): 413–422.

Tabak B. A., Teed A. R., Castle E. et al. (2019) Null results of oxytocin and vasopressin administration across a range of social cognitive and behavioral paradigms: Evidence from a randomized controlled trial. *Psychoneuroendocrinology,* 107: 124–132.

Taylor J. H., and French J. A. (2015) Oxytocin and vasopressin enhance responsiveness to infant stimuli in adult marmosets. *Hormones and Behavior*, 75: 154–159.

Terranova J. I., Song Z., Larkin T. E., 2nd, Hardcastle N., Norvelle A., Riaz A., and Albers H. E. (2016) Serotonin and arginine-vasopressin mediate sex differences in the regulation of dominance and aggression by the social brain. *Proceedings of the National Academy of Science USA*, 113(46): 13233–13238.

Tomasello M., and Carpenter M. (2007) Shared intentionality. *Developmental Science*, 10(1): 121–125.

Troisi A., Schino G., D'Antoni M., Pandolfi N., Aureli F., and D'Amato F. R. (1991) Scratching as a behavioral index of anxiety in macaque mothers. *Behavioral and Neural Biology*, 56: 307–313.

Uzefovsky F., Shalev I., Israel S. et al. (2015) Oxytocin receptor and vasopressin receptor 1a genes are respectively associated with emotional and cognitive empathy. *Hormones and Behavior*, 67: 60–65.

Uzefovsky F., Shalev I., Israel S., Knafo A., and Ebstein R. P. (2012) Vasopressin selectively impairs emotion recognition in men. *Psychoneuroendocrinology*, 37(4): 576–580.

Walum H., Westberg L., Henningsson S. et al. (2008) Genetic variation in the vasopressin receptor 1a gene (*AVPR1A*) associates with pair bonding behavior in humans. *Proceedings of the National Academy of Sciences USA*, 105(37): 14153–14156.

Warneken F., Chen F., and Tomasello M. (2006) Cooperative activities in young children and chimpanzees. *Child Development*, 77(3): 640–663.

Warneken F., and Tomasello M. (2009) The roots of human altruism. *British Journal of Psychology*, 100(Pt 3): 455–471.

Weiss A., King J. E., and Hopkins W. D. (2007) A cross-setting study of chimpanzee (*Pan troglodytes*) personality structure and development: Zoological parks and Yerkes National Primate Research Center. *American Journal of Primatology*, 69: 1264–1277.

Weiss A., King J. E., and Murray L. (2011) *Personality and Temperament in Nonhuman Primates*. New York: Springer.

Weiss A., King J. E., and Perkins L. (2006) Personality and subjective well-being in orangutans (*Pongo pygmaeus* and *Pongo abelii*). *Journal of Personality and Social Psychology*, 90(3): 501–511.

Weiss A., Staes N., Pereboom J. J., Inoue-Murayama M., Stevens J. M., and Eens M. (2015) Personality in Bonobos. *Psychological Science*, 26(9): 1430–1439.

Wilczynski W., Quispe M., Munoz M. I., and Penna M. (2017) Arginine vasotocin, the social neuropeptide of amphibians and reptiles. *Frontiers in Endocrinology*, 8: 186.

Wilson V. A. D., Inoue-Murayama M., and Weiss A. (2018) A comparison of personality in the common and Bolivian squirrel monkey (*Saimiri sciureus* and *Saimiri boliviensis*). *Journal of Comparative Psychology*, 132(1): 24–39.

Yirmiya N., Rosenberg C., Levi S. et al. (2006) Association between the arginine vasopressin 1a receptor (AVPR1a) gene and autism in a family-based study: Mediation by socialization skills. *Molecular Psychiatry*, 11(5): 488–494.

Zhang R., Zhang H. F., Han J. S., and Han S. P. (2017) Genes related to oxytocin and arginine-vasopressin pathways: Associations with autism spectrum disorders. *Neuroscience Bulletin*, 33(2): 238–246.

Ziegler T. E., and Crockford C. (2017) Neuroendocrine control in social relationships in non-human primates: Field based evidence. *Hormones and Behavior*, 91: 107–121.

Interim Summary

One of the foundational approaches to the evolution of behavior is the comparative approach. The underlying logic is to use similarities and differences across species to draw conclusions about the evolution of the trait in question, with the assumption that similar selective pressures lead to similar traits. When using this approach to understand humans, a natural starting place is the other primates, as we are primates ourselves. But this is not the only approach; we may also wish to know what the impact of a specific feature is, in which case we will focus on other species that have the same feature independent of phylogeny, or share a particular ecological or social niche. Convergences across disparate taxa may suggest the ways in which the trait in question is linked to a specific behavioral outcome.

Model organisms can also be used to better understand complex phenomena using a simpler system that provides better tractability for experimentation and fewer variables to hinder interpretation. These are typically very well-characterized species for which much is known about their genetics, anatomy, and physiology, including neurology and endocrinology, and their behavior, which allows scientists to choose a model species that is most appropriate for the question being asked. One example that is highly relevant for this book is the use of rodent models to understand the oxytocin and arginine vasopressin systems (for a review, see Chapter 5), particularly in the context of pair bonding. Much of this work was done with the prairie vole (*Microtus ochrogaster*) or the California mouse (*Perymyscus californicus*). A small rodent may seem like an unlikely candidate for understanding human behavior, but unlike nearly all other mammals, including approximately three-quarters of primates, prairie voles and California mice are monogamous, with the male and female in a mated pair forming long-term pair bonds. Even more importantly, for both species there are closely related non-monogamous species, which allows us to look for differences that are (largely) due to the species' mating systems. In addition, in rodents, scientists can use a variety of techniques to do things that are not possible with humans, such as look at brain function on the cellular level and manipulate the distribution of hormone receptors. Recently scientists have begun to explore other monogamous primates, such as titi monkeys (*Callicebus cupreus*) and callitrichids, to determine if there are differences in how these hormones interact in pair bonding in the primates as compared to rodents, which may be useful for understanding human pair bonding. One important point to remember is that while most work focuses on only a few model organisms, it is important to work with one that is a good choice for the question at hand, even if it is not the most commonly used organism. Indeed, we encourage the use of less commonly studied organisms to broaden our base of understanding.

In Part III of the book, the authors describe research that does just that. They take broadly comparative approaches to look at cooperation and competition in species that are more phylogenetically distant from humans. Despite this diversity, the same themes that have appeared throughout the book are echoed here. Although when we think of humans we typically consider complex social interactions (Chapter 1) and underlying cognitive mechanisms (Chapter 3), the chapters in this section highlight the degree to which studying other species uncovers situations in which complex behavior emerges without complex cognition. Two classic examples are outlined in Chapter 9, regarding cooperation in cleaner fish – a small, tropical reef fish – and Chapter 10, regarding bees and wasps. Both species show myriad traits that were long considered to be limited to the brainy "higher" animals, such as individual recognition, complex cooperation, reputation management, empathy, and impressive communication among individuals. This emerging recognition that smart brains do not need to look like ours is an important reminder that we should not assume that a specific mechanism is required for a particular behavior. Indeed, some of these "higher order" behaviors are likely much more widespread and phylogenetically ancient than assumed; think of the interplay between cooperation and conflict outlined by Ostrowski in Chapter 4. Aside from reshaping our understanding of what it means to be human, these emerging discoveries offer us the opportunity to test hypotheses about the origins and evolutionary function of some of these behaviors using the comparative approach.

This highlights another key point; although these species are all fairly distantly removed from humans, they are no less important for understanding human behavior. For instance, it does not take a large brain to evolve the ability to soothe a cooperation partner, as in the use of "massage" by cleaner fish (Chapter 9), which suggests both strong relationships and a knowledge of how to manipulate one's social partner. People who study humans are often used to assuming that any solution developed by humans must be cognitively complex and/or based on active decision-making, but many of these apparently complex solutions can be evolved strategies that do not require explicit decision-making. With respect to humans, it may well be that we solve some of these problems through explicit cognition, but it is important to consider these cognitively simpler strategies, which may function as alternative mechanisms, or work together with mechanisms requiring more complex cognition.

One particular challenge to working across disciplines is the differences in how terminology is used in different areas. This obscures communication at best, and at worst leads people to make inappropriate assumptions about what they are reading because they imbue a term with an understanding of it gained from a different discipline. Indeed, the main focus of Chapter 5 is a call to return to the original usage of the term "prosocial" and avoid using it in ways that are both misleading and cause researchers to overlook valuable model species. In Chapter 9, Bshary highlights the mutation of the term "mutualism" in over time, which obscures comparisons and leads to misinterpretation.

The key element that ties all of the chapters in this book together is the ubiquity of the tight interconnection between cooperation and conflict across species. Indeed, cooperation and conflict do not exist in isolation, but require one another. Ridley and Nelson-Flower (Chapter 8) make the case that conflict – both within and between groups – defines the cooperative breeding dynamics of the pied babbler. Rittschof and Grozinger argue in Chapter 10 that for social insects, the focus on cooperation has led to endemic conflict being overlooked, despite the fact that conflict defines the organization of eusocial insect societies just as much as cooperation. The fact that one seemingly cannot exist without the other tells us something fundamental about the evolution of social structure, and may be an essential lesson to those working to understand how to minimize conflict and improve cooperation in our own society.

Part III

Species Comparisons

8 Understanding the Trade-off between Cooperation and Conflict in Avian Societies

Amanda R. Ridley and Martha J. Nelson-Flower

Introduction

The evolutionary paradox of animals helping others in their social group, rather than living independently, has fascinated researchers for many decades. Ultimately, this cooperation is hypothesized to have evolved because the benefits of cooperation outweigh the costs (Hamilton, 1964), and considerable empirical evidence supports this hypothesis (see Koenig and Dickinson, 2016). However, one major cost of cooperation, particularly for territorial cooperative breeders, is conflict (Shen et al., 2017; Nelson-Flower et al., 2018a). This conflict may occur both within and between groups; such conflict includes access to reproductive opportunities, social status, territory, water, or food. We suggest that a consideration of the influence of *both* intergroup and intragroup levels of conflict in individual decisions to cooperate is essential, since intragroup conflict may lead to decisions to disperse if intergroup opportunities to mate are high, whereas high levels of intergroup conflict may promote intragroup cooperation in order to defend existing resources (see Chapter 10 for review of inter- vs. intragroup aggression in social insects). Conflict is therefore a natural outcome of cooperation, and the level of conflict may define the point at which cooperation is no longer a beneficial strategy for individuals. This possibility, that conflict at *both* inter- and intragroup levels define the stability of cooperation over time, remains relatively under-explored despite its potential importance in understanding the evolution of cooperation.

Avian species have contributed significantly to research into the evolution and maintenance of cooperative behavior (Koenig and Dickinson, 2016). Avian species have several attributes that make them suitable study species for research into the dynamics of cooperation: (a) a considerable proportion (more than 10%) of them breed cooperatively (Cockburn, 2006), (b) their reproductive attempts are often

We thank all past and current members of the Pied Babbler Research Project for their hard work in the field, and for insightful discussions into cooperative breeding behavior. Without their immense input and enthusiasm, we would never have been able to get such detailed long-term datasets. In particular, we thank Nichola Raihani for her incredible assistance in establishing the population and her input into interpreting pied babbler behavior. We thank the Percy FitzPatrick Institute of African Ornithology (University of Cape Town) for their unwavering logistical and scientific support over the entire duration of our long-term research project. Finally, we thank the Kalahari Meerkat Project, Professor Tim Clutton-Brock and Professor Marta Manser for their support of our research at the shared study site.

readily monitored (the presence of eggs in a nest can make it easier to monitor failed and attempted breeding compared to the internal gestation of a mammal, the fossorial breeding habits of many reptiles, and the broadcast spawning of fish), and (c) species with altricial young are often central-place foragers; direct provisioning of young is therefore easily observable (compared to lactation in a mammal, or precocial young in reptiles). Indeed, "helpers" in cooperative species were first formally documented by the famous ornithologist Alexander Skutch in 1935, who coined the term "helpers at the nest" for his observations of more than two adult birds provisioning a single brood (Skutch, 1935).

Evolutionary explanations for the occurrence of cooperation assume that individuals gain more benefit from cooperation than the costs incurred, and in many cases, cooperation is due to the presence of close relatives (Hamilton, 1964; Emlen, 1988; but see Riehl, 2013). This creates an evolutionarily stable strategy where individuals should continue to cooperate until the costs outweigh the benefits. Considerable empirical evidence has found a significant benefit of cooperation in terms of territory defense (reviewed in Christensen and Radford, 2018), predator detection and shared vigilance (Elgar, 1989; Ridley and Raihani, 2007a; Ridley et al., 2013), resource acquisition (foraging behavior – Hollén et al., 2008), body condition (Ridley et al., 2008), and reproductive success (Moehlman, 1979; Canestrari et al., 2008). However, the benefits of cooperation do not increase linearly with group size (Macedo and DuVal, 2018), nor does relatedness between individuals necessarily predict cooperation within a group (Riehl, 2013). At larger group sizes, the benefits of cooperation may begin to plateau and even decline because of rising conflict among group members. In addition, conflict may arise *despite relatedness* when the direct benefits of conflict outweigh the benefits to cooperating with kin (Shen et al., 2014). This conflict most commonly takes the form of conflict over resources, such as mating opportunities, foraging, and breeding sites (Box 8.1).

Pied Babblers

For the last 16 years, we have conducted research on the pied babbler (*Turdoides bicolor*) – a cooperatively breeding passerine bird (weighing 60–95 g) endemic to the semi-arid Kalahari savanna (Figure 8.1) – to investigate the trade-off between cooperation and conflict in animal societies (Ridley, 2016). The Pied Babbler Research Project was established in 2003, and has maintained long-term behavioral observations of individuals throughout their lifetime, facilitated through the process of habituation (Ridley, 2016). Habituation involves teaching wild animals that the observer is not a threat. With habituation, the birds tolerate observers walking with the group throughout the day, allowing detailed behavioral observations of interactions within and between groups. The study population typically comprises 12–18 habituated groups annually, with group size ranging between 3 and 11 adults (modal group size = 5 adults). Groups are highly territorial year-round (Golabek et al., 2012; Humphries, 2013), and there is one dominant breeding pair in each group who monopolize over 90% of breeding activity (Nelson-Flower et al., 2011). All other

---

Box 8.1 What Do We Mean by Conflict?

Conflict in avian societies can take many forms and is extremely common (Morales and Velando, 2013).

Forms of Conflict
Family Conflict

Parent–offspring conflict occurs when the interests of the parent and the offspring are not the same. For example, adults significantly reduce their feeding rate to nestlings to encourage them to fledge (Raihani and Ridley, 2007), and fledglings blackmail adults into feeding them by moving to dangerous locations (Thompson et al., 2013).

Sibling conflict can take the form of competition over access to provisioning adults or of competition over position in the social queue (Ekman et al, 2002; reviewed in Drummond, 2006).

Intragroup Conflict

Dominance conflict is where individuals compete over access to the dominant rank. Dominance can have several benefits, including access to a greater share of reproductive opportunities and food resources (Clutton-Brock, 2016; Alberts, 2019).

Reproductive conflict is where individuals compete over access to reproductive opportunities (Shen et al., 2017; Covas and Doutrelant, 2018).

Intergroup Conflict

Intergroup conflict is where groups compete against one another for access to a resource. Often, individuals within a group work together to compete against members of an opposing group. Territory defense is a form of intergroup conflict.

Interspecific Conflict

Interspecific conflict can take many forms, including conflict over access to nesting resources. For example, wattled starlings (*Creatophora cinerea*) eat pied babbler eggs so that they can lay their own eggs in the nest (Ridley and Thompson, 2012). Another example is brood parasitism, where one species lays egg/s in the nest of a heterospecific, and contributes no further parental care toward their young (Davies, 2000).

---

group members contribute to cooperative activities, which include brooding and provisioning young, sentinel behavior, predator mobbing, and territory defense (Ridley and Raihani, 2007b; Ridley, 2016).

Although cooperative behavior is prominent within groups, conflict between members of the same group is also a common occurrence. Overt conflict in pied babblers is readily observable through physical attacks, vocal displays, and posturing

Figure 8.1 An adult pied babbler (*Turdoides bicolor*) watching a fledgling forage. Fledgling pied babblers do not attain full adult plumage until approximately one year old, and develop foraging skills slowly, remaining dependent on adults for food for at least the first three months of life. During dry years, babblers must range widely over their large territories to find sufficient food for developing young. (Photo by A. Ridley.) (A black and white version of this figure will appear in some formats. For the color version, please refer to the plate section.)

(Raihani et al., 2010; Ridley, 2016). Because all individuals in the population are ringed with a unique color combination, dispersal to other groups is readily observed. Dispersal can be either voluntary or forced. Forced dispersal, also known as eviction, involves the continued attacking of a group member by one or several same-sex members of the group, until the attacked individual leaves the group (Raihani et al., 2010). The dominant breeding pair in each group maintains their position of dominance by aggressive displays toward subordinates. Such displays include jumping on a subordinate's back, pecking at them, puffing up their chest feathers while running toward them, or a low frequency vocal threat that we term an "aggressive purr." Subordinates typically react to such dominance displays by rolling onto their back with their feet in the air, opening their beak to beg, or crouching down close to the ground. These kinds of submissive gesture commonly serve to stop the attack of the dominant individual.

Intergroup conflict in pied babblers is readily observed, since it involves a conspicuous multimodal display. Groups typically chorus loudly while standing together in a prominent elevated position, facing the opposing group. Group members chorus synchronously, and the chorus is accompanied by piloerection of the head and body feathers, as well as rapid pumping of the wings. Occasionally, if a vocal display does not result in the retreat of one of the groups from the display area, physical conflict will ensue. This will involve members of opposing groups physically attacking one another either in air or on the ground. Physical attack can take various forms, but primarily involves pecking and leg grappling. Such attacks can result in injuries, but rarely death. Intergroup interactions are fairly common in pied babblers, occurring on average twice per week. Each pied babbler group is typically surrounded by three neighboring groups in the relatively homogenous semi-arid savanna they inhabit. In some cases, where groups are competing over access to a highly desired resource (such as access to water), intergroup interactions can occur several times a day. Intergroup interactions typically last 5–10 minutes, and fewer than 10% involve physical fighting.

Critical Group Size Effects

In cooperatively breeding species, small group sizes are often considered less beneficial to the individual group members. In black-backed jackals (*Canis mesomelas*), for example, reproductive success increases sharply as group size increases, and pairs without helpers rarely successfully raise young (Moehlman, 1979). Similarly, in wild dogs (*Lycaon pictus*), group sizes of less than five adults struggle to successfully defend their young and food resources against larger predators (Courchamp et al., 2002). The declining benefit of cooperation at small group sizes can lead to an Allee effect, where groups are less likely to reproduce successfully (or attract immigrants), and the likelihood of group extinction increases. Allee effects at small group sizes have been described for a number of cooperative species (reviewed in Angulo et al., 2018), including wild dogs (Angulo et al., 2009) and Arabian babblers (*Turdoides squamiceps*, Keynan and Ridley, 2016). However, the flip side of an Allee effect at small group sizes is that there can be an increased chance of poor reproductive success and group extinction at large group sizes; this has received relatively less attention in cooperative breeding species (Shen et al., 2014). Where membership of both very small and very large groups results in smaller benefits than membership of mid-sized groups, a quadratic relationship between group size and the cooperative benefit (e.g., reproductive success) occurs, known as the "group size paradox" (Olson, 1965; Esteban and Ray, 2001). The apex of this quadratic relationship is the critical group size, beyond which the benefits of cooperation (relative to the costs) begin to decline.

Critical group size effects are increasingly being recognized as an important aspect of intraspecific variation in dispersal decisions and helping behavior. While it has long been recognized that individual contributions to cooperative behaviors can decline with group size (e.g., personal contributions to vigilance Elgar, 1989; Beauchamp, 2009), how these declining contributions influence life history traits such as reproductive success and dispersal behavior is less clear. As group size increases, individuals are more likely to "free-ride" – where they reap the benefits of group membership without incurring the cost of investment in cooperative behaviors (Gavrilets, 2015). Such free-riding behavior can lead to the collective action problem (Willems et al., 2013), and lead to a critical group size effect – whereby the lowered investment in cooperation per individual makes membership of larger groups less beneficial. Many theoretical models support this idea (e.g., Gavrilets, 2015; Peña and Nöldeke, 2018); however empirical evidence of this effect in cooperatively breeding species is lacking.

Theoretically, a critical group size effect will be observed in all species at some point because resources are finite and will therefore become limiting (Shen et al., 2014). This effect might, at least in part, be an important factor in determining the balance between intragroup cooperation and conflict behaviors. If critical group size effects are present, we would expect to see increased rates of eviction, dispersal, and overt conflict as group size increases beyond a certain point. In contrast, below the critical group size point, we would expect to see groups become more willing to accept unrelated immigrants, and helpers being offered concessions to stay in the group (Vehrencamp, 1983; Reeve et al., 1998). An analysis of the 16-year pied babbler database confirms these trends. First, we found a quadratic relationship between reproductive success and group size, with group sizes of six adults having the highest reproductive success (Ridley, 2016). Second, we found a significant increase in aggression rates per individual as group size increased (Figure 8.2A and B). Third, we found individuals lower down the social queue were more likely to disperse (Nelson-Flower et al., 2018a), or *attempt* to disperse. And fourth, we found greater instability in large groups, with attempted dispersal rates (frequency of dispersal attempts per year on a per individual basis) higher for individuals in both large and small group sizes, but low in intermediate group sizes (see Figure 8.2B). We also found that small groups were more likely to non-aggressively accept immigrants, whereas large groups aggressively attack individuals attempting to immigrate (Ridley, 2016; Wiley and Ridley, 2018).

Combined, these demographic trends confirm the presence of a critical group size effect, making large groups highly unstable (expressed in high dispersal and eviction rates). Identifying a critical group size effect is significant because in species where the benefits of pair breeding versus cooperative breeding cannot be directly compared (due to the low frequency of pair breeding, for example), the factors that cause within-group cooperation to break down can give us important insights into the origin of cooperative breeding behaviors.

The relationship between group size and reproductive success can be misleading, because a simple increase in reproductive success with group size does not mean that individuals in large groups are receiving more benefits than individuals in small groups. To better determine the relationship between group size and reproductive success, it can be informative to look at reproductive success relative to the number of individuals helping to raise the young. This generates a different relationship, suggesting the indirect benefits of helping to raise related young are significantly smaller for helpers in larger groups (Figure 8.2A). Combined with an increase in conflict as group size increases, why do individuals stay in larger groups? There may be direct benefits of group size that outweigh the declining indirect fitness benefit, such as predator detection (Ridley et al., 2013), access to a high quality territory (Komdeur, 1992; Baglione et al., 2006), chances of attaining dominance/breeding access (Kokko and Johnstone, 1999), access to a high quality mate, and greater longevity (Kokko and Ekman, 2002; Beauchamp, 2014). Indeed, these benefits are likely to be the only benefits that would promote group membership for individuals unrelated to the young they help raise.

(A)

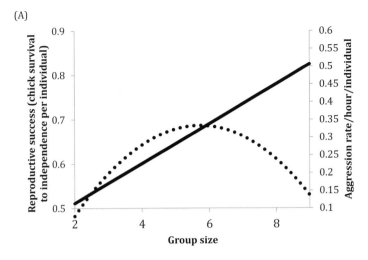

(B)

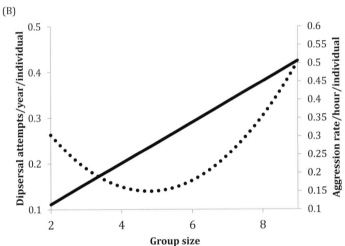

Figure 8.2 (A) The critical group size effect, showing the point where reproductive success declines (dashed line), at a group size of approximately six adults, while aggression rate continues to increase (solid line in graphs (A) and (B)). This should be the point at which dispersal increases, since the benefits of group membership become less, and indeed that is what we see: the probability of dispersal increases sharply at the critical group size of six adults (dashed line in Figure 8.2B). To understand this effect, it is important to examine aggression rates at the *individual* level. This is because in larger groups, there is likely to be a simple mathematical relationship between aggression rates and group size because when there are more individuals present, there are more opportunities to interact. To account for this effect, we have examined all rates presented above on a per individual basis.

We suggest that there are two possible causes of critical group size effects in avian societies: social conflict or ecological constraints. It is possible that the two are interrelated: social conflict could be exacerbated by limited resources due to prevailing ecological conditions (Shen et al., 2017; Nelson-Flower et al., 2018a). However, there may be a way to disentangle whether social conflict or ecological constraints are the primary influence on critical group size patterns. If ecological constraints were the primary cause of critical group size effects, we would expect to see a sliding scale of where this critical group size point was relative to prevailing ecological conditions. For example, in drought years where food resources are often more limited, we may expect to see a critical group size effect occur at a lower actual group size compared to a non-drought year. In contrast, if social influences were the primary cause of critical group size effects (for example, conflicts due to the length of the social queue and access to reproductive opportunities), then this critical group size point should not vary according to ecological conditions. It should, however, be expected to vary according to social group composition. For example, if access to reproductive opportunities was the primary driver of conflict according to group size, then we may expect to see a difference in the critical group size point in simple family groups versus complex groups, where adult helpers are not related to one or both members of the breeding pair. These are testable hypotheses which have not yet been investigated empirically in cooperative species. However, they could provide an important insight into the impact of social conflict in moderating group size and leading to a breakdown in cooperation. Indeed, Shen et al. (2014) developed theoretical models to predict the relationship among group size, productivity, and conflict – generating trends similar to those we hypothesized above for social influences as the primary cause of group size changes. Long-term datasets from a variety of different cooperative societies are now needed to formally test the ideas of Shen et al. (2017) empirically.

Kinship Effects

Relatedness within groups may have far-reaching effects. Groups consisting of close relatives are often more stable and productive, while the opposite is true for those that contain unrelated individuals (Lundy et al., 1998; Cooney and Bennett, 2000; Clutton-Brock et al., 2001). Individuals living in groups with relatives may experience higher survival rates than those living with non-relatives (Greisser et al., 2006) or may acquire better resources when they do establish independent territories (Dickinson et al., 2014).

As in many other cooperative breeders, intragroup conflict in southern pied babblers often revolves around the distribution of reproduction within the group, or reproductive skew. Typically, groups of pied babblers comprise the dominant pair and their adult offspring, which delay dispersal and help to raise their younger siblings (Nelson-Flower et al., 2011). Over time, one or both dominant individuals die or (rarely) leave the group, and a new, unrelated dominant enters the group, representing a breeding opportunity for the opposite-sex adult subordinates living therein. When adult subordinate females attempt to breed with their group's unrelated dominant male, this represents a cost to the dominant female and takes an appreciable toll on group productivity, as conflict among females takes the form of egg-eating behaviour (oophagy; Nelson-Flower et al., 2013). Dominant females in turn display significantly more aggression toward subordinate females that compete for reproduction (Nelson-Flower et al., 2013). Females that were close genetic relatives were just as likely as unrelated females to engage in this costly conflict (Nelson-Flower et al., 2018b), revealing that the benefits of a subordinate female's own reproduction are likely to outweigh the costs that this places on her genetic relatives.

Though the genetic relationship between females has no effect on their conflict, the genetic relationship between dominant and subordinate males appears to affect subordinate male reproductive behavior and dispersal (Nelson-Flower and Ridley, 2016; Nelson-Flower et al., 2018b). Subordinate male competition for dominance is not costly in and of itself (Nelson-Flower and Ridley, 2015). In fact, sons or brothers of dominant males are *more* likely to attempt to breed with unrelated dominant females (they are rarely successful); in contrast, stepsons or immigrant foreign subordinate males never compete to breed (Nelson-Flower et al., 2018b). In addition, these unrelated subordinate males are likely to disperse significantly earlier from their groups than those that are related to the dominant male, implying a role of genetic relatedness in tolerance (nepotistic tolerance, Nelson-Flower and Ridley, 2016; Nelson-Flower et al., 2018a). A similar role of nepotistic tolerance has been shown in other group-living species (Goldstein et al., 1998; Ekman and Griesser, 2002; Dickinson et al., 2014; Groenewoud et al., 2018) and implied in some (Clutton-Brock et al., 2010). In general, an understanding of the individual genetic relationships within social groups of cooperative breeders can illuminate a wide variety of drivers for important life-history events, such as dispersal and reproductive efforts within the group.

Intergroup Conflict

The study of cooperation tends to focus on intragroup dynamics that impact behavioral decisions to cooperate or disperse. (See long-term studies of cooperatively breeding species described in Koenig and Dickinson, 2016.) However, cooperative groups live within a wider population matrix, and intergroup interactions can be common (Thompson et al., 2017; Christensen and Radford, 2018; Mirville et al., 2018). In addition, recent theoretical models suggest that the costs of intergroup conflict can drive the evolution of cooperative behavior (Choi and Bowles, 2007; Rusch, 2014). Therefore, greater consideration should be given to the impact of intergroup interactions on intragroup dynamics. Recent research in several primate species further suggests that intergroup conflict has a strong influence on intragroup cooperative behaviors. For example, Crofoot (2013), Majolo et al. (2016), and Mirville (2018) found that aggressive intergroup interactions influenced intragroup affiliative behavior and movement patterns for several hours following an intergroup conflict. These findings suggest that territorial conflict influences intragroup dynamics beyond the immediate period of interaction and may therefore have a larger influence on group dynamics than is currently recognized.

Intergroup conflict can be costly, with individuals potentially suffering injuries, large energetic expenditure, loss of territory, loss of foraging time, and loss of body mass during the immediate period of conflict (Batchelor and Briffa, 2011; Humphries, 2013). In some cases, the costs of conflict can be large, such as the direct mortality of competing adults or infanticidal behavior (Packer and Pusey, 1983; van Schaik and Janson, 2000). However, the costs of intergroup conflict may have been significantly underestimated if the post-conflict impacts are not measured. The vast majority of research on intergroup conflict focusses on the costs of the interaction itself (reviewed in Mirville, 2018). However, a consideration of the change in a group's behavior post-conflict, and how long this change in behavior lasts, is an important measure of the impact of intergroup conflict (Thompson et al., 2017). The best way to determine the impact of an intergroup conflict may be to compare the behavior of individuals in the time period preceding the conflict, to the same time period post-conflict, such that a within-individual comparison of behavior can be made (Mirville, 2018). Although it has been recently recognized that inter-group interactions can have significant short-term costs for group members in the post-conflict period (Crofoot, 2013; Majolo et al., 2016; Mirville, 2018, reviewed in Christensen and Radford, 2018), the longer-term costs of intergroup conflict are more elusive.

We investigated the longer-term impacts of intergroup conflict in pied babblers, and found that groups that were involved in regular territorial conflicts during the breeding season had lower reproductive success (Figure 8.3). There could be several possible explanations for this effect: (a) territorial conflict is energetically costly (as

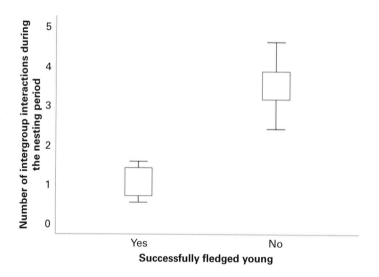

Figure 8.3 The relationship between frequency of aggressive intergroup interactions and outcome of a breeding attempt (mean ± s.e.). An attempt was considered successful if at least one nestling fledged. Data was based on 63 breeding attempts, of which 34 successfully fledged at least one young, and 29 did not.

shown in pied babblers by Humphries, 2013) and thus reduces the time available to incubate, brood, and provision young; (b) the loud and prolonged vocal displays that accompany intergroup interactions may attract predators to the nesting area (Ridley and Thompson, 2012); or (c) individuals may commit infanticide and destroy the eggs or nestlings of competing groups (Nelson-Flower et al., 2013). Oophagy has been observed in pied babblers, but our nest cameras only captured intragroup oophagy (Nelson-Flower et al., 2013). However, our nest camera coverage was limited to two breeding seasons and a select number of nests, and therefore we cannot rule out the possibility of intergroup oophagy in this species. Indeed, kidnapping of young has been observed in this species (Ridley, 2016), so intergroup infanticide remains a possibility.

In pied babblers, intergroup relatedness appears to affect intergroup conflict. Due to the short-distance dispersal commonly seen in this species, neighbors are often relatives (Nelson-Flower et al., 2012). Recent work has shown that close kinship between neighboring groups increases the proportion of shared territory between groups (Humphries, 2013). More shared territory means a greater number of costly territorial displays and conflicts, but kinship mitigates the length of the conflicts. Consequently, there is an overall decreased investment in conflict with related neighbors, especially where territorial overlap is mild to moderate.

Intergroup conflict may impact intragroup behavior in several ways, and can generate competing hypotheses. In many social species, not all group members participate in intergroup interactions (Nunn, 2000; reviewed in Christensen and Radford, 2018). Indeed, some group members may be considered "free riders" – where they benefit from the actions of other group members without expending any energy themselves (Gavrilets, 2015). If such free-riding reduces the collective strength of the group during an intergroup interaction, then free riders may be expected to receive punishment from defecting, a behavior that has been observed in several vertebrate societies, particularly primates (Willems et al., 2013). Alternatively, the intergroup threat may result in group members being more "cooperative," to induce group members to stay and help defend the territory. The idea of giving concessions to group members to encourage them to remain in the group was first introduced in early theoretical models of reproductive skew (e.g., Vehrencamp, 1983; Reeve et al., 1998). The idea behind the concessions model was that where group size was a significant predictor of fitness (e.g., through higher reproductive success or breeder longevity), helpers may be given parcels of "reproductive access" to induce them to stay and help (Reeve et al., 1998). The same concept can be applied to the threat posed by intergroup conflict: greater affiliative behavior may act as a concession to encourage participation in future intergroup interactions. Affiliative behaviors can be beneficial to individuals through the removal of ectoparasites (Hillegas et al., 2008) and lowering of stress levels through stronger social bonds (Wittig et al., 2008). Empirical research has found considerable evidence for intergroup conflict reducing intragroup aggression in woodhoopes (*Phoeniculus purpureus*, Radford, 2008) and mountain gorillas (*Gorilla beringei beringei*, Mirville, 2018), among others. In these species, intragroup affiliative behavior increases following an intergroup conflict, and aggressive behaviors decline. Thus, current evidence suggests that intergroup conflict may significantly influence intragroup dynamics, and we therefore encourage researchers to give greater consideration to the population-level effects that may impact group-level behaviors.

Group Size Considerations for Intergroup Conflict

Research on the patterns of conflict and cooperation in animal societies often focus primarily on the effects of group size, and we have done so ourselves in the figures in this chapter. However, we caution against a focus on *actual* group size alone. In some cases, *relative* group size, defined as the size of a group relative to the size of neighboring groups, may be more important in determining decisions to cooperate, compete, or disperse. A group size of 6 individuals may be considered large in the population as a whole, but not if it is surrounded by groups of size 9, 10, and 11, for example. In this theoretical case, although the group of size 6 is larger than the population average, it may be more likely to lose an intergroup interaction with its larger neighboring groups (but see Strong et al., 2017). In addition, where there are sex biases in dispersal and/or territory defense, the number of same-sex individuals present relative to neighboring groups may be a more informative predictor of contest outcome. For example, in Arabian babblers, individuals disperse in same-sex coalitions, and the number in the coalition relative to the number of same-sex individuals in the defending group is the primary predictor of dispersal success rather than absolute group size (Ridley, 2012). Group size and the number of individuals of each sex do not necessarily correlate, and therefore it is important to consider both parameters when considering the benefits of cooperative behaviors. Similarly, although group size and relative group size do correlate in some cases, there is often enough variation between the two to justify researchers to consider both as potential predictors of data patterns. Mirville et al.'s (2018) investigation into the factors influencing the intensity of intergroup interactions in mountain gorillas found that relative group size was a better predictor of interaction patterns than actual group size, and that, importantly, the relative number of males was a better predictor of escalation of intergroup interactions than actual group size.

The effect of habitat type on the maximum group size an area can support is also an important consideration when thinking about the effects of group size on patterns of cooperation and conflict. For species that live in heterogeneous landscapes, there are likely to be both high-quality and low-quality habitats. If a low-quality habitat cannot support the same group size as a high-quality habitat, then landscape-level (or between-site) comparisons of cooperation may become less meaningful when using *actual* group size as a predictor of the benefits of cooperation. We suggest a relative group size measure, where the reproductive success or the likelihood of winning a fight is considered relative to the size of other groups in that *habitat type*, may be the most pertinent measure of group size effects.

Conclusions

The benefits of cooperation, relative to the costs of conflict (Box 8.2) that group-living entails, determine the formation and stability of cooperative groups. Although cooperation tends to capture the imagination, it is really the outcome of *conflict* that defines the dynamics of cooperative breeding groups. Here, we have looked at the evidence for conflict as a driver of cooperation by discussing how conflict – more often defined as the "cost" of cooperation – can be empirically measured. It is only with true empirical measurement that we can fully test the prevailing theoretical models for the evolution of cooperation. Although cooperative breeding research has reached incredible levels of detail in impressive short- and long-term studies worldwide (see Koenig and Dickinson, 2016, for a summary of recent long-term studies), in this chapter we identify current gaps in our knowledge of what promotes cooperation, and what causes cooperation to break down. We argue that a consideration of critical group size effects is important because (a) the beneficial effect of group size is not linear, (b) the benefits of cooperation may be density-dependent, and (c) there is a point where group size is so large that conflict increases sharply, making cooperation no longer beneficial. Identifying these critical group size effects helps us to determine what is limiting individual decisions to cooperate or not. We also suggest that investigation of the effects of genetic relationships between same-sex individuals may provide new insights into the costs and benefits of reproductive conflict and delayed dispersal.

Box 8.2 How Do We Measure the *Costs* and *Benefits* of Cooperation?

Central to our understanding of how cooperation evolves is measuring the benefits of this strategy relative to the costs. While benefits and costs are often discussed in theoretical models of cooperation (Shen et al., 2017), quantitative measures of these are required to empirically support the theoretical models. Long-term studies of cooperation in vertebrates have found huge variation in the causes of cooperation (Koenig and Dickinson, 2016). This means that the costs and benefits of cooperation for one species cannot necessarily be generalized to another, and so an understanding of how to quantitatively measure these parameters in different species is required.

A *cost* can be considered a change in a behavior or life-history trait during cooperation that is negative relative to that same behavior or trait under a non-cooperative strategy. Some examples of this include the following:

- Lower reproductive success when multiple individuals compete to breed, such as when females eat one another's eggs (defined as oophagy, Nelson-Flower et al., 2013).
- Lower fitness-related benefits, for example, when helpers help to raise the young of non-relatives and thus get no indirect fitness benefits (Riehl, 2013).
- Loss of body condition and lower survival, for example, when individuals lose body mass when helping to raise young (Ridley and Raihani, 2007b), or when they receive injuries during inter- and intragroup conflict (Mirville, 2018).

Box 8.2 (*cont.*)

Measurable benefits of cooperation include the following:

- Access to higher-quality resources, resulting in higher foraging efficiency and body mass for individuals living in groups (Ridley et al., 2008).
- Higher survival rate, including greater predator detection and avoidance through shared vigilance and sentinel behavior (Elgar, 1989; Ridley et al., 2013).
- Greater chance of dispersing to a dominant breeding position (Ridley and Raihani, 2007b; Nelson-Flower et al., 2018b).
- Higher reproductive success in groups compared to when breeding alone (Ridley and van den Huevel, 2012).

We argue for a greater consideration of the population-level impacts on intragroup behavioral patterns. Too often cooperative breeding research focuses on parameters that influence cooperative behavior at the group level. In particular, we argue that intergroup conflict is likely to have a large impact on intragroup behavior, yet this possibility remains relatively unexplored. Using data from our own study species, we find that the frequency of intergroup conflict has a significant impact on reproductive success, and therefore can have long-term impacts on the benefits of group membership. Overall, we find that the field of cooperative breeding research has provided us with highly varied and detailed examples of why cooperation is beneficial, and how conflict moderates the benefits of cooperation. However, we feel that there is considerable scope for new avenues of research, and we look forward to the exciting new research discoveries to come.

References

Alberts S. C. (2019) Social influences on survival and reproduction: Insights from a long-term study of wild baboons. *Journal of Animal Ecology*, 88: 47–66.

Angulo E., Luque G. M., Gregory S. D., Wenzel J. W., Bessa-Gomes C., Berec L., and Courchamp F. (2018) Allee effects in social species. *Journal of Animal Ecology*, 87: 47–58.

Angulo E., Rasmussen G. S. A., MacDonald D. W., and Courchamp F. (2009) Do social groups prevent Allee effect related extinctions? The case of wild dogs. *Frontiers in Zoology*, 10: 11.

Baglione V., Canestrari D., Marcos J. M. 2006. Experimentally increased food resources in the natal territory promote offspring philopatry and helping in cooperatively breeding carrion crows. *Proceedings of the Royal Society London B Biological Science*, 273: 1529–1535.

Batchelor T. P., and Briffa M. (2011) Fight tactics in wood ants: Individuals in smaller groups fight harder but die faster. *Proceedings of the Royal Society London B Biological Science*, 278: 3242–3250.

Beauchamp G. (2009) How does food density influence vigilance in birds and mammals? *Animal Behaviour*, 78(2): 223–231.

Beauchamp G. (2014) Do avian cooperative breeders live longer? *Proceedings of the Royal Society London B Biological Science*, 281: 20140844.

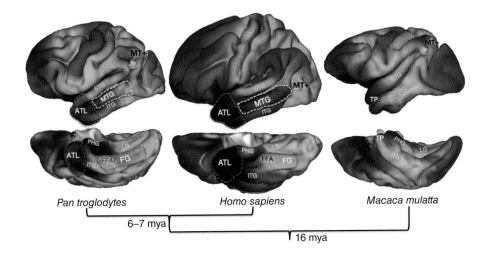

Figure 6.1 The organization of the human temporal lobe in evolutionary context. Divergence dates are from Steiper and Seiffert (2012). FG, Fusiform gyrus; MTG, Middle temporal gyrus; MT+, Visual motion area MT; ATL, Anterior temporal lobe. Dashed lines delineate evolutionarily novel cortical territories that have expanded in the hominid lineage. (A black and white version of this figure will appear in some formats.)

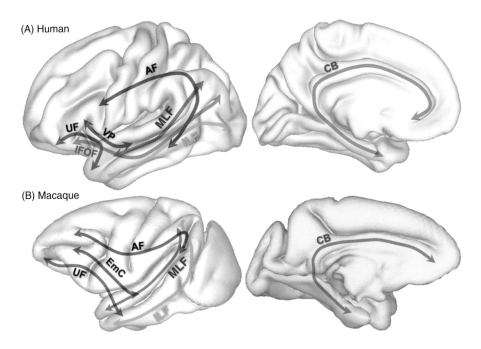

Figure 6.2 A simplified diagram of the major white matter tracts in humans and macaques that connect with and/or course through the temporal lobe. (A) Human fascicular trajectories are based on Catani and Thiebaut de Schotten (2008), Rilling et al. (2008), Saur et al. (2008), Frey et al. (2008), Makris et al. (2009), and Turken and Dronkers (2011). (B) Macaque fascicular trajectories are based on Schmahmann and Pandya (2009) and Schmahmann et al. (2007). AF, arcuate fasciculus; CB, cingulum bundle; EmC, extreme capsule; IFOF, inferior fronto-occipital fasciculus; ILF, inferior longitudinal fasciculus; MLF, middle longitudinal fasciculus; UF, uncinate fasciculus; VP, ventral pathway. Cortical reconstructions were generated using the Freesurfer image analysis suite, http://surfer.nmr.mgh.harvard.edu (Fischl, 2012). (A black and white version of this figure will appear in some formats.)

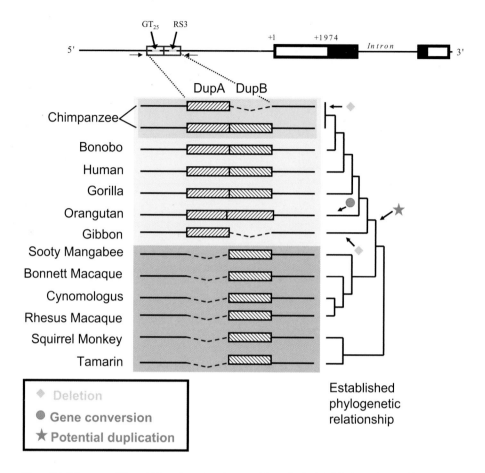

Figure 7.1 Diagram of DupA/B regions in primates. Note the indel deletion of the RS3 region in some chimpanzees which is absent in other great ape species including humans. (From Donaldson et al., 2008.) (A black and white version of this figure will appear in some formats.)

Figure 8.1 An adult pied babbler (*Turdoides bicolor*) watching a fledgling forage. Fledgling pied babblers do not attain full adult plumage until approximately one year old, and develop foraging skills slowly, remaining dependent on adults for food for at least the first three months of life. During dry years, babblers must range widely over their large territories to find sufficient food for developing young. (Photo by A. Ridley.) (A black and white version of this figure will appear in some formats.)

Figure 9.2 Cleaner fish *Labroides dimidiatus* cleans the mouth of a potato grouper (*Epinephelus tukula*), Great Barrier Reef, Australia. (Photo by Simon Gingins.) (A black and white version of this figure will appear in some formats.)

Figure 10.1 Model social insect systems. *Left panel:* Honeybees (*Apis mellifera*) live in perennial colonies with tens of thousands of female worker siblings who cooperate to rear the offspring of a single reproductive queen mother (Winston, 1991). Queens and workers are morphologically, physiologically, and behaviorally distinct, where workers are unable to mate or produce fertilized eggs, and queens specialize exclusively in egg production. The workers build the nest (which consists of a precisely built honeycomb with hexagonal cells) within a suitably sized cavity, produce specialized processed food which they feed to developing larvae (termed "brood"), provision the colony by foraging over vast areas (spanning hundreds of square kilometers), recruit nestmates to high-value forage sites through a symbolic dance language, defend the colony from predators and intruders (including robbing bees from other colonies), thermoregulate the colony during both hot and cold seasons, and search for new nest sites when the colony reproduces through a process of colony fission. (Photo by Nick Sloff, Penn State.) *Middle panel:* Bumblebees (*Bombus terrestris*) live in annual colonies that reach maximum sizes of a few hundred individuals, depending on the species (reviewed in Amsalem et al., 2015). Colonies are founded by a single queen (a foundress), who constructs the nest, forages, and rears the first cohort of worker brood. Once the workers have emerged, they take over foraging, brood care, and nest construction and defense, while the queen lays the eggs. Like honeybees, worker bumblebees are unmated. However, at the end of the season, the colony starts rearing the next generation of reproductive individuals (queens and males). At this point, the workers and queen compete with each other over reproduction of males, which develop from unfertilized eggs, and thus can be produced by unmated workers. The workers form a social and reproductive dominance hierarchy, with dominant workers behaving more aggressively and laying more male eggs. The males mate with the virgin queens from other nests, and the mated queens overwinter underground, while the males, workers, and old queen die. (Photo by Luca Franzini, Penn State.) *Right panel:* Wasps represent a diverse group of species, ranging from solitary to eusocial in their lifestyles (Jandt and Toth, 2015). Most work on wasp social behavior has focused on *Polistes* paper wasps. These wasps form annual colonies which are relatively small (~20 individuals) and are initiated in the spring by individual foundresses or by pairs of foundresses which establish a reproductive dominance hierarchy. As in the case of bumblebees, once the first cohort of workers is produced, they take over the colony tasks. Near the end of the colony life cycle, the next generation of males and queens are produced, and only the mated queens overwinter. *Polistes* workers have a clear social dominance hierarchy, and if the queen is lost, they will compete over reproductive dominance. (Photo by Jennifer Jandt, University of Otago.) (A black and white version of this figure will appear in some formats.)

Canestrari D., Marcos J. M., and Baglione V. (2008) Reproductive success increases with group size in cooperative carrion crows *Corvus corone corone*. *Animal Behaviour*, 75: 403–416.

Christensen C., and Radford A. N. (2018) Dear enemies or nasty neighbours? Causes and consequences of variation in the response of group-living species to territorial intrusions. *Behavioral Ecology*, 29: 1004–1013.

Choi J. K., and Bowles S. (2007) The coevolution of parochial altruism and war. *Science*, 318: 636–640.

Clutton-Brock T. H. (2016) *Mammal Societies*. West Sussex: John Wiley & Sons, Inc.

Clutton-Brock T. H., Brotherton P. N. M., Russell A. F. et al. (2001) Cooperation, control and concession in meerkat groups. *Science*, 291: 478–481.

Clutton-Brock T. H., Hodge S. J., Flower T. P., Spong G. F., and Young A. J. (2010) Adaptive suppression of subordinate reproduction in cooperative mammals. *American Naturalist*, 176: 664–673.

Cockburn A. (2006) Prevalence of different modes of parental care in birds. *Proceedings of the Royal Society London B Biological Science*, 273: 1375–1383.

Cooney R., and Bennett N. C. (2000) Inbreeding avoidance and reproductive skew in a cooperative mammal. *Proceedings of the Royal Society London B Biological Science*, 267: 801–806.

Courchamp F., Rasmussen G. S. A., and MacDonald D. W. (2002) Small pack size imposes a trade-off between hunting and pup-guarding in the painted hunting dog *Lycaon pictus*. *Behavioral Ecology*, 13: 20–27.

Covas R., and Doutrelant C. (2018) The sexual and social benefits of cooperation in animals. *Trends in Ecology and Evolution*, 34: 112–120.

Crofoot M. C. (2013) The cost of defeat: Capuchin groups travel further, faster and later after losing conflicts with neighbors. *American Journal of Physical Anthropology*, 152(1): 79–85, DOI: https://doi.org/10.1002/ajpa.22330.

Davies N. B. (2000) *Cuckoos, Cowbirds and Other Cheats*. London: T and AD Poyser Ltd.

Dickinson J. L., Ferree E. D., Stern C. A., Swift R., and Zuckerberg B. (2014) Delayed dispersal in western bluebirds: Teasing apart the importance of resources and parents. *Behavioral Ecology*, 25: 843–851.

Drummond H. (2006) Dominance in vertebrate broods and litters. *Quarterly Review of Biology*, 81: 3–32.

Ekman J., Eggers S., and Griesser M. (2002) Fighting to stay: The role of sibling rivalry for delayed dispersal. *Animal Behaviour*, 64: 453–459.

Ekman J., and Griesser M. (2002) Why offspring delay dispersal: Experimental evidence for a role of parental tolerance. *Proceedings of the Royal Society London B Biological Science*, 269: 1709–1713.

Elgar M. A. (1989) Predator vigilance and group size in mammals and birds – A critical review of the evidence. *Biology Review*, 64: 13–33.

Emlen S. T. (1988) The role of kinship in helping decisions among white-fronted bee-eaters. *Behavioral Ecology and Sociobiology*, 23: 305–315.

Esteban J., and Ray D. (2001) Collective action and the group size paradox. *American Political Science Review*, 95: 663–672.

Gavrilets S. (2015) The collective action problem in heterogeneous groups. *Philosophical Transactions of the Royal Society of London B*, 370: 20150016.

Golabek K. A., Ridley A. R., and Radford A. N. (2012) Food availability affects strength of seasonal territorial behaviour in a cooperatively breeding bird. *Animal Behaviour*, 83: 613–619.

Goldstein J. M., Woolfenden G. E., and Hailman J. P. (1998) A same-sex stepparent shortens a prebreeder's duration on the natal territory: Tests of two hypotheses in Florida scrub-jays. *Behavioral Ecology and Sociobiology*, 44: 15–22.

Griesser M., Nystrand M., and Ekman J. (2006) Reduced mortality selects for family cohesion in a social species. *Proceedings of the Royal Society London B Biological Science*, 273: 1881–1886.

Groenewoud F., Kingma S. A., Hammers M., Dugdale H. L., Burke T., Richardson D. S., and Komdeur J. (2018) Subordinate females in the cooperatively breeding Seychelles warbler obtain direct benefits by joining unrelated groups. *Journal of Animal Ecology*, 87: 1251–1263.

Hamilton W. D. (1964) The genetical evolution of social behavior I. *Journal of Theoretical Biology*, 7: 1–16.

Hillegas M. A., Waterman J. M., and Roth J. D. (2008) The influence of sex and sociality on parasite loads in an African ground squirrel. *Behavioral Ecology*, 19: 1006–1011.

Hollén L. I., Bell M. B.V., and Radford A. N. (2008) Cooperative sentinel calling? Foragers gain increased biomass intake. *Current Biology*, 18: 576–579, DOI: https://doi.org/10.1016/j.cub.2008.02.078.

Humphries D. (2013) The mechanisms and function of social recognition in the cooperatively breeding southern pied babler, *Turdoides bicolor*. PhD thesis, University of Western Australia.

Keynan O., and Ridley A. R. (2016) Component, group and demographic Allee effects in a cooperatively breeding bird species, the Arabian babbler (*Turdoides squamiceps*). *Oecologia*, 182: 153–161.

Koenig W. D., and Dickinson J. L. (2016) *Cooperative Breeding in Vertebrates: Studies of Ecology, Evolution and Behaviour*. Cambridge: Cambridge University Press.

Kokko H., and Ekman J. (2002) Delayed dispersal as a route of breeding: Territorial inheritance, safe havens, and ecological constraints. *American Naturalist,* 160: 468–484.

Kokko H., and Johnstone R. (1999) Social queuing in animal societies: A dynamic model of reproductive skew. *Proceedings of the Royal Society London B Biological Science*, 266: 571–578.

Komdeur J. (1992) Importance of habitat saturation and territory quality for the evolution of cooperative breeding in the Seychelles Warbler. *Nature*, 358: 493–495.

Lundy K. J., Parker P. G., and Zahavi A. (1998) Reproduction by subordinates in cooperatively breeding Arabian babblers is uncommon but predictable. *Behavioral Ecology and Sociobiology*, 43: 173–180.

Macedo R. H., and DuVal E. H. (2018) Friend or foe? The dynamics of social life. *Animal Behaviour*, 143: 139–143.

Majolo B., de Bortoli V. A., and Lehmann J. (2016) The effect of intergroup competition on intragroup affiliation in primates. *Animal Behaviour*, 114: 13–19.

Mirville M. O. (2018) The causes and consequences of intergroup interactions in mountain gorillas (*Gorillia beringei beringei*). PhD thesis, University of Western Australia.

Mirville M. O., Ridley A. R., Samedi J. P. M., Vecellio V., Ndagijimana F., Stoinski T. S., and Grueter C. C. (2018) Factors influencing individual participation during intergroup interactions in mountain gorillas. *Animal Behaviour*, 144: 75–86.

Moehlman P. D. (1979) Jackal helpers and pup survival. *Nature*, 277: 382–383.

Morales J., and Velando A. (2013) Signals in family groups. *Animal Behaviour*, 86: 11–16.

Nelson-Flower M. J., Hockey P. A. R., O'Ryan C., Raihani N. J., du Plessis M. A., and Ridley A. R. (2011) Monogamous dominant pairs monopolize reproduction in the cooperatively breeding pied babbler. *Behavioral Ecology*, 22: 559–565.

Nelson-Flower M. J., Hockey P. A. R., O'Ryan C., and Ridley A. R. (2012) Inbreeding avoidance mechanisms: Dispersal dynamics in cooperatively breeding southern pied babblers. *Journal of Animal Ecology*, 81: 876–883.

Nelson-Flower M. J., Hockey P. A. R., O'Ryan C. et al. (2013) Costly reproductive competition between females in a monogamous cooperatively breeding bird. *Proceedings of the Royal Society London B Biological Science*, 280: 20180728.

Nelson-Flower M. J., and Ridley A. R. (2015) Male–male competition is not costly to dominant males in a cooperatively breeding bird. *Behavioral Ecology and Sociobiology*, 69: 1997–2004.

Nelson-Flower M. J., and Ridley A. R. (2016) Nepotism and subordinate tenure in a cooperative breeder. *Biology Letters*, 12: 20160365.

Nelson-Flower M. J., Flower T. P., and Ridley A. R. (2018b) Sex differences in the drivers of reproductive skew in a cooperative breeder. *Molecular Ecology* 27: 2435–2446.

Nelson-Flower M. J., Wiley E. M., Flower T. P., and Ridley A. R. (2018a) Individual dispersal delays in a cooperative breeder: Ecological constraints, the benefits of philopatry and the social queue for dominance. *Journal of Animal Ecology*, 87: 1227–1238.

Nunn C. L. (2000) Collective benefits, free-riders, and male extra-group conflict. In Kappeler P. M., ed., *Primate Males: Causes and Consequences of Variation in Group Composition.* Cambridge: Cambridge University Press, pp. 192–204.

Olson M. (1965) *The Logic of Collective Action.* Cambridge, MA: Harvard University Press.

Packer C., and Pusey A. (1983) Adaptations of female lions to infanticide by incoming males. *American Naturalist,* 121: 716–728.

Peña J., and Nöldeke G. (2018) Group size effects in social evolution. *Journal of Theoretical Biology*, 457: 211–220.

Radford A. N. (2008) Duration and outcome of intergroup conflict influences intragroup affiliative behaviour. *Proceedings of the Royal Society London B Biological Science*, 275: 2787–2791.

Raihani N. J., Nelson-Flower M. J., Golabek K. A., and Ridley A. R. (2010) Routes to breeding in cooperatively breeding pied babblers *Turdoides bicolor. Journal of Avian Biology*, 41: 681–686.

Raihani N. J. and Ridley A. R. (2007) Adult vocalizations during provisioning: Offspring response and postfledging benefits in wild pied babblers. *Animal Behaviour*, 74(5): 1303–1309, DOI: https://doi.org/10.1016/j.anbehav.2007.02.025.

Reeve H. K., Emlen S. T., and Keller L. (1998) Reproductive sharing in animal societies: Reproductive incentives or incomplete control by dominant breeders. *Behavioral Ecology,* 9: 267–278.

Ridley A. R. (2012) Invading together: The benefits of coalition dispersal in a cooperative bird. *Behavioral Ecology and Sociobiology*, 66: 77–83.

Ridley A. R. (2016) Southern pied babblers: The dynamics of conflict and cooperation in a group-living society. In Koenig W. D., and Dickinson J. L., eds., *Cooperative Breeding in Vertebrates: Studies in Ecology, Evolution and Behaviour.* Cambridge: Cambridge University Press, pp. 115–132.

Ridley A. R., Nelson-Flower M. J., and Thompson A. M. (2013) Is sentinel behavior safe? An experimental investigation. *Animal Behaviour,* 85: 137–142.

Ridley A. R., and Raihani N. J. (2007a) Facultative response to a kleptoparasite by the cooperatively breeding pied babbler. *Behavioral Ecology*, 18: 324–330.

Ridley A. R., and Raihani N. J. (2007b) Variable postfledging care in a cooperative bird: Causes and consequences. *Behavioral Ecology*, 18: 994–1000.

Ridley A. R., Raihani N. J., and Nelson-Flower M. J. (2008) The cost of being alone: The fate of floaters in a population of cooperatively breeding pied babblers *Turdoides bicolor*. *Journal of Avian Biology*, 39: 389–392.

Ridley A. R., and van den Huevel I. M. (2012) Is there a difference in reproductive performance between cooperative and non-cooperative species? A southern African comparison. *Behaviour*, 149: 821–848.

Ridley A. R., and Thompson A. M. (2012) Heterospecific egg destruction by wattled starlings and the impact on pied babbler reproductive success. *Ostrich*, 82: 201–205.

Riehl C. (2013) Evolutionary routes to non-kin cooperative breeding in birds. *Proceedings of the Royal Society London B Biological Science*, 280: 20131445.

Rusch H. (2014) The evolutionary interplay of intergroup conflict and altruism in humans: A review of parochial altruism theory and prospects for its extension. *Proceedings of the Royal Society London B Biological Science,* 281: 20141539.

van Schaik C. P., and Janson C. H. (2000) *Infanticide by Males and Its Implications.* Cambridge: Cambridge University Press.

Shen S. F., Akçay E., and Rubenstein D. R. (2014) Group size and social conflict in complex societies. *American Naturalist*, 183: 301–310.

Shen S. F., Emlen S. T., Koenig W. D., and Rubenstein D. R. (2017) The ecology of cooperative breeding behaviour. *Ecology Letters*, 20: 708–720.

Skutch A. F. (1935) Helpers at the nest. *Auk*, 52: 257–273.

Strong M. J., Sherman B. L., and Riehl C. (2017) Home field advantage, not group size, predicts outcome of intergroup conflicts in a social bird. *Animal Behaviour*, 143: 205–213.

Thompson A. M., Raihani N. J., Hockey P. A. R., Britton A., Finch F. M., and Ridley A. R. (2013) The influence of fledgling location on adult provisioning: A test of the blackmail hypothesis. *Proceedings of the Royal Society London B Biological Science*, 280: 20130558.

Thompson F. J., Marshall H. H., Vitikainen E. I. K., and Cant M. A. (2017) Causes and consequences of intergroup conflict in cooperative banded mongooses. *Animal Behaviour*, 126: 31–40.

Vehrencamp S. L. (1983) A model for the evolution of despotic versus egalitarian societies. *Animal Behaviour*, 31: 667–682.

Wiley E. M., and Ridley A. R. (2018) The benefits of pair bond tenure in the cooperatively breeding pied babbler (*Turdoides bicolor*). *Ecology and Evolution*, 8: 7178–7185.

Willems E. P., Hellriegel B., and van Schaik C. P. (2013) The collective action problem in primate territory economics. *Proceedings of the Royal Society London B Biological Science*, 280: 20130081.

Wittig R. M., Crockford C., Lehmann J., Whitten P. L., Seyfarth R. M., and Cheney D. L. (2008) Focused grooming networks and stress alleviation in wild female baboons. *Hormones and Behavior*, 54: 170–177.

9 Cooperation and Conflict in Mutualisms with a Special Emphasis on Marine Cleaning Interactions

Redouan Bshary

Introduction

Take any ecology textbook and look up the chapter on interactions between species, and you will find that ecologists distinguish among three outcomes: mutualism, commensalism and parasitism/predation. Mutualisms are mutually beneficial (+/+), commensalisms are beneficial for one partner and neutral for the other (+/0), and parasitism/predation is beneficial for one and detrimental for the other (+/−). Mutualisms are at the core of the world as we know it; the evolution of the eukaryotic cell warranted the mutualistic integration of cell organelles (mitochondria and chloroplasts) into prokaryotic cells, and the radiation of flowering plants as a nutritional basis for the animal food chain is dependent on soil microorganisms for the fixation of nitrogen and phosphate as well as on pollinators (Bronstein, 2015). Therefore, studying mutualism is an integral part of ecological research and one that connects directly to understanding the evolution of cooperation. (See Chapter 4 for a discussion of mutualisms at the cell and genomic levels.)

Many partner species in mutualisms, such as plants and microbes, do not really behave. Nevertheless, mutual physiological adjustments can be described as conditional decisions in a game theoretic evolutionary framework. For example, leguminose plants grow new roots selectively where nitrogen is fixed for them by rhizobia bacteria (Kiers et al., 2003). As a consequence, rhizobia strains that do not fix nitrogen for plants have reduced access to plant resources such as carbohydrates and nodules (a sort of housing provided by roots). The example illustrates that research on mutualisms addresses the very same questions that dominates the literature on cooperation and conflict, including human cooperation and conflict: how can partners ensure that their investments yield return benefits, i.e., how did mutualisms evolve and how are they maintained? Indeed, research on mutualism borrows concepts from game theory, including human market analogies. Given that there is cooperation and conflict between partners in mutualisms, a key question is which side led to the origin of these interspecies interactions: did species A evolve mechanisms to seek the presence of species B to parasitise it or due to mutual benefits? The major part of this manuscript uses various examples to deal with these conceptual issues. A smaller part focuses on a particular system, the marine cleaning mutualism involving the cleaner fish *Labroides dimidiatus* and its client reef fishes. This mutualism illustrates how conflicts of interest may yield a great diversity of strategies employed by partners to tip payoff distributions in their own favor. Also, as this mutualism involves vertebrates, the cognitive and physiological mechanisms underlying interspecific social decision-making may be more readily compared to human and other primate social decision-making.

What Is Mutualism, What Is Cooperation?

As terminology differs both between and within disciplines, a proper understanding of any paper on social behavior warrants clear definitions. Key terms are summarized in Box 9.1.

Following Lehmann and Keller (2006), "helping" can be defined as the general phenomenon that warrants an explanation: Why should a focal individual perform an act that increases the direct fitness of a recipient? In evolutionary biology, the first

---

Box 9.1 Terminology Used in This Chapter

Helping: A behavior that increases the direct fitness of the recipient. Importantly, the behavior has been selected for at least in part because of this benefit.

Altruism: A helping behavior that increases the direct fitness of the recipient but reduces the direct fitness of the actor.

Cooperation: An interaction or a series of interactions between conspecific individuals that yields, on average, direct fitness benefits to all participants.

Mutualism: An interaction or a series of interactions between individuals belonging to different species that yields, on average, direct fitness benefits to all participants.

By-product benefit: An individual performs an immediately self-serving action that benefits another individual as a by-product.

Cheating/defecting: A behavior that increases the immediate payoff of the actor while reducing the immediate payoff of the recipient.

Partner control mechanism: A behavioral response to the partner cheating that reduces the partner's immediate or future payoff. For a mechanism to be efficient at promoting cooperation, the induced losses must annihilate the benefits of cheating.

Conflict of interest: A conflict of interest arises if partners differ with respect to their preferred combination of actions. In an interaction in which two players have the choice between two actions where the resulting four potential payoff combinations fit the standard prisoner's dilemma payoff matrix, each player benefits most from defecting against a cooperative partner.

Conflict: A conflict arises if an individual cheats/defects.

convincing answer was "biological altruism" (Hamilton, 1964a, b). Biological altruism involves a lifetime direct fitness loss of the helper. Such a loss might be more than compensated for if the recipient is a relative and therefore shares genes with the donor due to shared genealogy (Hamilton, 1964a, b), a phenomenon called "kin selection" (discussed in Chapter 4; see also Chapters 3, 8, and 10). As natural selection is about the propagation of genes in a population, helping relatives may increase the actor's inclusive fitness via indirect fitness benefits. Alternatively, the actor may obtain direct fitness benefits from helping. In this scenario, the helping act is either immediately self-serving, or it is not a lifetime cost but an investment that yields return benefits to the actor in the future. If interactions or chains of interactions yield, on average, direct fitness benefits to all involved individuals, the term "cooperation" can be used if partners belong to the same species (Lehmann and Keller, 2006; Bshary and Bergmüller, 2008; see also Chapters 8 and 10). In analogy, "mutualism" is defined by ecologists as an interaction between individuals belonging to different species, which yields, on average, direct fitness benefits to all partners involved (also discussed in Chapter 4). Defined in these ways, cooperation and mutualism describe the same fitness outcome (+/+) but distinguish between intraspecific and interspecific interactions. Ignoring the well-established definition for mutualism by ecologists, some behavioral biologists have recently caused quite some confusion with respect to terminology by calling conflict-free cases of intraspecific cooperation "mutualism." This is particularly unfortunate as most mutualisms are not conflict-free. Conflict-free cases of cooperation and mutualism are based on self-serving actions that produce by-product benefits for partners. For example, it has been argued that for a satiated meerkat (*Suricata suricatta*), the best self-serving behavioral option is to watch out for predators so that other group members can forage under low predation risk (Clutton-Brock et al., 1999). Being a watchman rather than resting benefits the satiated individual because its fitness is partly dependent on the size of its group ("group augmentation concept"; Kokko et al., 2001). The evolutionary origin and stability of helping based on by-product benefits is hence easy to understand.

A very essential question is what kind of data should be taken as evidence that there is a mutualistic relationship between individuals of two species. Ecologists and behavioral ecologists / evolutionary biologists study interspecific interactions on two different scales and hence use different criteria. Ecologists typically study population dynamics and hence use the term mutualism if the presence of species A causally increases the population density of species B, and vice versa. In contrast, evolutionarily minded biologists focus on the fitness consequences of interactions on the involved individuals. As long as interaction with an individual of species A gives an individual of species B a fitness edge relative to conspecifics, and vice versa, the relationship between the species is termed a mutualism. The second criterion is more suitable for the purpose of this chapter for two reasons. First, it is much easier to test, for example, whether a leguminous plant grows faster or has a higher seed set in the presence of nitrogen-fixating bacteria than to remove all those bacteria from natural habitats and to quantify the consequences on the leguminous plant population density. Second, the individual consequence definition takes into account that population size might be limited by factors that are unaffected by the mutualistic interaction. For example, the population size of a small damselfish on a coral reef might be limited by the abundance of crevices for shelter of a particular size, while the reduction of ectoparasite load through mutualistic interactions with cleaning organisms gives treated individuals a competitive advantage over untreated ones.

Advantages and Potential Pitfalls for the Use of Mutualisms to Understand Cooperation

In many social species like primates, the study of cooperation is rather challenging because social interactions occur between members that belong to the same meta-stable group, and hence may often involve relatives. Researchers should hence know genetic relationships in detail and consider the possibility that helping may be based on biological altruism rather than on direct benefits. Conversely, relatedness does not automatically imply that helping is based on biological altruism as the costs and benefits of helping may vary greatly, depending on the specifics of a species' life cycle (West et al., 2002; Lehmann and Rousset, 2010). In addition to genetic relatedness, researchers should know how kin is actually recognized to properly interpret acts of helping. As it stands, observing helping between non-relatives does not preclude that biological altruism is the underlying selective force: in any species in which individuals "recognize" kin based on having grown up together, rare mistakes may occur. Even in humans, incest avoidance is linked to growing up together rather than to genetic relatedness (Lieberman et al., 2007). The study of mutualisms avoids the problem that interaction partners are related. Moreover, the very same concepts proposed to explain intraspecific cooperation have also been used to explain mutualisms (Sachs et al., 2004; Bshary and Bronstein, 2011).

Having said that the same theories can be applied to both cooperation and mutualism, there are various features that are generally or often different. A general difference between mutualisms and intraspecific cooperation is that in mutualisms, interactants do not compete with each other over reproductive success but only with their respective conspecifics. Football lovers would compare interspecific interactions with a "three-point match," as when a top team plays against a low-ranking team. In contrast, intraspecific interactions would be described as "six-point match," as the three points one might lose would be gained by a direct competitor of similar strength. As a consequence, all else being equal, the conflict between partners is lower in mutualisms (Figure 9.1).

Another important difference is that partners in mutualisms play evolved roles, while partners in cooperation often play ontogenetic roles. Three basic services are commonly traded in mutualisms, namely protection, transport, and nutrition (Bronstein, 2001). In many cases, two different services are exchanged. Plants offer food to pollinators in exchange for transport of pollen. Various partner species offer food to ants in exchange for protection against predators. In some cases, the same services can be traded. Partners in mixed-species associations in ungulates

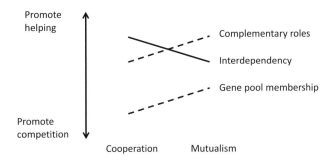

Figure 9.1 Three key factors that tend to promote or hinder helping for direct benefits to different degrees, depending on whether or not the interaction is between conspecifics (cooperation) or between members of different species (mutualism). In mutualisms, partners are more likely to have complementary roles (which promotes stable mutual helping), less likely to be strongly interdependent (which reduces self-serving effects of helping), and belong to different gene pools (which reduces competition). Note that there can be major deviations from general patterns in specific examples.

(Fitzgibbon, 1990) and in primates (Bshary and Noë, 1997) trade mutual protection, while plants and mycorrhizal fungi or rhizobia bacteria trade nutriments. But in these latter cases also, different evolved abilities are traded, such as abilities to detect different predator types (raptors versus ground predators) or different types of nutriments, i.e., carbohydrates versus nitrogen or phosphate. Evolved complementary roles are rare in cooperation. The one obvious exception is sexual reproduction between males and females. Trading sperm and eggs in a mating market is another classic application of biological market theory (Noë et al., 1991). Males and females may also play complementary roles during parental care, such as finding food versus incubating the eggs in various bird species. The latter case is unlikely to be based on evolved specializations, unless it could be shown that these differences are linked to the sex chromosomes. Otherwise, the same genes lead to different behavioral patterns depending on whether they are expressed in a male or in a female body. More generally, individuals of the same species tend to have the same general abilities. This similarity reduces the potential for mutual helping. For example, two neighboring plants are equally capable of using sunlight to produce carbohydrates and to collect water with their respective roots. Therefore, they have no services to exchange to create a surplus. A surplus can be obtained with microorganisms that provide plants with nitrogen or phosphate, which are otherwise inaccessible and which require only a cheap payment in the form of carbohydrates. In other words, mutualisms seem to be, on average, easier to evolve than cooperation because partners in mutualisms are more likely able to complement each other with respect to providing otherwise inaccessible goods/services (Figure 9.1).

Finally, the interaction patterns may often differ between mutualisms and cooperation, though there is overlap. It is reasonably easy to choose to study a mutualism involving interaction partners whose future success is rather independent of each other and independent of the success of conspecifics. In the marine cleaning mutualism described later in more detail, cleaner fish and most client species have free access to several potential partners, and if one partner dies it is typically replaced by a new member of equal value. Also, how a cleaner and a client interact will typically not have any side effects on their respective relatives as most reef fish species are open-water spawners with pelagic egg and larval stages, leading to a lack of kin structure in the populations. Note, however, that population structure and hence relatedness within each partner species may nevertheless be of potential importance in other mutualisms, for example when an ant colony of related individuals protects clones of aphids in exchange for the sugary excretions.

Some mutualisms are symbioses. A symbiosis is the intricate living-together of organisms in time and space (so it includes cases of commensalism and parasitism). In such cases, the interests between partners become more aligned and conflicts of interest reduced, as the fall of one partner will negatively affect the fitness of the other partner. In other words, the fitness of interaction partners becomes interdependent (Box 9.2). To give a few examples, ants living on acacia trees, zooxanthellae living with polyps to form corals, as well as algae living with fungi to form lichens are mutualisms with high interdependence between partners (see Bronstein, 2015). In the most extreme cases, hosts transfer symbionts together with own offspring to the next generation; i.e., there is vertical transmission of partners (Herre et al., 1999). Therefore, symbionts are very much like cell organelles (mitochondria and chloroplasts) in the sense that host and symbiont fitness are strongly interdependent: one cannot live without the other, which largely removes any potential for conflict. Interdependency arises in any meta-stable social group. It has been argued to be of particular importance to understand the evolution of the high levels of cooperation observed between unrelated human individuals (Tomasello et al., 2012). The fact that partners in mutualisms are always unrelated to each other allows us to study causes of interdependency and consequences of cooperation without having to consider potentially confounding effects of altruism based on kin selection. However, interdependencies occur more frequently in cases of intraspecific helping, as helping typically takes place within groups of relatively stable composition. In such groups, interaction partners may not only be genetically related to each other ("genetic interdependence") but their future successes might also be interdependent, which can easily become a game changer (Box 9.2). Both genetic and social interdependencies hence favor the evolution and stability of cooperation relative to mutualisms (Figure 9.1).

Similarities between Human Cooperation and Mutualisms

Given the three main differences between mutualisms and cooperation listed above, it emerges that if we want to compare the diversity of human forms of cooperation with other species, some forms are better compared to other examples of cooperation while some forms are better compared to mutualisms. Most notably, trading in human markets and in mutualisms feature the exchange of different currencies between

Box 9.2 Interdependence As a Factor That Stabilizes Cooperation and Mutualism

Even very simple game theoretic models make a considerable list of assumptions regarding the social interaction that is analyzed. If one studies a game with a prisoner's dilemma payoff matrix (see Box 3.1 in Chapter 3 for a review of games), it is typically assumed that the fitness of the interaction partners is a function of one's own cumulative gains, independently of the partner's gains. They play a certain number of rounds and then reproduce independently as a function of accumulated gains. Under these assumptions, it pays to exploit an unconditionally cooperative partner by always defecting. Now take the example of two female monkeys. They can only be expected to cooperate if they are members of the same group. Group living itself is an adaptation to reduce predation pressure due to various mechanisms, some of which are inherent features of group life and hence not subject to the question of cooperating versus cheating: the reduction of attack frequencies corrected for the number of individuals, the dilution of risk of being killed should a predator be successful, many eyes and ears for timely detection of ambush predators, and confusion effects during flights. The magnitude of these benefits is a function of group size: the larger the better. Groups also compete with neighboring groups over space and the resources therein. Sometimes, competition over resources can also be interspecific, like between lions and hyenas over a carcass. Again, the magnitude of benefits in the form of winning contests is a function of group size. As a consequence, a selfish individual that exploits cooperative group members may lose out, i.e., may lower its own direct fitness, if its actions lead to a smaller group size. Group members are stakeholders (Eshel and Shaked, 2001; Roberts, 2005), and contributions to group benefits may be self-serving due to the beneficial effects of group augmentation (Kokko et al., 2001). Apparently costly behaviors such as sentinel duties in meerkats, where one individual watches out for predators while the rest of the group forages, are self-serving when viewed in this framework (Clutton-Brock et al., 1999). In species such as baboons (*Papio cynocephalus*) in which a stable social network is important for survival and reproduction (Silk et al., 2003), bonded individual pairs are even more "trapped" in mutual interdependence. Exploiting a friend and causing its downfall means that there is less support for self in the future. Therefore, investing into a friend readily becomes self-serving ("pseudoreciprocity"; Connor, 1986), at least up to a point.

A prisoner's dilemma payoff matrix that simply summarizes immediate material benefits, such as being groomed, and immediate costs, such as the time invested in grooming, fails to account for the future fitness consequences due to interdependency. Bshary et al. (2016) explicitly presented various scenarios of how material payoffs translate into fitness as a function of the degree of interdependency between interaction partners. In the most extreme case, an apparent prisoner's dilemma turns into a by-product benefits game in which cooperating is self-serving and defecting is, hence, under negative selection. More likely, a prisoner's dilemma material payoff matrix may turn into a so-called snowdrift game fitness matrix, in which individuals cooperate with a certain probability even in one-shot interactions. Iteration can still lead to high levels of reciprocal cooperation (Bshary et al., 2016).

specialized traders who are otherwise rather independent of each other – in the most extreme case between strangers with no perspective of shared future interactions. As a consequence, various concepts of human markets can be applied nicely to mutualisms. Foremost, biological market theory proposes a key role of partner choice in determining exchange rates based on supply and demand ratios (Noë et al., 1991; Noë and Hammerstein, 1995; Hammerstein and Noë, 2016). A diversity of studies on mutualisms has successfully tested how changes in supply or demand affect services. To give one classic example in more detail: the larvae of various lycaenid butterfly species are tended by ants. The larvae have special glands to produce a sugar-rich solution for the ants, which in turn protects the larvae against predators and parasitoid wasps that would lay eggs inside the larvae (Pierce et al., 2002). Axen et al. (1996) experimentally manipulated the number of ants (*Lasius niger and L. flavus*) tending various subject larvae (*Polyommatus icarus*). They could show that low ant presence led to increased nectar production as well as increased production of an ant-attracting pheromone. Mutualism markets can be highly complex for at least two reasons. First, many mutualisms are characterized by two trader classes where each class consists of several species. A single ant species may simultaneously protect a variety of partner species, and a single partner species may be attended by various ant species (Palmer et al., 2003). Best known to the public should be the fact that many flowering plants offer nectar to a variety of unspecialized pollinators that, in turn, visit a diversity of plant species (Olesen et al., 2007). A second factor that introduces complexity is that a group of individuals may interact with a partner and must coordinate their trading strategies, as described for nutrient exchanges between arbuscular mycorrhizal fungi (AMF) and their plant hosts via communal structures (Noë and Kiers, 2018). AMF appears to be a single large organism but it consists of polymorphic nuclei (i.e., cells are genetically different from each other). In order to explain the exchange characteristics of this mutualism, the authors draw analogies to exchanges between divisions of firms and to human co-ops, something that has never been used to explain cooperation in non-human species.

Another feature of human markets that fits mutualisms is the specialization of traders and the resulting obligate dependence on finding partners to trade. Some plant–pollinator relationships form large multispecies networks (Olesen et al., 2007), while others like the fig–fig wasp species complex often involve highly specific species pairs (Machado et al., 2005). More generally, there is the ecological concept of compensatory trait loss: own capacities are reduced and, instead, the feature is outsourced to a partner species (Ellers et al., 2012). Typical cases involve protection against predators: grass roots would be highly susceptible to herbivores without fungal endophytes providing chemical defenses (Müller and Krauss, 2005), and some acacia trees partly rely on ants defending them physically against herbivores (Janzen, 1966). Similarly, defenses against pathogens or competitors can also be reduced due to protection by mutualistic partners, as described for fungus protected by ants (Currie and Stuart, 2001) and garden algae protected by damselfish (Hata and Kato, 2006).

As long as the interdependency resulting from compensatory trait loss is mutual, the potential for conflict is reduced on the level of partner species. However, the same cannot be said for competition between individuals within a species over access to individuals of the partner species. The prisoner's dilemma game logic emphasizes competition/conflict between the interaction partners because each partner is tempted to cheat (see Box 3.1 in Chapter 3 for a description of various games). In contrast, market theory shifts the emphasis to competition/conflict between members of the same trader class over access to partners. If partner choice is easy for the members of the trader class that offers a good in high demand, the resulting competition through outbidding may virtually equal the benefits of trading in the chosen class. This is best explained in the so-called veto game, in which two or few individuals offering one good compete over access to a single individual offering a complementary good for a coalitionary action (Noë, 1990). Under these conditions, the single individual is in the veto position and can obtain the lion's share of the benefits out of the exchange, as potential partners have to outbid each other in order to maintain the possibility of being chosen.

Exchanges between traders of the same species are less easily achieved because all traders are typically quite similar with respect to the goods or services they could potentially provide, with the major exception of sexual reproduction. Therefore, many forms of cooperation are based on the principle "more is better." Living in groups, the most general form of cooperation (often combined with kin selection), may yield benefits due to safety in numbers, the combined effects of dilution of risk, many eyes and ears for the early detection of predators, and confusion effects preventing predators to single out prey (Kenward, 1978; Landeau and Terborgh, 1986; De Vos and O'Riain, 2009). Similarly, coalition formation is useful as joining the same forces increases fighting abilities, either as predators to capture prey (Stander, 1992, Boesch, 1994), as prey to defend against predators (Curio, 1978), or as a team beating a stronger individual (Noë, 1990). Allogrooming, mostly studied in primates, is a peculiar case of cooperation as it is based on a universal shortcoming, i.e., individuals cannot reach all parts of their own body, which makes the help of another individual valuable. Therefore, grooming can and is traded against grooming (Schino and Aureli, 2008). Interestingly, once the commodity of grooming is established, it can also be traded against other commodities that are asymmetrically distributed between individuals, typically linked to the rank in a hierarchy. Dominants can offer coalitionary support or tolerance in exchange for grooming (Seyfarth, 1977; Cheney et al., 2010; Borgeaud and Bshary, 2015; Wubs et al., 2018). More generally, hierarchies may lead to various forms of specializations and trading of different services. In cooperatively breeding cichlid fish species such as *Neolamprologus pulcher*, the youngest helpers may remove sand from nests and care for the eggs of the territory owners, while older helpers contribute to territory and anti-predator defense (Taborsky, 1984). It is thus a widespread feature of cooperation that the same individual may play different roles and hence cooperate in different ways during its life. The position in a hierarchy may be predictable if it is linked to age/size but it may also be unpredictable if an individual's rank is inherited from its mother's rank, as found in so-called matrilineal societies (described for primates, hyenas [*Crocuta Crocuta*], and cetaceans). Age-related transitions may be prepared with genetic programs, while socially transmitted roles warrant individuals to assess their environment and to adjust flexibly. The latter provides a starting point to link cooperation with cognition, an option that is rather understudied in mutualisms. I will return to this topic at the end of this chapter.

Evolutionary Origins and Stability of Mutualisms

Given that conflicts of interest between mutualism partners are widespread, how did mutualisms evolve under the assumption that not-helping was the ancestral state? There are actually two opposing potential starting points for mutualisms. A competition-based hypothesis proposes that host–parasite or predator–prey relationships may have been at the beginning of various mutualisms. Take the evolution of eukaryote cells. Prokaryotic cells may have initially preyed on mitochondria, which then adapted to avoid being digested, leading eventually to division of labor and full interdependence with the host (Gray and Doolittle, 1982). In this scenario, the cooperative interactions behind mutualism are (evolutionary) reactions to conflict between the interacting members.

The alternative hypothesis for the evolution of mutualisms is that one partner produced benefits for another species as a by-product of self-serving actions. For example, aphids produce sugary excretions not primarily to attract tending ants but because feeding on plant material leads to a surplus of sugar (Leimar and Connor, 2003). Because of this waste product, ants benefit from defending aphids against predators and parasitoids, and aphids may, in response, improve the nutritional value of the excretions to become a better partner (Leimar and Connor, 2003). The repeated independent evolution of marine cleaning organisms is also most likely due to by-product benefits (Barbu et al., 2011): Eating small crustaceans off substrates may include ectoparasites on living, where self-serving foraging creates benefits to clients as a by-product. Distinguishing between a competitive and a by-product benefit starting point is not necessarily straightforward. A recent reconstruction of the evolution of ant-flowering plant mutualisms concludes that ants colonized plants as a foraging strategy, long before plants provided extrafloral nectaries as food sources and shelter for ant nests (domatia) (Nelsen et al., 2018). The initial diet of the ants may have been phytophageous insects, yielding by-product benefits to the plant, but also plant products such as nectar, which is a form of predation/parasitism.

No matter what the evolutionary origins, the vast majority of mutualisms involve investments from at least one side. Where there is investment, the question arises as to why mutants that produce less or no investment do not thrive, leading eventually to the extinction of the mutualism. The danger of an erosion of the mutualism between partners can be considered to be typically higher than for intraspecific cooperation, i.e., for all cases in which partners are not interdependent. Furthermore, various authors pointed out that mutualisms may be at risk because of third-party species that exploit the service offered by one partner without giving anything in return. For example, a plant that produces nectar may be used as a food source not only by pollinators but also by many other species. Given the diversity of mutualisms, it is not easy to provide all-inclusive answers to the question why mutualisms appear to be rather evolutionarily stable against erosion, despite investments and the resulting conflicts of interest. A few mutualisms show interdependence between partners (Box 9.1), most obviously in cases in which symbionts are vertically transferred to the next host generation, as are mitochondria. Furthermore, scientists have struggled to find interaction patterns where the payoff matrix corresponds to an iterated prisoner's dilemma, which would create maximal conflicts of interest between partners (for a possible exception, see Kiers et al., 2011). Instead, many mutualisms appear to follow the following pattern: Species A invests in a commodity for species B, which makes it self-serving for species B to provide a service to species A in return. In other words, as long as species A delivers, cheating by species B is under negative selection. Here are a few examples: as long as various partner species (acacia plants, aphids, lycaenid butterfly larvae) of ants deliver housing and/or food, it is in the interest of ants to defend their resources. As long as cleaner wrasse remove ectoparasites, it is in the self-interest of predatory reef fish clients to let them live. As long as plants deliver nectar, it is self-serving for pollinators to visit the flowers. As long as rhizobia bacteria fix nitrogen in the soil, it is in the self-interest of plants to grow roots with nodules (little capsules that provide shelter for bacteria).

A peculiar form of asymmetric strategic option in mutualism is farming. The farming species invests in the partner species, which uses the investment in a selfish way: to grow and to reproduce, without helping the investor. Instead, the investor takes its share without asking the farmed partner. Humans farm plants and animals for food as well as for other products and services (Diamond, 2002). The farmed plants and animals have no say in what happens to them, but as species, our farmed animals have been highly successful, both with respect to numbers and worldwide distributions. While humans are the only vertebrates known to farm, there are various examples of farming in other species. Prominent examples include ants, termites, and ambrosia beetles farming fungus as crops (Aanen et al., 2002; Currie et al., 2003; Hulcr and Cognato, 2010), and even amoeba and fungi farming bacteria (Brock et al., 2011; Pion et al., 2013).

Even in mutualisms in which services like defense by one partner are self-serving *if* the other partner cooperates through investments, it is still important to investigate how continuous investment is stable against mutants that invest less or nothing. As argued by Jones et al. (2015), such cheaters would gain, as a consequence, higher payoffs than the population average if unchecked. The issue of controlling the partner's action to cause positive selection on investment seems to be particularly pertinent in mutualisms as many involved species lack central nervous systems that would allow them to play complicated strategies based on memory and flexible behavior. The advantage of the asymmetries in payoff structure between partners is that the controlling partner's decision can be best described with optimality theory. As it turns out, very simple rules may hence select against cheating. For example, a pollinator's optimal probing time depends on the amount of nectar provided by a flower. A plant mutant that produces less nectar will hence face shorter probing duration, which in turn reduces pollen deposition, leading to an overall lower seed set, and hence to negative selection on reduced nectar production (Brandenburg et al., 2012). The previously mentioned case of plants growing new roots primarily on existing roots with a high nitrogen uptake, i.e., in the vicinity of cooperative bacteria that fix a lot of nitrogen is another good example (Kiers et al., 2003).

The observation that partner control mechanisms used in mutualisms are typically straightforward does not mean that selection on partners to win the conflict may not sometimes yield adaptive mutations that cause erosion toward parasitism. A striking example involves a few lycaenid butterfly species that have secondarily become parasites of ants. A phylogenetic reconstruction shows that the production of a sugar-rich solution by the butterfly larvae caused the origin of the interspecific interactions with ants (Pierce et al., 2002). In the derived parasitic butterfly species, the butterfly larvae produce an odor that mimics the odor of ant larvae. As a consequence, they are carried into the ants' nest, where they feed on ant eggs and larvae (Pierce et al., 2002). Apart from such clear-cut examples of evolved parasitism, there are many gray zone cases where both the cooperative side and the conflict side are displayed in obvious manners. A peculiar example involves drongos (*Dicrurus adsimilis*) and various partner species (Flower, 2011). Drongos are birds that are extremely good at learning vocalizations, including the alarm calls of various partner species. Drongos benefit their partner by giving proper alarm calls when predators approach. However, drongos may also give false alarm calls when they see an individual of the partner species handling an attractive food item. That individual may drop the food in response, which is then stolen by the drongo (Flower, 2011). Another case of complex interspecific relationships displaying mutual interest and conflict involves *Wolbachia* bacteria and their various invertebrate hosts (Werren, 1997). *Wolbachia* are the closest free-living relatives of mitochondria, and they share various features with those. The key shared feature of interest is the vertical transmission to the next host generation via the female germ line only. For a *Wolbachia*, being inside a male is therefore a dead-end street. One way to be, nevertheless, still useful is to kill any egg that does not contain the same line of *Wolbachia*. The system works because of a poison–antidote mechanism: the same *Wolbachia* line that fills male sperm with poison provides the antidote to female eggs (Bossan et al., 2011). As a consequence, females containing *Wolbachia* are better off than females without *Wolbachia* because their zygotes survive no matter whether the father was *Wolbachia* positive or negative. It is like the mafia: provide the threat and the solution; in the *Wolbachia* case at the expense of male carrier reproductive success (as all fertilizations of *Wolbachia*-negative eggs are killed off). (See Chapter 4 for another example of a poison–antidote system.)

A Specific Case: Marine Cleaning Mutualism

Generally speaking, cleaning mutualisms involve one partner that "cleans" by removing ectoparasites and dead or infected tissue from a "client" partner that typically cooperates by staying put to allow for efficient foraging by the cleaner. While the mutual benefits hence appear to be straightforward, i.e., food for the cleaner and parasite removal for the client, there are conflicts of interests between clients and some cleaner species that prefer to eat healthy client tissue (mucus, scales, or blood), where such consumption constitutes cheating. For example, oxpeckers (*Buphagus erythrorhynchus*), small birds found on African savannahs, do not actively wound ungulate clients but try to keep already existing wounds open to maintain easy access to blood (Weeks, 2000). The oxpeckers' preferences between blood and ectoparasites have not been assessed, but it appears that their preferences resemble those of cleaner fish of the genus *Labroides*, which have been shown to prefer the protective mucus of client reef fish over client ectoparasites (Grutter and Bshary, 2003). In contrast, cleaning gobies (*Elacatinus sp.*) from the Caribbean prefer to eat ectoparasites over client mucus, which means that their mutualism with clients is stable due to simple by-product benefits (Soares et al., 2010). Crucial for observational studies, a cleaner cheating by taking a bite of mucus correlates with the client performing a short jolt of the body in response, while clients typically remain still in response to cleaners removing an ectoparasite (Bshary and Grutter, 2002a). Thus, the frequency of client jolts, or, even more precisely, the frequency of client jolts relative to the total number of times a cleaner's mouth touches the client, provides estimates of a cleaner's service quality, i.e., its level of cooperating versus defecting.

Overall, more than 100 species have been described to gain food from cleaning at least at some stage of their life. While there are exotic observations such as gulls cleaning a sunfish, the vast majority of described cleaner species are fishes and shrimps that occur on coral reefs (Côté, 2000; Vaughan et al., 2017). Of the fishes, only species belonging to two clades, namely wrasses of the genus *Labroides* (Figure 9.2) and gobies of the genus *Elacatinus*, are fully dependent on cleaning for their diet during their entire life. All other cleaner fish only gain small parts of their food from cleaning, typically only as juveniles (Côté, 2000; Vaughan et al., 2017).

While none of these facultative species has explicitly been subject to preference tests, my prediction would be that all of them prefer ectoparasites over mucus, and that cleaning largely occurs as a by-product of them including living substrates in their foraging habits. In fact, there is one feature of *Labroides* cleaners that sets them apart from all other cleaners: They evolved from a clade that is specialized on feeding on coral mucus (Huertas and Bellwood, 2018). This peculiarity might explain why *Labroides* cleaners prefer client mucus, and why other cleaner species might be indifferent to mucus consumption. In line with this view, facultative cleaners from the Red Sea caused fewer jolts per time unit than *L. dimidiatus* individuals did (Barbu et al., 2011). *Labroides* cleaners have evolved a specific jaw anatomy adapted to cleaning (Wainwright et al., 2004), allowing them to lift off ectoparasites attached to the client's surface with the incisors on their lower jaw. In other words, they can remove ectoparasites efficiently without biting into the mucus. In contrast, facultative cleaners may bite accidentally into the mucus when taking an ectoparasite with both lower and upper jaws. It is hence conceivable that the *Labroides* cleaners sometimes decide to eat mucus and cause client jolts "on purpose," while facultative cleaners only lack the precision to reliably remove ectoparasites without an additional piece of mucus.

Figure 9.2 Cleaner fish *Labroides dimidiatus* cleans the mouth of a potato grouper (*Epinephelus tukula*), Great Barrier Reef, Australia. (Photo by Simon Gingins.) (A black and white version of this figure will appear in some formats. For the color version, please refer to the plate section.)

The result that facultative cleaners cheat less frequently may appear counterintuitive at first sight. If facultative cleaners cheat less, why do clients not preferentially seek their service, making them more obligate cleaners? One likely factor is that *Labroides* cleaners are more efficient at spotting and removing ectoparasites, as their jaw anatomy allows them to swim in parallel to the client's surface, while facultative cleaners inspect in position that is perpendicular to the client. This hypothesis has not been tested yet. A second factor known to positively affect client decisions is that only cleaners of the genus *Labroides* are known to provide tactile stimulation with the pectoral and especially pelvic fins to clients. Clients typically stop coordinated swimming movements in response. Access to such stimulation reduces client (surgeonfish – *Ctenochaetus striatus*) basal and acute stress levels, as measured with cortisol concentrations in their blood (Soares et al., 2011). This result is interesting for two reasons. First, the assessment of cortisol was based on studies on human massage therapy, where success is inferred if patients show a reduction in cortisol levels (Field et al., 2005). As in the human case, clients do not obtain any material benefit (such as reduced predation risk, reduced parasite load, extra food, access to a partner, etc.) from receiving a tactile stimulation. Thus, the benefit of stress hormone reduction is apparently based on a psychological effect. Second, *Labroides* cleaners have evolved a behavior that allows them to pay and make up for their cheating. They can reconcile with clients they just cheated, and they can manipulate client decisions about inter-action duration by providing tactile stimulation (Bshary and Würth, 2001).

The usage of tactile stimulation is only one of various impressive aspects of strategic sophistication, as mostly revealed in studies on *L. dimidiatus*, that sets the *Labroides* mutualism apart from other well-studied mutualisms that involve invest-ments and hence potential for conflict. The other systems typically involve plants, microbes, and insects, i.e., species with no or small brains. Those systems are hence highly suitable to study the genetic evolution underlying mutualism, as different phenotypes are highly correlated with different genotypes. Selection lines that differ in their offers may be produced in relatively short time, and also genetic evolution under different environmental conditions may be achieved with relative ease. Studying these systems in detail allows addressing the question of what ecological circumstances lead to stable exchange of services with limited or even no cognition involved. Unfortunately, selection line methods cannot be used on *Labroides* cleaners. These fish are open-water spawners with pelagic egg and larval stages, and their life cycle cannot be completed in the laboratory. The positive aspect of studying the cleaning mutualism is that vertebrates interact with each other, leading to interaction patterns that become much more comparable to cases of cooperation studied in other vertebrates, including monkeys, apes, and humans. Indeed, various game theoretic concepts that have been used to explain high levels of cooperation humans – and hence implicitly assume that advanced cognitive processes are involved – have arguably been applied more successfully to cleaning mutualism than to cooperation in nonhuman species.

In order to appreciate the diversity of games and corresponding strategies played in the *Labroides* cleaning mutualism, it is important to note that client species differ in their strategic options for how to respond to cheating by cleaners. A first distinction has to be made between predatory and non-predatory clients. Predatory clients have the "threat of reciprocity" (Bshary and Bronstein, 2004): a cleaner that cheats is not a cleaner but a potential food item; i.e., predators may respond to being cheated with cheating and trying to eat the cleaner. This strategic option of predators is difficult to study as cleaners typically refrain from cheating them (Bshary, 2001). In contrast, non-predatory clients lack any option to cheat a cheating cleaner in return. They must thus use other partner control mechanisms, defined as behavioral responses that reduce the payoff of a cheating cleaner, to keep cleaner service quality high. For non-predatory clients, a second distinction becomes important. Resident client species occupy small territories or home ranges that include only one cleaning territory. They have thus access to the local cleaner only; the local cleaner has a veto position concerning the question of whether a resident is cleaned at all or not. In contrast, visitor client species occupy larger territories or home ranges that include several cleaning stations. They have thus the potential to choose between cleaners. Intuitively, one might expect that visitors receive a better service than residents do. However, this is only partly true. Residents change the predictions based on market theory by introducing punishment of cheating cleaners through aggressive chasing. If one corrects for body size (smaller clients tend to jolt more frequently than larger clients do), it appears that the threat of being punished and the threat of losing future interactions with a client to another cleaner yield about the same service quality (Bshary, 2001). There is only one aspect of cleaning service where visitors benefit from their choice options: The cleaner gives them priority if they seek service simultaneously with a resident (Soares et al., 2013). This is because the resident has to wait for inspection, while the visitor can swim off and visit another cleaner instead. If visitors do not exert their choice options, they lose this priority (Triki et al., 2018).

An additional layer of complexity in cleaner–client interactions is introduced by the fact that just considering two-player interactions is not sufficient. Instead, as cleaners have typically about 2,000 interactions per day (Grutter, 1995; Wismer et al., 2014), these interactions often take place in a communication network (McGregor, 1993), in which potential clients swimming toward a cleaner may be a bystander to the cleaner's current interaction. The bystanders extract information about the cleaner's current performance, i.e., they attribute an image score to the cleaner, evidenced by their response to what they saw: if the cleaner cooperated, bystanders are likely to invite them for inspection, while they tend to avoid cleaners they have seen cheating their current client (Bshary, 2002). As a response to image-scoring bystanders, cleaners behave more cooperatively when they are observed (Bshary and Grutter, 2006; Pinto et al., 2011; see Chapter 3 for a discussion of the role of observation and reputation in humans). Some cleaners may even attract image-scoring bystanders by being nice to small residents by providing tactile stimulation, only to cheat the new client (Bshary, 2002). The usage of tactile stimulation outside of its normal context qualifies as functional tactical deception (Hauser, 1997), without implying that cleaners have any causal understanding why the deception works.

Apart from reputation management, cleaners also face the challenge that the male cleaners may sometimes co-inspect clients with a female member of their harem. Typically, this situation occurs when a large visitor client invites for inspection, i.e., a client that swims off in response to being cheated. The situation between the cleaners is therefore equivalent to an iterated prisoner's dilemma, as only the cleaner that cheats first obtains the benefits of feeding on mucus, while both cleaners share the cost of the client leaving prematurely (Bshary et al., 2008). The solution is that males provide pretty much the same service quality as when they inspect alone, while females become more cooperative. This is because the larger males punish female partners for cheating clients through aggressive chasing, which makes the females behave more cooperatively during future interactions (Raihani et al., 2010). To fully understand this solution, one has to consider the life history of the cleaners. All individuals start their reproductive career as females and only change sex and become male when they are large enough to defend a harem of females against other males (Robertson, 1973). As a consequence, a male sees his female harem members both as partners for reproduction and as future rivals over access to females. If a female cheats a client, this is good for the male only if she invests the extra energy in the production of eggs. If the energy is used for growth, this increases the risk that the male loses her sooner. Males indeed adjust their punishment to various key factors: they are apparently more tolerant toward females cheating during spawning days

(Bshary and D'Souza, 2005), and they are more likely to punish larger females than smaller females (Raihani et al., 2012).

The Cognitive Toolbox Underlying the Cleaners' Sophisticated Decision Rules

Marine cleaning mutualism involving the cleaner wrasse *L. dimidiatus* has become a textbook example for sophisticated decision rules in fishes. It has contributed to a change within the scientific community regarding the perception of cognitive abilities of fishes (Brown, 2015; Bshary et al., 2014). Indeed, the list of phenomena of interest presented above is impressive: reconciliation, tactical deception, punishment, reputation management, and cooperative solution to an iterated prisoner's dilemma. The strategic sophistication is clearly linked to the conflict between cleaner and client due to the cleaners' food preferences. If cleaners prefer ectoparasites over mucus, as in the Caribbean cleaning gobies cleaning mutualism, clients do not punish and do not switch partners even though gobies also cause jolts (Soares et al., 2008).

Research on the cognitive toolbox of cleaner wrasse revealed various more or less surprising results. Potentially less surprising results include individual recognition of clients (Tebbich et al., 2002), remembrance of the "what" and "when" of interactions (Salwiczek and Bshary, 2011), and the ability of self-restraint, expressed as the ability to feed against preference to prolong interactions (Bshary and Grutter, 2005). More surprising might be their use of generalized rule learning in the context of conflict with resident clients: When a resident chases a cleaner, the latter may seek the interaction with a passing predatory client. This effectively ends the chasing, which means that predators are used as social tools against punishing clients (Bshary et al., 2002). Experiments using laminated pictures of predators and non-predators revealed that once a cleaner learned that a predator picture yields safety while a non-predatory fish picture does not, they can generalize this information to the next combination of pictures, even if the shape of the predator changes in extreme ways, like from a grouper (*Pessuliferus*) to a moray eel (*Gymnothorax javanicus*; Wismer et al., 2016). The most spectacular recent finding has been the evidence that cleaners pass the mirror task (Kohda et al., 2018). As the authors note, the resulting open question is whether cleaners have indeed self-consciousness or whether the task fails to provide definite conclusions about self-consciousness. Finally, of particular relevance for the following part of this article is an ability of cleaners for which the underlying mechanisms are still unclear: Cleaners outperform a variety of primates and other vertebrates in the "biological market task" or "ephemeral choice task." The task was designed to capture the challenge to give priority to a visitor (ephemeral) client over a resident (permanent) client (Bshary and Grutter, 2002b). In its "standard" form, two Plexiglas plates of equal shape and size, distinguishable through differences in color patterns, offering the same amount of food, are presented simultaneously to the subject. The subject must make a choice from which plate it eats. If it chooses the resident plate, the visitor plate is removed before the subject can access it. The subject hence obtains a single unit of food. If it chooses the visitor plate, the resident plate remains and hence the subject can obtain double the amount of food. Initial studies documented that adult cleaners can learn to solve the initial task, as well as the reversal of the roles of the two plates, typically within 200 trials, while capuchins, chimpanzees, orangutans, rats, and pigeons fail (Salwiczek et al., 2012; Zentall et al., 2016, 2017). Only African gray parrots also perform well in this task (Pepperberg and Hartsfield 2014). Recent modeling proposes that the task can only be solved if the two presented plates are seen as a unit (a "chunk") rather than as two separate entities (Quiñones et al., 2020). Normally, chunking is studied in the context of human language, where the identification of sub-structures is essential for information-processing by listeners (Kolodny et al., 2015).

Shifts in the Cleaning Market and Shifts in Strategic Sophistication

More recent research on cleaners revealed that the ability to solve the biological market task is not ubiquitous, and neither is their ability to manage their reputation (Wismer et al., 2014; Binning et al., 2017; Triki et al., 2018). Instead, the performance of individual cleaners in their respective experiments or in nature is highly consistent within a site in a given year but differs between sites or within a site between years. The studied sites are all part of the same reef system that surrounds Lizard Island, Great Barrier Reef. Hence all the cleaners belong to the same population. Moreover, cleaners are open-water spawners with pelagic egg and larval stages. Finally, cleaners take about one year to become adults and then live for another two to three years. As a consequence, it is unlikely that the observed differences between sites and between years are due to genetically coded differences in strategies. Instead, ontogenetic adjustments to the local interspecific social environment are the likely reason for the observed variation. Indeed, sites harboring cleaners with low strategic sophistication in market and reputation tasks are typically socially simple, characterized by low cleaner and (visitor) client densities. Changes within a site between years correlate with severe environmental perturbations: the northern part of Lizard Island was struck by two cyclones in 2014 and 2015, while the entire reefs around the island were affected by severe coral bleaching due to the 2016 El Niño event (Triki et al., 2018). As a consequence of these perturbations, previously socially complex sites lost about 80% of cleaners and about 60% of visitor clients, rendering the fish densities very similar to the socially simple sites (Triki et al., 2018).

The causes and consequences of the intraspecific variation in strategic sophistication are currently being investigated. The focus is on effects of the social environment on brain development (all data still unpublished). With respect to the topic of this chapter, it is important to note that a lack of reputation management affects both cooperation and conflict in the system. First, if cleaners do not adjust service quality to the presence of bystanders, then service quality diminishes. Second, reputation management is also essential for tactical deception. Cleaners from socially simple sites are apparently not capable to selectively give tactile stimulation to small clients and cheat large clients (Binning et al., 2017). Thus, this source of conflict is removed in socially simple sites. Finally, the apparent lack of ability to prioritize an ephemeral food source in the market experiment has potential implications for biological market theory. Current thinking emphasizes how shifts in supply and demand affect exchange rates, while the cost of partner choice is typically considered to be constant (Noë and Hammerstein, 1994; Johnstone and Bshary, 2008). In the cleaning mutualism, lower cleaner densities clearly increase the visitor clients' cost of switching between partners in terms of time budget ("opportunity losses") and energy, independently of whether or not supply and demand ratios change. As a consequence, markets may change from "free markets" to oligopolies, and the visitor clients' loss of receiving priority might be a consequence (Triki et al., 2019).

Conclusions

As the various examples throughout this chapter illustrated, many mutualisms involve investments by one or both partner species. The resulting momentary payoff reduction must yield future benefits in order to be under positive selection. As soon as benefits are contingent on the partner's actions, conflicts of interest arise. In a cleaner's ideal world, it can feed on client mucus without the client terminating the interaction in response. Partner species of ants / flowering plants would ideally obtain the ants' protection/ pollination without providing food resources to the partner. Therefore, most mutualistic systems provide ample opportunities to study how conflicts of interest are resolved so that the overall outcome is mutually beneficial. Indeed, mutualisms offer great opportunities to study the interplay between helping and cheating from diverse angles. Cases in which mutualisms have eroded into parasitism over evolutionary time provide additional perspectives to the question of what factors normally yield stable mutual benefits. How interdependency between unrelated individuals reduces most potential for conflict is arguably better considered in mutualisms than in cooperation studies, as the latter face the challenge to disentangle direct benefits from altruism based on kin selection. On the other extreme, it is easier to find examples of mutualism in which partners have independent futures and hence may focus on personal payoffs than it is to find such cases in cooperation. Thus, mutualisms in their totality resemble the diverse forms of cooperation found in humans.

For various mutualisms, the life cycles of each partner species as well as key ecological variables are reasonably well understood. For mutualisms involving plants and microbes, such knowledge allows us to determine what conditions yield stable mutual helping in the absence of advanced and even rather basic cognitive abilities like individual recognition or memory-based decisions (Brosnan et al., 2010). In other words, various mutualisms provide null-models of mutual helping based on investments, i.e., despite conflicts of interest, allowing us to ask how alternative ecological conditions warrant more advanced cognitive abilities, to make more specific predictions under which conditions such as social behavior and cognitive abilities might coevolve.

Interestingly, few mutualisms are currently used to study more sophisticated cognition. The cleaner fish mutualism is largely an exception in this respect. Other mutualisms that also involve vertebrates have, at least in theory, great potential.

Cooperative hunting between groupers of the genus *Pessuliferus* and partner species (moray eels, Napoleon wrasse [*Cheilinus undulatus*], and octopus [*Octopus cyanea*]) relies on groupers giving referential gestures to guide partners toward a potentially vulnerable prey (Vail et al., 2013). Furthermore, groupers know when they need a partner and when they can hunt alone, and they can select the best collaborator (Vail et al., 2014), just as chimpanzees do (Melis et al., 2006). These results provide evidence that sophisticated cognition may be found in mutualisms, even if both partners act in a self-serving way, i.e., trying to maximize their own prey capture rate, with benefits for the partner accruing as a by-product of the coordinated movements. Two other interesting mutualisms involve humans as partners of wild animals: Honeyguides (*Indicator indicator*) are African birds that show humans the location of bee nests, and dolphins (*Tursiops truncatus*) may herd fish into nets. Both mutualisms occur only within a limited range of the honeyguides' or dolphins' geographic range, indicating that participation is socially transmitted (Zapes et al., 2011; Spottiswoode et al., 2016). In both cases, human hunters and fishermen seem to know what amount of reward (wax for birds, fish for the dolphins) suffices to keep partners motivated to participate again during future encounters. In principle, the involvement of humans would provide ideal conditions to test the decision rules of partner individuals in the wild. For example, how would a honeyguide react to a hunter not providing any wax? We would finally have a system in which we can introduce cooperators and cheaters in a predetermined counterbalanced way by instructing human subjects what to do. Unfortunately, as local people depend on these interactions for their living, such experiments are most likely not feasible.

Finally, mutualism complexes, in which each class of traders is represented by several partner species, deserve continued detailed attention. Ecologists have been fascinated by those for a long time (Bronstein, 1994; Palmer et al., 2003) but the complexity of such systems prevents easy answers. These mutualism complexes raise the same questions that anybody going shopping in a supermarket must ask her-/himself on a regular basis, especially in the USA: how is it possible that competing products (salad dressing, yogurt, etc.) produced by a diversity of companies can fill several meters of shelves? How can these products coexist? In mutualisms, it appears very clear that some partner species provide more benefits than others. If this is so, why does partner choice not eliminate the other species? How should we classify species that provide below-average benefits? Are they cheaters if they obtain higher-than-average benefits from the interactions (Jones et al., 2015), or should the term "cheater" be reserved to cases in which the interaction yields a lower payoff to the partner species than the outside option "no interaction at all"? Understanding the details of all species involved in a mutualism complex will be very challenging, especially as non-participating species may add indirect effects (Guimarães et al., 2017). Nevertheless, tools exist and networks are described (Olesen et al., 2007; Vázquez et al., 2009). By focusing on systems that involve insects, plants, and microbes, we can study the coexistence of evolved offers and ask what ecological conditions promote such coexistence and compare the results to explanations for coexistence of products in human markets.

References

Aanen D. K., Eggleton P., Rouland-Lefèvre C., Guldberg-Frøslev T., Rosendahl S., and Boomsma, J. J. (2002) The evolution of fungus-growing termites and their mutualistic fungal symbionts. *Proceedings of the National Academy of Sciences USA*, 99(23): 14887–14892.

Axen A. H., Leimar O., and Hoffman V. (1996) Signalling in a mutualistic interaction. *Animal Behaviour*, 52: 321–333.

Barbu L., Guinand C., Alvarez N., Bergmüller R., and Bshary R. (2011) Cleaning wrasse species vary with respect to dependency on the mutualism and behavioural adaptations in interactions. *Animal Behaviour*, 82: 1067–1074.

Binning S. A., Rey O., Wismer S., Triki Z., Glauser G., Soares M. C., and Bshary R. (2017) Reputation management promotes strategic adjustment of service quality in cleaner wrasse. *Scientific Reports*, 7(1): 8425.

Boesch C. (1994) Cooperative hunting in wild chimpanzees. *Animal Behaviour*, 48: 653–667.

Borgeaud C., and Bshary R. (2015) Wild vervet monkeys trade tolerance and specific coalitionary support for grooming in experimentally induced conflicts. *Current Biology*, 25: 3011–3016.

Bossan B., Koehncke A., and Hammerstein P. (2011) A new model and method for understanding Wolbachia-induced cytoplasmic incompatibility. *PLoS ONE*, 6(5): e19757.

Brandenburg A., Kuhlemeier C., and Bshary R. (2012) Hawkmoth pollinators decrease seed set of a low nectar *Petunia axillaris* line through reduced probing time. *Current Biology*, 22:1635–1639.

Brock D. A., Douglas T. E., Queller D. C., and Strassmann J. E. (2011) Primitive agriculture in a social amoeba. *Nature*, 469(7330): 393–396.

Bronstein, J. L. (1994) Conditional outcomes in mutualistic interactions. *Trends in Ecology and Evolution*, 9(6): 214–217.

Bronstein, J. L. (2001) Mutualisms. In Fox C., Fairbairn D., and Roff D., eds., *Evolutionary Ecology: Perspectives and Synthesis*. Oxford University Press, pp. 315–330.

Bronstein, J. L., ed. (2015) *Mutualism*. Oxford: Oxford University Press.

Brosnan S. F., Salwiczek L., and Bshary R. (2010) The interplay of cognition and cooperation. *Philosophical Transactions of the Royal Society B: Biological Sciences*, 365: 2699–2710.

Brown C. (2015) Fish intelligence, sentience and ethics. *Animal Cognition*, 18: 1–17.

Bshary R. (2001) The cleaner fish market. In Noë R., Van Hooff J. A. R. A. M., and Hammerstein P., eds., *Economics in Nature: Social Dilemmas, Mate Choice and Biological Markets*. Cambridge: Cambridge University Press, pp. 146–172.

Bshary R. (2002) Biting cleaner fish use altruism to deceive image-scoring client reef fish. *Proceedings of the Royal Society London B Biological Sciences*, 269: 2087–2093.

Bshary R., and Bergmüller R. (2008) Distinguishing four fundamental approaches to the evolution of helping. *Journal of Evolutionary Biology*, 21: 405–420.

Bshary R., and Bronstein J. L. (2004) Game structures in mutualisms: What can the evidence tell us about the kind of models we need? *Advances in the Study of Behaviour*, 34: 59–101.

Bshary R., and Bronstein J. S. (2011) A general scheme to predict partner control mechanisms in pairwise cooperative interactions between unrelated individuals. *Ethology*, 117: 271–283.

Bshary R., and D'Souza A. (2005) Indirect reciprocity in interactions between cleaner fish and client reef fish. In P. McGregor, ed., *Communication Networks*. Cambridge: Cambridge University Press, pp. 521–539.

Bshary R., Gingins S., and Vail A. L. (2014) Social cognition in fishes. *Trends in Cognitive Sciences*, 8(9): 465–471.

Bshary R., and Grutter A. S. (2002a) Asymmetric cheating opportunities and partner control in a cleaner fish mutualism. *Animal Behaviour*, 63: 547–555.

Bshary R., and Grutter A. S. (2002b) Experimental evidence that partner choice is a driving force in the payoff distribution among cooperators or mutualists: The cleaner fish case. *Ecology Letters*, 5(1): 130–136.

Bshary R., and Grutter A. S. (2005) Punishment and partner switching cause cooperative behaviour in a cleaning mutualism. *Biology Letters*, 1(4): 396–399.

Bshary R., and Grutter A. S. (2006) Image scoring and cooperation in a cleaner fish mutualism. *Nature*, 441: 975–978.

Bshary R., Grutter A. S., Willener A. S. T., and Leimar O. (2008) Pairs of cooperating cleaner fish provide better service quality than singletons. *Nature*, 455(7215): 964–966.

Bshary R., and Noë R. (1997) Red colobus and diana monkeys provide mutual protection against predators. *Animal Behaviour*, 54: 1461–1474.

Bshary R., and Schäffer D. (2002) Choosy reef fish select cleaner fish that provide high-quality service. *Animal Behaviour*, 63: 557–564.

Bshary R., Wickler W., and Fricke H. (2002) Fish cognition: A primate's eye view. *Animal Cognition*, 5(1): 1–13.

Bshary R., and Würth M. (2001) Cleaner fish *Labroides dimidiatus* manipulate client reef fish by providing tactile stimulation. *Proceedings of the Royal Society of London B Biological Science*, 268: 1495–1501.

Bshary R., Zuberbühler K., and van Schaik C. P. (2016) Why mutual helping in most natural systems is neither conflict-free nor based on maximal conflict. *Philosophical Transactions of the Royal Society B*, 371: 20150091.

Cheney D. L., Moscovice L. R., Heesen M., Mundry R., and Seyfarth R. M. (2010) Contingent cooperation between wild female baboons. *Proceedings of the National Academy of Sciences USA*, 107(21): 9562–9566.

Clutton-Brock T. H., O'Riain M. J., Brotherton P. N. M., Gaynor D., Kansky R., Griffin A. S., and Manser M. (1999) Selfish sentinels in cooperative mammals. *Science*, 284: 1640–1644.

Connor R. C. (1986) Pseudo-reciprocity: Investing in altruism. *Animal Behaviour*, 34: 1562–1566.

Côté I. M. (2000) Evolution and ecology of cleaning symbioses in the sea. *Oceanography and Marine Biology Annual Review*, 38: 311–355.

Curio E. (1978) The adaptive significance of avian mobbing. I. Teleonomic hypotheses and predictions. *Zeitschrift fur Tierpsychology*, 48: 175–183.

Currie C. R., and Stuart A. E. (2001) Weeding and grooming of pathogens in agriculture by ants. *Proceedings of the Royal Society of London B Biological Science*, 268: 1033–1039.

Currie, C. R., Wong, B., Stuart, A. E. et al. (2003) Ancient tripartite coevolution in the attine ant–microbe symbiosis. *Science*, 299(5605): 386–388.

De Vos A., and O'Riain M. J. (2009) Sharks shape the geometry of a selfish seal herd: Experimental evidence from seal decoys. *Biology Letters*, 6(1): 48–50.

Diamond J. (2002) Evolution, consequences and future of plant and animal domestication. *Nature*, 418: 700–707.

Ellers J., Toby K. E., Currie C. R., McDonald B. R., and Visser B. (2012) Ecological interactions drive evolutionary loss of traits. *Ecology Letters*, 15: 1071–1082.

Eshel I., and Shaked, A. (2002) Partnership. *Journal of Theoretical Biology*, 208: 457–474.

Field T., Hernandez-Reif M., and Diego M. (2005) Cortisol decreases and serotonin and dopamine increase following massage therapy. *International Journal of Neuroscience*, 115: 1397–1413.

FitzGibbon C. D. (1990) Mixed-species grouping in Thomson's and Grant's gazelles: The antipredator benefits. *Animal Behaviour*, 39: 1116–1126.

Flower T. (2011) Fork-tailed drongos use deceptive mimicked alarm calls to steal food. *Proceedings of the Royal Society of London B: Biological Sciences*, 278: 1548–1555.

Gray M. W., and Doolittle W. F. (1982) Has the endosymbiont hypothesis been proven?. *Microbiological Reviews*, 46(1): 1–42.

Grutter A. S. (1995) Relationship between cleaning rates and ectoparasite loads in coral reef fishes. *Marine Ecology Progress Series*, 118: 51–58.

Grutter A. S., and Bshary R. (2003) Cleaner wrasse prefer client mucus: Support for partner control mechanisms in cleaning interactions. *Proceedings of the Royal Society of London B Biological Science*, 270: S242–S244.

Guimarães P. R., Jr., Pires M. M., Jordano P., Bascompte J., and Thompson J. N. (2017) Indirect effects drive coevolution in mutualistic networks. *Nature*, 550(7677): 511.

Hamilton W. D. (1964a) The genetical evolution of social behaviour. I *Journal of Theoretical Biology*, 7(1): 1–16.

Hamilton W. D. (1964b) The genetical evolution of social behaviour. II. *Journal of Theoretical Biology*, 7(1): 17–52.

Hammerstein P., and Noë R. (2016) Biological trade and markets. *Philosophical Transactions of the Royal Society B*, 371(1687): 20150101.

Hata H., and Kato M. (2006) A novel obligate cultivation mutualism between damselfish and Polysiphonia algae. *Biology Letters*, 2: 593–596.

Hauser M. D. (1997) Minding the behaviour of deception. In Byrne R. W., and Whiten A., eds., *Machiavellian Intelligence II: Extensions and Evaluations*. Cambridge: Cambridge University Press, pp. 112–143.

Herre E. A., Knowlton N., Mueller U. G., and Rehner S. A. (1999) The evolution of mutualisms: Exploring the paths between conflict and cooperation. *Trends in Ecology and Evolution*, 14: 49–53.

Huertas V., and Bellwood D. R. (2018) Feeding innovations and the first coral-feeding fishes. *Coral Reefs*, 37(3): 649–658.

Hulcr J., and Cognato A. I. (2010) Repeated evolution of crop theft in fungus-farming ambrosia beetles. *Evolution*, 64: 3205–3212.

Janzen D. H. (1966) Coevolution of mutualism between ants and acacias in Central America. *Evolution*, 20: 249–275.

Johnstone R. A., and Bshary R. (2008) Mutualism, market effects and partner control. *Journal of Evolutionary Biology*, 21(3): 879–888.

Jones, E. I., Afkhami, M. E., Akçay, E. et al. (2015) Cheaters must prosper: Reconciling theoretical and empirical perspectives on cheating in mutualism. *Ecology Letters*, 18(11): 1270–1284.

Kenward R. E. (1978) Hawks and doves: Factors affecting success and selection in goshawk attacks on woodpigeons. *Journal of Animal Ecology*, 47: 449–460.

Kiers E. T., Duhamel M., Beesetty Y. et al. (2011) Reciprocal rewards stabilize cooperation in the mycorrhizal symbiosis. *Science*, 333(6044): 880–882.

Kiers E. T., Rousseau R. A., West S. A., and Denison R. F. (2003) Host sanctions and the legume–rhizobium mutualism. *Nature*, 425: 78–81.

Kohda M., Takashi H., Takeyama T., Awata S., Tanaka H., Asai J., and Jordan A. (2018) Cleaner wrasse pass the mark test. What are the implications for consciousness and self-awareness testing in animals? *BioRxiv*, 397067.

Kokko H., Johnstone R. A., and Clutton-Brock T. H. (2001) The evolution of cooperative breeding through group augmentation. *Proceedings of the Royal Society of London B Biological Science*, 268: 187–196.

Kolodny O., Lotem A., and Edelman S. (2015) Learning a generative probabilistic grammar of experience: A process-level model of language acquisition. *Cognitive Science*, 39(2): 227–267.

Landeau L., and Terborgh J. (1986) Oddity and the "confusion effect" in predation. *Animal Behaviour*, 34: 1372–1380.

Lehmann L., and Keller L. (2006) The evolution of cooperation and altruism – A general framework and a classification of models. *Journal of Evolutionary Biology*, 19: 1365–1376.

Lehmann L., and Rousset F. (2010) How life history and demography promote or inhibit the evolution of helping behaviours. *Philosophical Transactions of the Royal Society of London B*, 365(1553): 2599–2617.

Leimar O., and Connor R. C. (2003) By-product benefits, reciprocity, and pseudoreciprocity in mutualism. In Hammerstein P., ed., *Genetic and Cultural Evolution of Cooperation*. Cambridge, MA: MIT Press, pp. 203–222.

Lieberman D., Tooby J., and Cosmides L. (2007) The architecture of human kin detection. *Nature*, 445(7129): 727.

Machado C. A., Robbins N., Gilbert M. T. P., and Herre E. A. (2005) Critical review of host specificity and its coevolutionary implications in the fig/fig-wasp mutualism. *Proceedings of the National Academy of Sciences USA*, 102(Suppl 1): 6558–6565.

McGregor P. K. (1993) Signalling in territorial systems: A context for individual identification, ranging and eavesdropping. *Philosophical Transactions of the Royal Society London B*, 340(1292): 237–244.

Melis A. P., Hare B., and Tomasello M. (2006) Chimpanzees recruit the best collaborators. *Science*, 311(5765): 1297–1300.

Müller C. B., and Krauss J. (2005) Symbiosis between grasses and asexual fungal endophytes. *Current Opinion in Plant Biology*, 8: 450–456.

Nelsen M. P., Ree R. H., and Moreau C. S. (2018) Ant–plant interactions evolved through increasing interdependence. *Proceedings of the National Academy of Sciences USA*, 115 (48): 12253–12258.

Noë R. (1990) A veto game played by baboons: A challenge to the use of the Prisoner's Dilemma as a paradigm for reciprocity and cooperation. *Animal Behaviour*, 39(1): 78–90.

Noë R., and Hammerstein P. (1994) Biological markets: Supply and demand determine the effect of partner choice in cooperation, mutualism and mating. *Behavioral Ecology and Sociobiology*, 35(1): 1–11.

Noë R., and Hammerstein P. (1995) Biological markets. *Trends in Ecology and Evolution*, 10: 336–339.

Noë R., and Kiers E. T. (2018) Mycorrhizal markets, firms, and co-ops. *Trends in Ecology and Evolution*, 33(10): 777–789.

Noë R., van Schaik C. P., and van Hooff J. A. R. A. M. (1991) The market effect: An explanation for pay-off asymmetries among collaborating animals. *Ethology*, 87: 97–118.

Olesen J. M., Bascompte J., Dupont Y. L., and Jordano P. (2007) The modularity of pollination networks. *Proceedings of the National Academy of Sciences USA*, 104(50): 19891–19896.

Palmer T. M., Stanton M. L., and Young T. P. (2003) Competition and coexistence: Exploring mechanisms that restrict and maintain diversity within mutualist guilds. *The American Naturalist*, 162(S4): S63–S79.

Pepperberg I. M., and Hartsfield L. A. (2014) Can grey parrots (*Psittacus erithacus*) succeed on a "complex" foraging task failed by nonhuman primates (*Pan troglodytes, Pongo abelii, Sapajus paella*) but solved by wrasse fish (*Labroides dimidiatus*)? *Journal of Comparative Psychology,* 128(3): 298–306.

Pierce N. E., Braby M. F., Heath A., Lohman D. J., Mathew J., Rand D. B., and Travassos M. A. (2002) The ecology and evolution of ant association in the Lycaenidae (Lepidoptera) *Annual Review of Entomology*, 47: 733–771.

Pinto A., Oates J., Grutter A., and Bshary R. (2011) Cleaner wrasses *Labroides dimidiatus* are more cooperative in the presence of an audience. *Current Biology*, 21(13): 1140–1144.

Pion M., Spangenberg J. E., Simon, A. et al. (2013) Bacterial farming by the fungus *Morchella crassipes. Proceedings of the Royal Society of London B: Biological Sciences*, 280(1773): 20132242.

Quiñones A., Lotem A., Leimar O., and Bshary R (2020) Reinforcement learning theory reveals the cognitive requirements for solving the cleaner fish market task. *American Naturalist*, 195(4): 664–677.

Raihani N. J., Grutter A. S., and Bshary R. (2010) Punishers benefit from third-party punishment in fish. *Science*, 327(5962): 171–171.

Raihani N. J., Thornton A., and Bshary R. (2012) Punishment and cooperation in nature. *Trends in Ecology and Evolution*, 27(5): 288–295.

Roberts G. (2005) Cooperation through interdependence. *Animal Behaviour*, 70(4): 901–908.

Robertson D. R. (1973) Field observations on the reproductive behaviour of a pomacentrid fish, *Acanthochromis polyacanthus. Zeitschrift Für Tierpsychologie*, 32(3): 319–324.

Sachs J. L., Mueller U. G., Wilcox T. P., and Bull J. J. (2004) The evolution of cooperation. *Quarterly Review of Biology*, 79(2): 135–160.

Salwiczek L. H., and Bshary R. (2011) Cleaner wrasses keep track of the 'when' and 'what' in a foraging task 1. *Ethology*, 117(11): 939–948.

Salwiczek L. H., Prétôt L., Demarta L. et al. (2012) Adult cleaner wrasse outperform capuchin monkeys, chimpanzees and orang-utans in a complex foraging task derived from cleaner – client reef fish cooperation. *PLoS ONE*, 7(11): e49068.

Schino G., and Aureli F. (2008) Grooming reciprocation among female primates: A meta-analysis. *Biology Letters*, 4: 9–11.

Seyfarth R. M. (1977) A model of social grooming among adult female monkeys. *Journal of Theoretical Biology*, 65(4): 671–698.

Silk J. B., Alberts S. C., and Altmann J. (2003) Social bonds of female baboons enhance infant survival. *Science*, 302: 1231–1234.

Soares M. C., Cardoso S. C., Nicolet K. J., Côté I. M., and Bshary R. (2013) Indo-Pacific parrotfish exert partner choice in interactions with cleanerfish but Caribbean parrotfish do not. *Animal Behaviour,* 86: 611–615.

Soares M. C., Côté I. M., Cardoso S. C., Oliveira R. F., and Bshary R. (2010) Caribbean cleaning gobies prefer client ectoparasites over mucus. *Ethology*, 116: 1244–1248.

Soares M. C., Côté I. M., Cardoso S. C., and Bshary R. (2008) The cleaning goby mutualism: A system without punishment, partner switching or tactile stimulation: Choice options and partner control. *Journal of Zoology*, 276(3): 306–312.

Soares M. S., Oliveira R. F., Ros A. F. H., Grutter A. S., and Bshary R. (2011) Tactile stimulation lowers stress in fish. *Nature Communications,* 2: 534.

Spottiswoode C. N., Begg K. S., and Begg C. M. (2016) Reciprocal signaling in honeyguide-human mutualism. *Science*, 353(6297): 387–389.

Stander P. E. (1992) Foraging dynamics of lions in a semi-arid environment. *Canadian Journal of Zoology*, 70: 8–21.

Taborsky M. (1984) Broodcare helpers in the cichlid fish *Lamprologus brichardi*: Their costs and benefits. *Animal Behaviour*, 32(4): 1236–1252.

Tebbich S., Bshary R., and Grutter A. S. (2002) Cleaner fish *Labroides dimidiatus* recognise familiar clients. *Animal Cognition*, 5(3): 139–145.

Tomasello, M., Melis, A. P., Tennie, C., Wyman E., Herrmann, E. (2012) Two key steps in the evolution of human cooperation: The interdependence hypothesis. *Current Anthropology*, 53(6), 000–000.

Triki Z., Wismer S., Levorato E., and Bshary R. (2018) A decrease in the abundance and strategic sophistication of cleaner fish after environmental perturbations. *Global Change Biology*, 24: 481–489.

Triki Z., Wismer S., Rey O., Binning S. A., Levorato E., and Bshary R. (2019) Biological market effects predict cleaner fish strategic sophistication. *Behavioral Ecology*, online https:// doi. org/10.1093/beheco/arz111.

Vail A. L., Manica A., and Bshary R. (2013) Referential gestures in fish collaborative hunting. *Nature Communications*, 4: 1765.

Vail A. L., Manica A., and Bshary R. (2014) Fish choose appropriately when and with whom to collaborate. *Current Biology,* 24: R791–R793.

Vaughan D. B., Grutter A. S., Costello, M. J., and Hutson K. S. (2017) Cleaner fishes and shrimp diversity and a re-evaluation of cleaning symbioses. *Fish and Fisheries*, 18(4): 698–716.

Vázquez D. P., Blüthgen N., Cagnolo L., and Chacoff N. P. (2009) Uniting pattern and process in plant–animal mutualistic networks: A review. *Annals of Botany,* 103: 1445–1457.

Wainwright P. C., Bellwood D. R., Westneat M. W., Grubich J. R., and Hoey A. S. (2004) A functional morphospace for the skull of labrid fishes: Patterns of diversity in a complex biomechanical system. *Biological Journal of the Linnean Society*, 82(1): 1–25.

Weeks P. (2000) Red-billed oxpeckers: Vampires or tickbirds? *Behavioral Ecology*, 11: 154–160.

Werren J. H. (1997) Biology of Wolbachia. *Annual Review of Entomology*, 42: 587–609.

West S. A., Pen I., and Griffin A. S. (2002) Cooperation and competition between relatives. *Science*, 296: 72–75.

Wismer S., Grutter A., and Bshary R. (2016) Generalized rule application in bluestreak cleaner wrasse (*Labroides dimidiatus*): Using predator species as social tools to reduce punishment. *Animal Cognition*, 19(4): 769–778.

Wismer S., Pinto A. I., Vail A. L., Grutter A. S., and Bshary R. (2014) Variation in cleaner wrasse cooperation and cognition: Influence of the developmental environment? *Ethology*, 120(6): 519–531.

Wubs M., Bshary R., and Lehmann L. (2018) A reinforcement learning model for grooming up the hierarchy in primates. *Animal Behaviour,* 138: 165–185.

Zappes C. A., Andriolo A., Simões-Lopes P. C., and Di Beneditto A. P. M. (2011) "Human–dolphin (*Tursiops truncatus* Montagu, 1821) cooperative fishery" and its influence on cast net fishing activities in Barra de Imbé/Tramandaí, Southern Brazil. *Ocean and Coastal Management*, 54(5): 427–432.

Zentall T. R., Case J. P., and Berry J. R. (2017) Rats' acquisition of the ephemeral reward task. *Animal Cognition*, 20(3): 419–425.

Zentall T. R., Case J. P., and Luong J. (2016) Pigeon's (*Columba livia*) paradoxical preference for the suboptimal alternative in a complex foraging task. *Journal of Comparative Psychology*, 130(2): 138–144.

10 The Fundamental Role of Aggression and Conflict in the Evolution and Organization of Social Groups

Clare C. Rittschof and Christina M. Grozinger

Introduction

Research on the evolution of social groups has focused substantially on processes that increase cooperative behaviors (Bourke, 2011). Indeed, each major evolutionary transition (the evolution of eukaryotes from prokaryotes, the evolution of multicellular organisms from single-celled organisms, the evolution of eusocial insect societies from solitary species, etc.) involves increasing cooperative behavior to the point where previously independent units now must interact to successfully replicate (Szathmary and Smith, 1995). With the focus on cooperation, the importance of aggression and conflict in societies is typically overlooked, despite the fundamental nature of these processes in giving rise to and maintaining social structures (see Chapters 2, 8, and 9 for other examples). For instance, maternal antipredator aggression is the critical antecedent to a stable social unit that exists in a central place (e.g., a nest or burrow) (Brunton, 1990; Groom, 1992). Within social groups, aggression among individuals establishes dominance hierarchies, manages conflict, and leads to division of labor in contexts of reproduction and offspring care (Ratnieks et al., 2006; Wittemyer and Getz, 2007). Here we discuss the possible roles of aggression and conflict in the evolution and organization of social groups, and explore the underlying molecular and physiological mechanisms associated with these drivers of sociality. We define aggression as behaviors that carry a potential physical cost or otherwise could reduce the direct fitness of the individuals involved, and we define conflict as situations in which the fitness interests of interacting individuals diverge, regardless of the behavioral outcome.

For this examination of aggression, conflict, and social groups, we focus on insect societies, which best represent eusocial societies, the pinnacle of cooperative behavior

We would like to thank the Münster Graduate School of Evolution and its Evolutionary Think Tank for providing support and stimulating discussions during the development of this chapter. We would like to thank members of the Grozinger and Rittschof laboratories, Walter Wilczynski, and Sarah Brosnan for critical reading of the text. Rittschof is supported by the National Institute of Food and Agriculture, US Department of Agriculture Hatch Program under accession number 1012993, and the Foundation for Food and Agricultural Research Pollinator Health Fund (Grant ID: 549049). Grozinger is supported by the National Science Foundation, United States Department of Agriculture National Institute of Food and Agriculture, Agriculture and Food Research Initiative (USDA-NIFA-AFRI) Program, and the US Israel Binational Science Foundation.

> **Box 10.1** Organization of Eusocial Insect Societies
>
> Eusocial societies are defined as having cooperative brood care (where multiple adults provide care for young), overlapping generations (where parents and adult offspring remain together on the nest), and, most critically, a reproductive division of labor, where only one or a few individuals in the colony reproduce (e.g., female queens), while the others (workers) remain sterile and help rear their close relatives' offspring (Robinson et al., 2008). In the most advanced eusocial systems (e.g., honeybees and most ants), queens and workers represent distinct and stable "castes," with dramatic differences in morphology, physiology, and behavior, as a result of distinct developmental trajectories (reviewed in Bourke, 2011). There are also more primitively social species where worker reproductive potential is seasonally dynamic and constrained to producing only males (e.g., bumblebees), and others where multiple individuals retain full reproductive potential, and dominance hierarchies are established to determine who will reproduce at a given point in time (e.g., wasps). In many cases, insect societies also exhibit a division of labor among workers, where different colony tasks (e.g., collecting food, building, cleaning, and defending the nest) are carried out by different subsets of individuals (Smith et al., 2008). Workers may segregate to different tasks according to morphological or physiological features, some of which are a function of adult age (known as age or temporal polyethism). Both caste differentiation and worker division of labor are the result of interactions between genetic variation and environmental conditions, with nutrition playing a key role (Smith et al., 2008).

and social evolution in the animal kingdom (Wilson, 2000; Bourke, 2011) (see Box 10.1 and Figure 10.1). Insects are an outstanding system in which to study the evolution of complex social behavior, since related species can span the range from solitary to eusocial behavior, thus allowing us to explore the role of aggression and conflict in shaping increasingly complex social units (Figure 10.2). The organization of insect societies is based on an elaborated division of labor among females, encompassing not only differential reproduction but also sterile worker division of labor, where individuals specialize on caring for the developing larvae (brood), foraging for resources, and defending the nest. These reproductive and worker divisions of labor systems provide unique contexts to address how aggression and conflict shape social structure, the mechanistic bases of these processes, and how these systems evolved.

Role of Defensive Aggression in Social Group Formation and Organization

Role of Maternal Aggression in the Evolution of Social Groups

Post-zygotic maternal care, in which a mother stays to care for eggs and young at the expense of additional reproductive opportunities, is a costly form of care that is rare

Figure 10.1 Model social insect systems. *Left panel:* Honeybees (*Apis mellifera*) live in perennial colonies with tens of thousands of female worker siblings who cooperate to rear the offspring of a single reproductive queen mother (Winston, 1991). Queens and workers are morphologically, physiologically, and behaviorally distinct, where workers are unable to mate or produce fertilized eggs, and queens specialize exclusively in egg production. The workers build the nest (which consists of a precisely built honeycomb with hexagonal cells) within a suitably sized cavity, produce specialized processed food which they feed to developing larvae (termed "brood"), provision the colony by foraging over vast areas (spanning hundreds of square kilometers), recruit nestmates to high-value forage sites through a symbolic dance language, defend the colony from predators and intruders (including robbing bees from other colonies), thermoregulate the colony during both hot and cold seasons, and search for new nest sites when the colony reproduces through a process of colony fission. (Photo by Nick Sloff, Penn State.) *Middle panel:* Bumblebees (*Bombus terrestris*) live in annual colonies that reach maximum sizes of a few hundred individuals, depending on the species (reviewed in Amsalem et al., 2015). Colonies are founded by a single queen (a foundress), who constructs the nest, forages, and rears the first cohort of worker brood. Once the workers have emerged, they take over foraging, brood care, and nest construction and defense, while the queen lays the eggs. Like honeybees, worker bumblebees are unmated. However, at the end of the season, the colony starts rearing the next generation of reproductive individuals (queens and males). At this point, the workers and queen compete with each other over production of males, which develop from unfertilized eggs, and thus can be produced by unmated workers. The workers form a social and reproductive dominance hierarchy, with dominant workers behaving more aggressively and laying more male eggs. The males mate with the virgin queens from other nests, and the mated queens overwinter underground, while the males, workers, and old queen die. (Photo by Luca Franzini, Penn State.) *Right panel:* Wasps represent a diverse group of species, ranging from solitary to eusocial in their lifestyles (Jandt and Toth, 2015). Most work on wasp social behavior has focused on *Polistes* paper wasps. These wasps form annual colonies which are relatively small (~20 individuals) and are initiated in the spring by individual foundresses or by pairs of foundresses which establish a reproductive dominance hierarchy. As in the case of bumblebees, once the first cohort of workers is produced, they take over the colony tasks. Near the end of the colony life cycle, the next generation of males and queens are produced, and only the mated queens overwinter. *Polistes* workers have a clear social dominance hierarchy, and if the queen is lost, they will compete over reproductive dominance. (Photo by Jennifer Jandt, University of Otago.) (A black and white version of this figure will appear in some formats. For the color version, please refer to the plate section.)

among insects, but evolves when offspring care improves parental fitness (Bourke, 2011). Maternal care behavior is defined as any activity that improves the fitness of offspring, including building and maintaining a nest, providing food to altricial offspring, and defending the nest against predators and parasites (Tallamy and

Maternal defense
of nest

Social defense
of nest

Dominance hierachies
leading to reproductive
division of labor

Aggression towards
juvenile offspring leading
to nutritional constraints
and caste differentiation

Large social groups
with elaborated worker
caste can have individuals
specialized for defense

Figure 10.2 Different contexts and modes of aggression and conflict associated with the evolution of eusociality. *Top panel:* Maternal antipredator behavior is fundamental for the establishment of a nest, and over evolutionary time, a social group made up of a mother and offspring. If adult offspring delay dispersal, the social group increases in size and stability, and coordinated defense against predators and conspecifics becomes increasingly important. *Middle panel:* In social groups with multiple adults, there can be conflict over reproduction. Physical aggression among individuals mediates this conflict, and leads to the establishment of dominance hierarchies and reproductive division of labor. Furthermore, often the subordinate, non-reproductive individuals engage in nest defense. Conflict over reproduction can be further mediated by more nuanced aggressive behaviors that constrain individuals' ability to compete for dominance and/or reproduce. For example, adults in the nest can nutritionally deprive individuals during development, resulting in a sterile adult worker caste with distinct morphological, physiological, and behavioral traits compared to the well-nourished reproductives. *Lower panel:* In social groups with a worker caste, further division of labor among workers generates distinct groups that specialize in nest defense. (Bear image from IconsPNG; Graphic design by Harland Patch, Penn State.)

Brown, 1999). Most research on social insects has focused on the food-provisioning aspect of maternal care in the evolution of sociality, particularly the mechanistic underpinnings that explain how this suite of behavioral traits became distinct from the egg-laying role (Amdam et al., 2006; Toth et al., 2007). The focus on food provisioning could reflect the fact that some of the most elaborate forms of sociality involve large, conspicuous nests with high levels of food storage and even food cultivation (De Fine Licht et al., 2014). Food provisioning is also related to other evolutionary drivers of sociality, e.g., an ability to take advantage collectively of rare or ephemeral food resources, and prolonged feeding interactions with offspring (O'Dowd and Hay, 1980; Scott, 1998; Seeley, 2012), which may give rise to overlapping adult generations (Nowak et al., 2010) (Box 10.1). However, food provisioning is not a unique feature of higher-level sociality, and it is not a sufficient precondition for its evolution (Carey, 2001). For example, in all bee species, including solitary species, the females construct nests for their eggs, and maternal provisioning is the norm, presumably because relatively immobile larvae are the ancestral state in hymenoptera (bees, ants, and wasps; Tallamy and Brown, 1999).

Strong selection for nest defense and adequate defensive capabilities are less well-studied but critically important criteria for social evolution (Andersson, 1984). Maternal nest defense is empirically well-supported as an evolutionary driver of maternal care evolution (Crespi, 1994; Nakahira and Kudo, 2008). In some eusocial species, nest defense may outweigh worker food provisioning in terms of significance for group survival (Wong et al., 2013), and in other cases it is the primary adaptive outcome of social grouping (Strassmann, 1981; Strassmann et al., 1988). The hymenoptera, an insect group in which sociality evolved multiple times (Figure 10.1), take advantage of the myriad benefits of social living at a reduced cost due to the evolution of a poisonous stinger, which makes them particularly effective at defense (Andersson, 1984; Machado, 2002). Thus, antipredator behavior is important in both the establishment and success of social groups (Figure 10.2).

Role of Defense against Predators and Conspecifics in Social Structure and Organization

The need to repel predators (including conspecifics) influences the structure and function of social groups. For example, predation pressure can cause or respond to variation in colony size, over ecological or evolutionary time frames. Honeybee (*Apis mellifera*) defensive response increases in intensity with colony size, with larger, more mature colonies typically exhibiting higher levels of antipredator aggression (Hunt et al., 2003; Kastberger et al., 2008). In the social paper wasp *Polistes bellicosus*, complete nest destruction from vertebrate predators (e.g., birds, raccoons, and opossums) is so common that only large groups of females are able to successfully defend against total nest failure (Strassmann et al., 1988). In contrast, in a congener, *Polistes exclamans*, bird predation is a common source of nest failure, but this threat, in combination with high levels of infestation by several species of parasitoid wasp, leads to the formation of smaller, more defensible nests (Strassmann et al., 1988). Phenotypic responses to predator threat may reflect plasticity (in the case of the honeybee colony), genetically encoded variation, or some combination (Giray et al., 2000). For example, *Polistes metricus* wasp foundresses that are genetically "bolder" – defined as less likely to leave the nest when prodded repeatedly with a plastic brush – produce larger colonies that are less responsive to predators (Wright et al., 2017). Thus overall, predation threat can impact both social group size and the trade-offs associated with cooperating with conspecifics.

Communication strategies have evolved to facilitate the defense of nests, by organizing coordinated attacks on predators (Figure 10.2) or identification of invading conspecifics from other nests through chemical cues used for nestmate recognition (Breed, 1983; Breed et al., 1990, 2004). For example, in honeybees, workers will emit alarm pheromones if the nest is threatened. This is a multi-component chemical blend, which serves to attract nestmates and stimulate defensive responses. The gland that produces the main pheromone components is associated with the sting sheath, and thus the pheromone is released in large quantities if the individual stings a target. Several other hymenopteran species produce alarm pheromones, and in some cases, these pheromones are included in the venom (Bruschini et al., 2006). Alarm pheromone enables a rapid, collective behavioral response from numerous individuals simultaneously. The result is an enhanced ability to repel large predators that become greater threats as colonies grow in size and food stores. Thus, colony size and predator threat are factors that interact to shape social structure, organization, and sophistication over evolutionary time.

Conspecific competition can take a variety of forms, but it is usually associated with competition for nutritional resources, and thus can vary with environmental conditions. In animal species with high food-storage needs or ephemeral resources, robbing conspecific nests is a common practice (Cardinal et al., 2010). For example, in honeybees, robbing activity is common at times of nectar dearth, particularly prior to the onset of winter, when colonies must store enough food to survive until the following spring (Doke et al., 2015). During a robbing bout, large numbers of foragers enter a nearby (usually weaker) colony, kill worker bees, and steal honey and bring it back to their home hive. A subset of middle-aged honeybee workers specializes in "guarding" behavior (Breed, 1983; Breed et al., 1990, 2004). Guarding activity (in terms of the number of guards and the threshold for conspecific acceptance versus rejection) changes with environmental conditions that make robbing more or less prevalent (Couvillon et al., 2008); thus, honeybee colonies have clearly evolved the ability to adjust investment in conspecific defensive resources according to the external environmental pressures. Specific defensive mechanisms and behaviors that target conspecific predators are also observed in other social insect groups, e.g., termites (Thorne et al., 2003).

Unlike in the antipredator context, conspecifics have the ability to communicate with one another through honest signaling, and in some cases, ritualized aggression has evolved to resolve conflict among colonies with minimal physical injury. One iconic example is the honey ant (*Myrmecocystus mimicus*). In this species, neighboring colonies typically make an effort to avoid one another when foraging. However, periodically, when high-quality food resources are located nearby, they do invade one another's territory to gain access. Elaborate aggressive "tournaments," ritualized displays among workers from neighboring colonies that convey both individual and colony strength, prevent the food-seeking colony from entering the conspecific ant nest and stealing colony members and nectar as food resources (Hölldobler, 1981). Tournaments involve stilting displays, abdominal inflation, antennal drumming, and standing on rocks and debris to appear larger (Hölldobler and Wilson, 1994).

In some cases, competition among colonies of conspecifics can give way to colony mergers, a form of cooperation across social groups. In the case of *Zootermopsis nevadensis* termites, this can weaken the reproductive dominance of the original kings and queens (Thorne et al., 2003). In other cases, super colonies increase the competitive ability of individual colony constituents. Unicolonial ant species (e.g., Argentine ants) present an interesting balance between aggression and cooperation among different colonies. These species exist as polygynous (multi-queen) super colonies, where individuals freely move among unrelated nests. While within–super colony aggression is low, individuals recognize and attack individuals from other super colonies as well as other ant species. The ability to form super colonies gives these species a competitive advantage and may facilitate invasion and range expansion (Vásquez and Silverman, 2008).

Role of Within-Group Aggression in the Evolution of Social Behavior

A hallmark of eusocial insect societies is reproductive division of labor, where some individuals produce more offspring than others (Bourke, 2011). This reproductive division of labor would appear to reduce the benefits of the cooperative behavior and increase conflict by reducing the fitness of the cooperating individual (but see Box 10.1). Aggressive behaviors are integral to the establishment and maintenance of reproductive dominance hierarchies because they increase the costs of non-cooperative behavior and thus reduce conflict within groups (reviewed in Ratnieks et al., 2006; Ratnieks and Wenseleers, 2008). Similar to intraspecific interactions across social groups (see second section of this chapter), within social groups, non-damaging aggressive interactions have evolved to reduce the costs of physical acts of aggression, which carry risks for both the actor and the recipient. Aggressive behaviors include honest/reliable signaling of dominance (which can include ritualized aggression, visual or chemical signals), coercion (which reduces the benefit of selfish action, such as policing to remove subordinate-laid eggs), or constraint (where the behavior generates individuals that simply do not have the capacity to act selfishly, as is observed in pre-adult caste differentiation). It is important to note that in the context of dominance interactions within a social group, the cost of aggression is often not evaluated, and in some cases it may be less costly to be the aggressive dominant than to perform subordinate behaviors (Weiner et al., 2009). Because aggressive behaviors shape social structures among adult females within the colony, and ultimately lead to caste differentiation between queens and workers, these relative costs may also change or become difficult to directly measure over evolutionary time as other mechanisms (e.g., morphological differentiation) weaken individuals' ability to transition among dominant and subordinate behaviors.

Aggression Shapes Social Dominance and Reproductive Dominance within Social Groups

In social species where multiple adults in the colony have the ability to reproduce, individuals can form social and reproductive dominance hierarchies (Figure 10.2). Aggression level (social dominance) is generally positively correlated with reproductive potential (reproductive dominance; West-Eberhard, 1967; Amsalem et al., 2015). This link between social dominance and reproductive dominance ensures that competitive interactions result in the most fecund individual reproducing, which can maximize helpers' inclusive fitness in groups of relatives (West-Eberhard, 1967). Fitness returns for subordinates include waiting to inherit the nest or helping rear relatives (Keller and Nonacs, 1993; Keller and Chapuiset, 1999; Cant, 2011).

In many species, honest or reliable signaling mechanisms have evolved that accurately reflect physical abilities, allowing individuals in the group to avoid potentially damaging physical aggression and maximize their fitness returns. Ritualized aggression is the most obvious form of honest/reliable signaling of dominance, and is observed in co-founding reproductive females known as gynes (e.g., in wasps, West-Eberhard, 1967), and among queens and workers or within queenless worker groups (e.g., bumblebees, Amsalem et al., 2015). However, individuals can also engage in a "pheromonal competition," where members of a group compete to produce the greatest amount of fertility-signaling pheromones (Moritz et al., 2004). In *Polistes dominulus* paper wasps, facial color patterns are associated with dominance rank, and individuals use this information to structure their aggressive interactions: individuals prefer to challenge those whose facial patterns have been altered to indicate a low rank and physical condition (Tibbetts and Lindsay, 2008).

Social group dynamics, most notably loss of the dominant reproductive individual, are characterized by temporary shifts in aggression among group members. In many species of *Polistes* wasps, if the dominant egg-layer is removed, the most aggressive worker with the largest ovaries and the best reproductive potential takes over (West-Eberhard, 1967). In honeybees (*A. mellifera*), if the queen is lost and there are insufficient female eggs and larvae available to produce a new one, a subset of the workers activate their ovaries and lay unfertilized eggs that develop into males (Naeger et al., 2013). Since there is limited time to produce males before the colony collapses (Page and Erickson, 1988), and laying workers can pheromonally suppress ovary activation of their sisters (Sakagami, 1958), there should be selection for workers to activate their ovaries as rapidly as possible. Indeed, paired queenless workers will compete with each other to produce the greatest amount of the pheromone that indicates fertility (Moritz et al., 2004).

As with antipredator behavior, aggression within social groups also is impacted by group size. In bumblebees (*Bombus terrestris*), when group size increases, the dominant individual (the α female) engages in more aggressive interactions, and directs most of her aggression at the next dominant female (β female) in the hierarchy (Amsalem and Hefetz, 2011). Despite the increased overall aggression, the β females in larger groups showed greater levels of ovary activation (an indicator of active reproduction). Furthermore, lower-ranked workers in the group showed activated ovaries, presumably because the α individual is unable to engage in sufficient aggressive interactions with all the members of the group (Amsalem and Hefetz, 2011).

Aggression during Development Leads to Caste Differentiation

Social and reproductive dominance hierarchies are formed when multiple individuals in the colony compete over reproduction. However, many social insect species manipulate the developmental trajectories of developing larvae (or "brood") within the nest, to generate adults (subordinates) with reduced ability to compete with the dominant individuals and with overall reduced reproductive potential, which then reduces conflict and facilitates cooperative behaviors (Figure 10.2; Ratnieks et al., 2006; Ratnieks and Wenseleers, 2008). Subordinate individuals cooperate by caring for the brood of the dominant reproductives, collecting food, and defending the nest. Manipulation of the developmental trajectories of larvae is an example of a developmental constraint, and, as it reduces the fitness of the recipient, it is thus considered an aggressive behavior.

In some societies, subordinate and dominant individuals are morphologically similar, but differences during development has resulted in physiological and behavioral differences in the adults. For example, lower-quality diets during larval development in the subsocial bee *Ceratina calcarata* produce a "dwarf" daughter, who subsequently aids the mother in rearing her sisters and brothers and forgoes reproduction (Lawson et al., 2017). Similarly, in *Polistes* wasps, it is hypothesized that nutritional deprivation during larval development leads to adults with reduced nutritional stores and reproductive capacity, and limited ability to survive the winter – thus a subordinate worker phenotype rather than a dominant queen phenotype (Hunt et al., 2007; Jandt and Toth, 2015; Jandt et al., 2017). *Polistes fuscatus* appears to have developed a ritualized, non-nutritional signal that manipulates larval development; vibrations of larval cells by antennal drumming from the queen generates adult females with reduced nutritional stores resembling a subordinate worker-like physiological state, though mechanisms underpinning this process remain to be determined (Suryanarayanan et al., 2011).

In advanced eusocial species, rather than generating weaker or less reproductively capable adults, brood manipulation (which is often, but not exclusively, nutritional) can activate unique developmental pathways leading to alternative phenotypes (queen-versus-worker castes), which vary dramatically in morphology, physiology, and behavior. Nutritional switch points are common among insect species, perhaps owing to the unusually significant contribution of larval nutrition to adult phenotypes in general

(Moczek, 2010). In honeybees, differential feeding of female larvae (differences in diet quantity and quality, and perhaps the presence/absence of plant phytochemicals or honeybee-produced proteins) results in profound caste differences mediated by changes in epigenetic, transcriptional, metabolic, and hormonal signaling pathways (Mao et al., 2015; Maleszka, 2018). Workers can no longer mate or store sperm, and their ovaries are dramatically smaller than those of queens, while queens have lost the structures for collecting pollen. In several ant species, differences in developmental trajectories produce workers that cannot produce viable eggs because of mislocalization of maternally inherited factors in those eggs (Khila and Abouheif, 2008).

Are Aggression and Reproduction Physiologically Coupled?

Though the link between aggression and reproductive dominance is well-established, a fascinating and outstanding question is whether aggressive behaviors facilitate – or are perhaps even necessary for – reproduction and ovary activation, particularly in societies without developmentally determined caste differentiation. For example, in multiple species, bumblebee nest-founding queens that are placed with a second queen are much more likely to lay eggs than if they are alone (Plowright and Jay, 1966). Moreover, in bumblebee worker groups where the queen is lost, individual aggressive interactions both precede and predict the degree of ovary activation and egg-laying in workers (Amsalem and Hefetz, 2010; Sibbald and Plowright, 2012). Similarly, in *Ropalidia marginata* wasps, where a subordinate worker can rise to queen status if the queen is removed, one of the workers (the "potential queen") increases her aggressive behaviors substantially for a brief period before she becomes the queen, after which she ceases aggressing her nestmates (Skukla et al., 2014). Surprisingly, there is little evidence that aggression from the potential queen suppresses reproduction in nestmates. Instead, this behavior seems to trigger ovary activation in the aggressor: solitary potential queens, with no one to aggress, take significantly longer to lay their first eggs.

The physiological mechanism by which aggression and reproduction are coupled remains to be fully understood. Juvenile hormone (JH) signaling or regulation is likely involved, as it plays a role in mediating both reproduction and aggression in insects (Hartfelder, 2000; Tibbetts and Crocker, 2014). In insect species, JH generally acts as a gonadotropin and serves as a critical regulator of female reproduction, with increased hemolymph (blood) titers of JH during pre-adult stages associated with increased ovary size, and increased titers in adult stages associated with ovary activation and egg-laying (Chapman, 2012). In a species of solitary wasp where females compete over beetle larvae, which serve as hosts in which wasp mothers lay their eggs, JH treatments result in increased egg production, which is associated with increased aggression toward other females (Mathiron et al., 2019). In *Polistes* wasps, which live in small social groups with one reproductive queen and a small number of sterile workers (Figure 10.1), increasing JH levels increases ovary activation and social dominance in queens, but does not increase ovary activation in workers and instead causes them to show increased nest-guarding (aggression) and foraging (Giray et al., 2005; Tibbetts and Sheehan, 2012; Jandt and Toth, 2015). Mechanistically, this variation in the response to JH may be mediated by the nutritional status of the individual the queen, with great nutritional resources and higher reproductive potential, responds to JH with increased reproductive output, while the workers, with reduced nutritional status, respond to JH with increased helping behaviors. In honeybees, JH continues to be associated with aggression in workers: higher JH levels are correlated with increased nest-guarding effort (Huang et al., 1994; Hartfelder, 2000); the effects of JH on queen aggression remain to be determined. Thus, JH mediates both reproductive and aggressive behaviors, but these relationships vary depending on the degree and type of social organization and with the physiological and social contexts of the individual. (Chapters 5 and 7 discuss a similar sex- and context-dependent action of the peptide vasopressin in mammals, which similarly modulates both reproductive and aggressive behaviors.)

Are Aggression, Reproduction, and Brood Care Physiologically Uncoupled?

In vertebrate species, the hormone testosterone mediates a trade-off between reproductive potential and parental care. The "challenge hypothesis" is a conceptual framework used to explain social and reproductive modulation of testosterone levels (Wingfield et al., 1990). Under the challenge hypothesis, high testosterone is associated with high reproductive output (both spermatogenesis and the potential to gain access to mates through aggression), but a decrease in the care component of reproduction (Wingfield et al., 1990; Hirschenhauser and Oliveira, 2006; Tibbetts and Crocker, 2014).

Interestingly, a very similar negative correlation is observed between aggression/reproductive potential and brood care in social insects, potentially mediated by JH (Tibbetts and Crocker 2014). However, because tasks are distributed across colony members in social insect groups, and thus are expressed simultaneously, it is unlikely that trade-offs between care and aggression maintain this mechanistic coupling. In social insects, brood care is defined by direct interactions between adult individuals and developing larvae, typically involving food-provisioning and thermoregulatory activities. Dominant, high-aggression individuals (with higher JH levels) produce more brood and provide low levels of brood care. Brood care is instead provided by low-aggression, sterile subordinates. In honeybees, reproduction, aggression, and brood care are even further separated, as queens engage in reproduction while both nest defense (aggression) and brood care are performed by workers. Fascinatingly, worker differences in JH still reflect a trade-off between aggression and care behaviors – high-JH, high-aggression, older worker bees perform nest defense and non-care tasks outside of the hive, while younger bees with lower JH levels care for the brood (Hartfelder, 2000). Thus, aggression (and JH) seems to essentially separate adults in a social group along a continuum of reproduction and brood-care potential in social insects. Subordinate individuals with diminished JH (either as a cause or effect of variation in aggressive encounters) are already physiologically primed to perform care behaviors. This physiological state, coupled with an incentive to provide care (see Box 10.1), allows the primary reproductive individuals to maintain high reproductive output with minimal cost. The sum total is an efficient division of labor that results in high colony growth potential.

Role of Aggression in Facilitating Worker Division of Labor in Social Groups

Aggression may have facilitated the evolution of subspecialized worker behaviors, and ultimately worker division of labor (Figure 10.2). Nest defense is increasingly important as colonies become larger, more stable, and more likely to store food. In turn, larger colony size is associated with the evolution of subspecialized defensive worker castes (Fjerdingstad and Crozier, 2006). In species in which individual workers are specialized for nest defense, the relative proportions of subcastes are altered by interspecific aggression, which is dynamic across environments (Passera et al., 1996); in these cases, subspecialization allows the colony to fine tune its worker effort to match its needs (Yang et al., 2004). For example, honeybee colonies will increase their investment in guard bees during periods of nutritional dearth, when robbing by other honeybee colonies becomes more likely (discussed earlier).

In some cases, worker-defense specialization is accompanied by distinct morphological features, including larger body size (Wheeler and Nijhout, 1981; Yang et al., 2004), large heads (Passera et al., 1996; Rajakumar et al., 2012), and well-developed jaw muscles (Muscedere et al., 2011). Morphological subspecialization of workers into clearly distinct subcastes occurs in ~15% of ant species, most termite species, only one known bee species, and no wasps (Robinson, 1992; Gruter et al., 2012). In ants, morphological subspecialization is more common in species with greater morphological distinction between queen and worker castes (Fjerdingstad and Crozier, 2006). This suggests that as workers become more distinct from reproductives, they may suffer less of a reproductive penalty from further specialization. This pattern is reflected at the mechanistic level: in ants, subcaste development into the soldier phenotype is a result of nutritional differences and is mediated by JH exposure at the later larval stages, after the queen–worker dimorphism has been set (Wheeler and Nijhout, 1981; Rajakumar et al., 2012).

Most social insect species do not evolve a morphologically distinct soldier caste; rather, workers show age-related task specialization (age polyethism), which can be modified by external environmental factors and social cues in the nest (Mcdonald and Topoff, 1985; Robinson, 1987, 1992; Breed et al., 1990; Seid and Traniello, 2006). Despite the lack of external morphological variation, there can be major physiological differences among subcastes of workers, including changes in muscle structure (Muscedere et al., 2011), brain size (Farris et al., 2001; Fahrbach et al., 2003), and fat stores (Toth and Robinson, 2005). As with morphological caste determination during larval development, age polyethism has dietary correlates and is mediated by

JH (Robinson, 1987; Huang and Robinson, 1992; Shorter and Tibbetts, 2008). Thus, morphological specialization and age polyethism share many features, and as opposed to representing a more extreme form of worker subspecialization, the existence of morphological worker subcastes may simply be a different evolutionary route to a similar functional outcome.

Even in species with well-established worker subspecialization, ongoing conflict between queens and workers over reproduction can facilitate worker division of labor. In honeybees, there is no overt physical aggression between queens and workers, and queen reproductive dominance is maintained through queen pheromones (Kocher and Grozinger, 2011). Exposure to queen pheromone also slows down temporal changes in worker subspecialization, specifically the transition from nursing to foraging behaviors (Robinson et al., 1998). Workers show individual variation in their degree of contact with the queen, where workers with larger ovaries, a measure of reproductive potential, have reduced contact with the queen (Kocher et al., 2010; Galbraith et al., 2015). Relatively low levels of contact with the queen may enhance worker reproduction in the event that the queen is lost by allowing individuals to more rapidly activate their ovaries and initiate egg-laying (Galbraith et al., 2015). This differential contact, however, may also impact the timing of the transition from nursing to foraging, and other elements of task subspecialization that exist under normal, queen-right colony conditions. Indeed, several studies have demonstrated that ovary size, which predicts queen contact, also predicts likelihood to specialize on pollen (rather than nectar) foraging (Amdam et al., 2006). Thus, in honeybees, ongoing conflict between the queen and workers appears to play an important role in overall social organization.

Molecular Mechanisms Mediating Aggression

In social insects, genomics approaches have been broadly applied to identify genes associated with the expression and evolution of various social behaviors, including aggression (Zayed and Robinson, 2012). These approaches find evidence that similar genes regulate aggression across very diverse contexts for behavioral expression. For example, in honeybees, genes showing brain expression differences across subspecies that differ in aggressive behavior overlap with genes that are differentially activated in the brain within a subspecies as a function of aggressive experience (Alaux et al., 2009). There is evidence for a common neuromolecular signature of conspecific intruder aggression across distantly related vertebrate and invertebrate species (Rittschof et al., 2014). Interestingly, some of the same genes that mediate defense against predators also mediate within-nest reproductive dominance: suites of genes that respond to intruder threat in *Drosophila* flies, honeybees, and mice, are also differentially regulated between dominant and subordinate paper wasps (Toth et al., 2014). Thus, aggressive phenotypes from different contexts are not easily distinguished on a molecular level. This makes it difficult to use molecular data to infer how physiological mechanisms have been modified over evolutionary time to give rise to elaborated and context-specific aggressive phenotypes (commonly observed in social groups, as discussed in the second section of this chapter), or to explain variable relationships between aggression and reproductive activities among individuals (a defining feature of advanced societies, as discussed in the third section of this chapter). It is easier to define similarities in molecular data than differences, and thus future studies focused on identifying key molecular differences across aggressive phenotypes and contexts could be an avenue forward in this area of behavioral genomics (Rittschof and Robinson, 2016). Another option is to link gene expression patterns to higher orders of biological organization that may distinguish phenotypes (Rittschof et al., 2018).

Because aggression is a high-energy, performance-related phenotype, and genes associated with variation in aggression are pleiotropic, variation in aggression may be a by-product of selection on other phenotypes. For example, in the honeybee, there is evidence that high aggression is correlated with positive health outcomes, including decreased parasitic mite loads and increased survival following pesticide exposure (Rittschof et al., 2015), and greater overall colony survival (Wray et al., 2011). Similarly, older adult worker bees are the most aggressive, but also perform the other most energetically costly behavior in the hive – foraging for food. Thus, variation within and among colonies in aggressive response could reflect selection for resilience to myriad health stressors, or an adaptive response to ecological conditions that shape colony nest-defense strategy, foraging effort, or both.

The strong coupling between aggression and foraging activity in honeybee workers manifests at the molecular level. A study comparing the brain neuromolecular signature of high aggression for bees responding to an acute aggression-inducing stimulus and bees showing high aggression as a function of adult age found evidence of molecular similarity (Alaux et al., 2009). One of the major brain changes associated with high aggression at the molecular level, for bees responding to a predator or conspecific threat, is a shift in oxidative phosphorylation activity, a metabolic pathway localized to the mitochondria (Alaux et al., 2009; Rittschof et al., 2014; Chandrasekaran et al., 2015). However, this change also occurs with worker age, tracking changes in both aggression and foraging behaviors (Alaux et al., 2009). Physiological data decouple this apparent mechanistic similarity in foraging and aggressive behavior: Measures of brain mitochondrial bioenergetics confirm that both aggression and foraging are associated with variation in brain energy metabolism, but that foraging activity modulates the brain energetic response to aggression-inducing cues (Rittschof et al., 2018, 2019). Thus, the physiological scale distinguishes apparent similarities that manifest at the level of brain gene-expression patterns, and functional studies are needed to interpret evidence of molecular similarities across behavioral contexts for aggression.

The ubiquity of aggression, in terms of the individuals who express it and the contexts in which it is expressed, its complex regulation, and its pleiotropic effects presents a significant challenge to interpreting its role in social evolution. Unlike the evolution of care behaviors, which are completely excluded from certain castes over evolutionary time, it could be that over the course of social evolution, aggressive responses to particular social contexts have been excluded from some castes, while aggression in other contexts has been maintained. The mechanistic framework to facilitate this partitioning may already be present due to the modular nature of aggressive responses, which are highly flexible and context-dependent in nonsocial and social species alike (Bee, 2003; Fawcett and Johnstone, 2010; Ishikawa and Miura, 2012; Li-Byarlay et al., 2014; Asahina, 2017). Thus, it may be valuable to consider "aggression modules" (Barron and Robinson, 2008), which are behavioral modules (presumably with molecular or physiological correlates; see Bloch and Grozinger, 2011) regulated by distinct social and environmental cues in different individuals, species, and contexts.

Conclusions

While most studies and discussions of the evolution and maintenance of social insect societies have focused on ultimate and proximate mechanisms that have enhanced cooperation and reduced conflict, it is clear that aggression and conflict play direct roles in these processes (Figure 10.2; see Chapters 5 and 8 for similar perspectives in vertebrates). Defense of offspring and the nest is critical to the formation of social groups, and it lays the groundwork for the development of larger and more stable nests with abundant food resources. Subsequently, larger nests experience increased predation risk and threats from robbing conspecifics, leading to the development of more elaborated social structures, including communication systems to support kin recognition and mediate social defense. Aggression and conflict among nestmates facilitate the evolution of reproductive division of labor, ultimately laying the groundwork for caste differentiation. Caste differentiation allows for more efficient worker division of labor and larger colony size, which both necessitates and facilitates the evolution of worker behavioral subspecialization on defense.

Elucidating the molecular mechanisms underlying aggression and defense, and how these may have influenced the evolution of sociality, is a fascinating and fundamental area of for future research. Reproductive division of labor, caste differentiation, and worker division of labor have been hypothesized to result from the uncoupling of key molecular and physiological pathways, such that certain behaviors and physiological states are expressed only at certain times, within specific individuals, or under particular contexts (Rehan and Toth, 2015; Kapheim, 2017). However, comparative transcriptomic studies indicate that the same suite of genes mediate intruder defense across invertebrates and vertebrates (Rittschof et al., 2014), and that within insects, the same genes associated with intruder defense in subordinates are also associated with reproductive dominance between dominants and subordinates (Toth et al., 2014). These findings suggest that similar molecular mechanisms and pathways underlie defense and aggression in both reproductive and non-reproductive phenotypes alike. Thus, it may be that it is the stimuli and sensory pathways triggering activation of these aggression pathways that differ across individuals and contexts (Bloch and Grozinger, 2011). Moreover, variation in aggression is associated with variation in brain metabolic processes and has been linked pleiotropically with many traits associated with social insect health, and thus future studies are needed to elucidate how selective, context-dependent regulation of aggressive behaviors evolved. With such a high degree of pleiotropy, there are undoubtedly constraints and trade-offs in changes in regulation of these genes. It remains to be determined if the trade-offs observed in social species (where aggressive behaviors are positively associated with reproduction and negatively associated with brood care) are ancestral to solitary species or established during the course of social evolution. With the advent of technologies that allow us to readily map variation in genomes, transcriptomes, and metabolomes, and an increasing interest in developing detailed behavior studies and assays in nontraditional model systems (Robinson et al., 2008; Rubenstein and Hofmann, 2015), it is now possible to comprehensively study the role of aggression and its evolution in a diverse array of social insect species and contexts, and furthermore, to determine to what extent behavioral, molecular, and physiological modules are conserved in vertebrate species. Further molecular studies, comparing aggression in different contexts, castes, and species, are needed to fully unravel the mechanisms underpinning these behaviors and how they are modified during – and how they may contribute to – social evolution.

References

Alaux C., Sinha S., Hasadsri L. et al. (2009) Honey bee aggression supports a link between gene regulation and behavioral evolution. *Proceedings of the National Academy of Science USA,* 106(36): 15400–15405.

Amdam G. V., Csondes A., Fondrk M. K., and Page R. E., Jr. (2006) Complex social behaviour derived from maternal reproductive traits. *Nature,* 439(7072): 76–78.

Amsalem E., Grozinger C. M., Padilla M., and Hefetz A. (2015) The physiological and genomic bases of bumble bee social behaviour. *Advances in Insect Physiology: Genomics, Physiology and Behavior of Social Insects,* 48: 37–94.

Amsalem E., and Hefetz A. (2010) The appeasement effect of sterility signaling in dominance contests among *Bombus terrestris* workers. *Behavioral Ecology and Sociobiology,* 64(10): 1685–1694.

Amsalem E., and Hefetz A. (2011) The effect of group size on the interplay between dominance and reproduction in *Bombus terrestris. PLoS ONE,* 6(3): e18238.

Andersson M. (1984) The evolution of eusociality. *Annual Review of Ecology and Systematics,* 15(1): 165–189.

Asahina K. (2017) Neuromodulation and strategic action choice in *Drosophila* aggression. *Annual Review of Neuroscience,* 40: 51–75.

Barron A. B., and Robinson G. E. (2008) The utility of behavioral models and modules in molecular analyses of social behavior. *Genes Brain and Behavior,* 7(3): 257–265.

Bee M. A. (2003) Experience-based plasticity of acoustically evoked aggression in a territorial frog. *Journal of Comparative Physiology A,* 189(6): 485–496.

Bloch G., and Grozinger C. M. (2011) Social molecular pathways and the evolution of bee societies. *Philosophical Transactions of the Royal Society B Biological Science,* 366(1574): 2155–2170.

Bourke A. F. G. (2011) *Principles of Social Evolution*. Oxford: Oxford University Press.

Breed M. D. (1983) Nestmate recognition in honey bees. *Animal Behaviour*, 31: 86–91.

Breed M. D., Guzman-Novoa E., and Hunt G. J. (2004) Defensive behavior of honey bees: Organization, genetics, and comparisons with other bees. *Annual Review of Entomology*, 49: 271–298.

Breed M. D., Robinson G. E., and Page R. E. (1990) Division of labor during honey bee colony defense. *Behavioral Ecology and Sociobiology*, 27: 395–401.

Brunton D. H. (1990) The effects of nesting stage, sex, and type of predator on parental defense by Killdeer (Charadrius-Vociferus) – Testing models of avian parental defense. *Behavioral Ecology and Sociobiology*, 26(3): 181–190.

Bruschini C., Cervo R., and Turillazzi S. (2006) Evidence of alarm pheromones in the venom of *Polistes dominulus* workers (Hymenoptera: Vespidae). *Physiological Entomology*, 31(3): 286–293.

Cant M. A. (2011) The role of threats in animal cooperation. *Proceedings of the Royal Society B Biological Science*, 278: 170–178.

Cardinal S., Straka J., and Danforth B. N. (2010) Comprehensive phylogeny of apid bees reveals evolutionary origins and antiquity of cleptoparasitism. *Proceedings of the National Academy of Sciences USA*, 107(37): 16207–16211.

Carey J. R. (2001) Demographic mechanisms for the evolution of long life in social insects. *Experimental Gerontology*, 36: 713–722.

Chandrasekaran S., Rittschof C. C., Djukovic D., Gu H., Raftery D., Price N. D., and Robinson G. E. (2015) Aggression is associated with aerobic glycolysis in the honey bee brain. *Genes Brain and Behavior*, 14(2): 158–166.

Chapman R. F. (2012) *The Insects: Structure and Function*. Cambridge, UK: Cambridge University Press.

Couvillon M. J., Robinson E. J. H., Atkinson B., Child L., Dent K. R., and Ratnieks F. L. W. (2008) En garde: Rapid shifts in honeybee, *Apis mellifera*, guarding behaviour are triggered by onslaught of conspecific intruders. *Animal Behaviour*, 76(5): 1653–1658.

Crespi B. J. (1994) Three conditions for the evolution of eusociality: Are they sufficient? *Insectes Sociaux*, 41: 395–400.

De Fine Licht H. H., Boomsma J. J., and Tunlid A. (2014) Symbiotic adaptations in the fungal cultivar of leaf-cutting ants. *Nature Communications*, 5: 5675.

Doke M. A., Frazier M., and Grozinger C. M. (2015) Overwintering honey bees: Biology and management. *Current Opinion in Insect Science*, 10: 185–193.

Fahrbach S. E., Farris S. M., Sullivan J. P., and Robinson G. E. (2003) Limits on volume changes in the mushroom bodies of the honey bee brain. *Journal of Neurobiology*, 57(2): 141–151.

Farris S. M., Robinson G. E., and Fahrbach S. E. (2001) Experience- and age-related outgrowth of intrinsic neurons in the mushroom bodies of the adult worker honeybee. *The Journal of Neuroscience*, 21(16): 6395–6404.

Fawcett T. W., and Johnstone R. A. (2010) Learning your own strength: Winner and loser effects should change with age and experience. *Proceedings of the Royal Society B Biological Science*, 277(1686): 1427–1434.

Fjerdingstad E. J., and Crozier R. H. (2006) The evolution of worker caste diversity in social insects. *The American Naturalist*, 167(3): 390–400.

Galbraith D. A., Wang Y., Amdam G. V., Page R. E., and Grozinger C. M. (2015) Reproductive physiology mediates honey bee (*Apis mellifera*) worker responses to social cues. *Behavioral Ecology and Sociobiology*, 69(9): 1511–1518.

Giray T., Giovanetti M., and West-Eberhard M. J. (2005) Juvenile hormone, reproduction, and worker behavior in the neotropical social wasp *Polistes canadensis*. *Proceedings of the National Academy of Science USA*, 102(9): 3330–3335.

Giray T., Guzman-Novoa E., Aron C. W., Zelinsky B., Fahrbach S. E., and Robinson G. E. (2000) Genetic variation in worker temporal polyethism and colony defensiveness in the honey bee *Apis mellifera*. *Behavioral Ecology*, 11(1): 44–55.

Groom M. J. (1992) Sand-colored Nighthawks parasitize the antipredator behavior of three nesting bird species. *Ecology*, 73(3): 785–793.

Gruter C., Menezes C., Imperatriz-Fonseca V. L., and Ratnieks F. L. W. (2012) A morphologically specialized soldier caste improves colony defense in a neotropical eusocial bee. *Proceedings of the National Academy of Sciences USA*, 109(4): 1182–1186.

Hartfelder K. (2000) Insect juvenile hormone: From "status quo" to high society. *Brazilian Journal of Medical and Biological Research*, 33: 157–177.

Hirschenhauser K., and Oliveira R. F. (2006) Social modulation of androgens in male vertebrates: Meta-analyses of the challenge hypothesis. *Animal Behaviour*, 71(2): 265–277.

Hölldobler B. (1981) Foraging and spatiotemporal territories in the honey ant *Myrmecocystus mimicus* (Hymenoptera: Formicidae). *Behavioral Ecology and Sociobiology*, 9: 301–314.

Hölldobler B., and Wilson E. O. (1994) *War and Foreign Policy. Journey to the Ants*. Cambridge, MA: Harvard University Press, p. 70.

Huang Z., and Robinson G. E. (1992) Honeybee colony integration: Worker–worker interactions mediate hormonally regulated plasticity in division of labor. *Proceedings of the National Academy of Sciences USA*, 89: 11726–11729.

Huang Z. Y., Robinson G. E., and Borst D. W. (1994) Physiological correlates of division of labor among similarly aged honey bees. *Journal of Comparative Physiology A*, 174(6): 731–739.

Hunt G. J., Guzmán-Novoa E., Uribe-Rubio J. L., and Prieto-Merlos D. (2003) Genotype–environment interactions in honeybee guarding behaviour. *Animal Behaviour*, 66(3): 459–467.

Hunt J. H., Kensinger B. J., Kossuth J. A., Henshaw M. T., Norberg K., Wolschin F., and Amdam G. V. (2007) A diapause pathway underlies the gyne phenotype in *Polistes* wasps, revealing an evolutionary route to caste-containing insect societies. *Proceedings of the National Academy of Science USA*, 104(35): 14020–14025.

Ishikawa Y., and Miura T. (2012) Hidden aggression in termite workers: Plastic defensive behaviour dependent upon social context. *Animal Behaviour*, 83(3): 737–745.

Jandt J. M., Suryanarayanan S., Hermanson J. C., Jeanne R. L., and Toth A. L. (2017) Maternal and nourishment factors interact to influence offspring developmental trajectories in social wasps. *Proceedings of the Royal Society London B Biological Science*, 284(1857): 1–9.

Jandt J. M., and Toth A. L. (2015) Physiological and genomic mechanisms of social organization in wasps (Family: Vespidae). *Advances in Insect Physiology: Genomics, Physiology and Behavior of Social Insects*, 48: 95–130.

Kapheim K. M. (2017) Nutritional, endocrine, and social influences on reproductive physiology at the origins of social behavior. *Current Opinion in Insect Science*, 22: 62–70.

Kastberger G., Thenius R., Stabentheiner A., and Hepburn R. (2008) Aggressive and docile colony defence patterns in *Apis mellifera*: A retreater–releaser concept. *Journal of Insect Behavior*, 22(1): 65–85.

Keller L., and Chapuiset M. (1999) Cooperation among selfish individuals in insect societies. *BioScience*, 49(11): 899–909.

Keller L., and Nonacs P. (1993) The role of queen pheromones in social insects: Queen control or queen signal? *Animal Behavior,* 45(4): 787–794.

Khila A., and Abouheif E. (2008) Reproductive constraint is a developmental mechanism that maintains social harmony in advanced ant societies. *Proceedings of the National Academy of Science USA,* 105(46): 17884–17889.

Kocher S. D., Ayroles J. F., Stone E. A., and Grozinger C. M. (2010) Individual variation in pheromone response correlates with reproductive traits and brain gene expression in worker honey bees. *PLoS ONE,* 5(2): e9116.

Kocher S. D., and Grozinger C. M. (2011) Cooperation, conflict, and the evolution of queen pheromones. *Journal of Chemical Ecology,* 37(11): 1263–1275.

Lawson S. P., Helmreich S. L., and Rehan S. M. (2017) Effects of nutritional deprivation on development and behavior in the subsocial bee *Ceratina calcarata* (Hymenoptera: Xylocopinae). *Journal of Experimental Biology,* 220(Pt 23): 4456–4462.

Li-Byarlay H., Rittschof C. C., Massey J. H., Pittendrigh B. R., and Robinson G. E. (2014) Socially responsive effects of brain oxidative metabolism on aggression. *Proceedings of the National Academy of Science USA,* 111(34): 12533–12537.

Machado G. (2002) Maternal care, defensive behavior, and sociality in neotropical *Goniosoma* harvestmen (Arachnida, Opiliones). *Insectes Sociaux,* 49: 388–393.

Maleszka R. (2018) Beyond Royalactin and a master inducer explanation of phenotypic plasticity in honey bees. *Communications Biology,* 1: 8.

Mao W., Schuler M. A., and Berenbaum M. R. (2015) A dietary phytochemical alters caste-associated gene expression in honey bees. *Science Advances,* 1(7): e1500795.

Mathiron A. G. E., Earley R. L., and Goubault M. (2019) Juvenile hormone manipulation affects female reproductive status and aggressiveness in a non-social parasitoid wasp. *General and Comparative Endocrinology,* 274: 80–86.

Mcdonald P., and Topoff H. H. (1985) Social regulation of behavioral development in the ant, *Novomessor albisetosus* (Mayr). *Journal of Comparative Psychology,* 99(1): 3–14.

Moczek A. P. (2010) Phenotypic plasticity and diversity in insects. *Philosophical Transactions of the Royal Society London B Biological Science,* 365(1540): 593–603.

Moritz R. F., Lattorff H. M., and Crewe R. M. (2004) Honeybee workers (*Apis mellifera capensis*) compete for producing queen-like pheromone signals. *Proceedings of the Royal Society London B Biological Science,* 271 (Suppl 3): S98–S100.

Muscedere M. L., Traniello J. F. A., and Gronenberg W. (2011) Coming of age in an ant colony: Cephalic muscle maturation accompanies behavioral development in *Pheidole dentata.* *Naturwissenschaften,* 98(9): 783–793.

Naeger N. L., Peso M., Even N., Barron A. B., and Robinson G. E. (2013) Altruistic behavior by egg-laying worker honeybees. *Current Biology,* 23(16): 1574–1578.

Nakahira T., and Kudo S.-I. (2008) Maternal care in the burrower bug *Adomerus triguttulus*: Defensive behavior. *Journal of Insect Behavior,* 21(4): 306–316.

Nowak M. A., Tarnita C. E., and Wilson E. O. (2010) The evolution of eusociality. *Nature,* 466 (7310): 1057–1062.

O'Dowd D. J., and Hay M. E. (1980) Mutualism between harvester ants and a desert ephemeral: Seed escape from rodents. *Ecology,* 61(3): 531–540.

Page R. E., and Erickson E. H. (1988) Reproduction by worker honey bees (*Apis mellifera* L.). *Behavioral Ecology and Sociobiology,* 23: 117–126.

Passera L., Roncin E., Kaufman B. A., and Keller L. (1996) Increased soldier production in ant colonies exposed to intraspecific competition. *Nature,* 379: 630–631.

Plowright R. C., and Jay S. C. (1966) Rearing bumble bee colonies in captivity. *Journal of Apicultural Research*, 5(3): 155–165.

Rajakumar R., San Mauro D., Dijkstra M. B. et al. (2012) Ancestral developmental potential facilitates parallel evolution in ants. *Science*, 335: 79–82.

Ratnieks F. L., Foster K. R., and Wenseleers T. (2006) Conflict resolution in insect societies. *Annual Review of Entomology*, 51: 581–608.

Ratnieks F. L., and Wenseleers T. (2008) Altruism in insect societies and beyond: Voluntary or enforced? *Trends in Ecology and Evolution*, 23(1): 45–52.

Rehan S. M., and Toth A. L. (2015) Climbing the social ladder: The molecular evolution of sociality. *Trends in Ecology and Evolution*, 30(7): 426–433.

Rittschof C. C., Bukhari S. A., Sloofman L. G. et al. (2014) Neuromolecular responses to social challenge: Common mechanisms across mouse, stickleback fish, and honey bee. *Proceedings of the National Academy of Science USA*, 111(50): 17929–17934.

Rittschof C. C., Coombs C. B., Frazier M., Grozinger C. M., and Robinson G. E. (2015) Early-life experience affects honey bee aggression and resilience to immune challenge. *Science Reports*, 5: 15572.

Rittschof C. C., and Robinson G. E. (2016) Behavioral genetic toolkits: Toward the evolutionary origins of complex phenotypes. *Current Topics in Developmental Biology*, 119: 157–204.

Rittschof C. C., Vekaria H. J., Palmer J. H., and Sullivan P. G. (2018) Brain mitochondrial bioenergetics change with rapid and prolonged shifts in aggression in the honey bee, *Apis mellifera*. *Journal of Experimental Biology*, 221(Pt 8): jeb176917, DOI: 10.1242/jeb.176917.

Rittschof C. C., Vekaria H. J., Palmer J. H., and Sullivan P. G. (2019) Biogenic amines and activity levels alter the neural energetic response to aggressive social cues in the honey bee *Apis mellifera*. *Journal of Neuroscience Research*, 97: 991–1003.

Robinson G. E. (1987) Modulation of alarm pheromone perception in the honey bee: Evidence for division of labor based on hormonally regulated response thresholds. *Journal of Comparative Physiology A*, 160: 613–619.

Robinson G. E. (1992) Regulation of division of labor in insect societies. *Annual Review of Entomology*, 37: 637–665.

Robinson G. E., Fernald R. D., and Clayton D. F. (2008) Genes and social behavior. *Science*, 322(5903): 896–900.

Robinson G. E., Winston M. L., Huang Z., and Pankiw T. (1998) Queen mandibular gland pheromone influences worker honey bee (*Apis mellifera* L.) foraging ontogeny and juvenile hormone titers. *Journal of Insect Physiology*, 44(7–8): 685–692.

Rubenstein D. R., and Hofmann H. A. (2015) Editorial overview: The integrative study of animal behavior. *Current Opinion in Behavioral Sciences*, 6: v–viii.

Sakagami S. F. (1958) The false queen: Fourth adjustive response in dequeened honeybee colonies. *Behaviour*, 13: 280–296.

Scott M. P. (1998) The ecology and behavior of burying beetles. *Annual Review of Entomology*, 43: 595–618.

Seeley T. D. (2012) Progress in understanding how the waggle dance improves the foraging efficiency of honey bee colonies. In Galizia C., Eisenhardt D., Giurfa M., eds., *Honeybee Neurobiology and Behavior*. Dordrecht: Springer, pp. 77–87.

Seid M. A., and Traniello J. F. A. (2006) Age-related repertoire expansion and division of labor in *Pheidole dentata* (Hymenoptera: Formicidae): A new perspective on temporal polyethism and behavioral plasticity in ants. *Behavioral Ecology and Sociobiology*, 60(5): 631–644.

Shorter J. R., and Tibbetts E. A. (2008) The effect of juvenile hormone on temporal polyethism in the paper wasp *Polistes dominulus*. *Insectes Sociaux*, 56(1): 7–13.

Sibbald E. D., and Plowright C. M. S. (2012) On the relationship between aggression and reproduction in pairs of orphaned worker bumblebees (*Bombus impatiens*). *Insectes Sociaux*, 60(1): 23–30.

Smith C., Toth A., Suarez A. et al. (2008) Genetic and genomic analyses of the division of labour in insect societies. *Nature Reviews Genetics*, 9: 735–748, DOI: https://doi.org/10.1038/nrg2429.

Skukla S., Pareek V., and Gadagkar R. (2014) Ovarian development in a primitively eusocial wasp: Social interactions affect behaviorally dominant and subordinate wasps in opposite directions relative to solitary females *Behavioral Processes*, 106: 22026.

Strassmann J. E. (1981) Parasitoids, predators, and group size in the paper wasp *Polistes exclamans*. *Ecology*, 62(5): 1225–1233.

Strassmann J. E., Queller D. C., and Hughes C. R. (1988) Predation and the evolution of sociality in the paper wasp *Polistes bellicosus*. *Ecology*, 69(5): 1497–1505.

Suryanarayanan S., Hermanson J. C., and Jeanne R. L. (2011) A mechanical signal biases caste development in a social wasp. *Current Biology*, 21(3): 231–235.

Szathmary E., and Smith J. M. (1995) The major evolutionary transitions. *Nature*, 374(6519): 227–232.

Tallamy D. W., and Brown W. P. (1999) Semelparity and the evolution of maternal care in insects. *Animal Behaviour*, 57: 727–730.

Thorne B. L., Breisch N. L., and Muscedere M. L. (2003) Evolution of eusociality and the soldier caste in termites: Influence of intraspecific competition and accelerated inheritance. *Proceedings of the National Academy of Sciences USA*, 100(22): 12808–12813.

Tibbetts E. A., and Crocker K. C. (2014) The challenge hypothesis across taxa: Social modulation of hormone titres in vertebrates and insects. *Animal Behaviour*, 92: 281–290.

Tibbetts E. A., and Lindsay R. (2008) Visual signals of status and rival assessment in *Polistes dominulus* paper wasps. *Biology Letters*, 4(3): 237–239.

Tibbetts E. A., and Sheehan M. J. (2012) The effect of juvenile hormone on Polistes wasp fertility varies with cooperative behavior. *Hormones and Behavior*, 61(4): 559–564.

Toth A. L., and Robinson G. E. (2005) Worker nutrition and division of labour in honeybees. *Animal Behaviour*, 69(2): 427–435.

Toth A. L., Tooker J. F., Radhakrishnan S., Minard R., Henshaw M. T., and Grozinger C. M. (2014) Shared genes related to aggression, rather than chemical communication, are associated with reproductive dominance in paper wasps (*Polistes metricus*) *BMC Genomics*, 15: 75.

Toth A. L., Varala K., Newman T. C. et al. (2007) Wasp gene expression supports an evolutionary link between maternal behavior and eusociality. *Science*, 318(5849): 441–444.

Vásquez G. M., and Silverman J. (2008) Intraspecific aggression and colony fusion in the Argentine ant. *Animal Behaviour*, 75(2): 583–593.

Weiner S. A., Woods W. A., Jr., and Starks P. T. (2009) The energetic costs of stereotyped behavior in the paper wasp, Polistes dominulus. *Naturwissenschaften*, 96(2): 297–302.

West-Eberhard M. J. (1967) Foundress associations in Polistine wasps: Dominance hierarchies and the evolution of social behavior. *Science*, 157(3796): 1584–1585.

Wheeler D. E., and Nijhout H. F. (1981) Soldier determination in ants: New role for juvenile hormone. *Science*, 213(4505): 361–363.

Wilson E. O. (2000) *Sociobiology: The New Synthesis*. Cambridge, MA: The Belknap Press of Harvard University Press.

Wingfield J. C., Hegner R. E., Dufty A. M., and Ball G. F. (1990) The "Challenge Hypothesis": Theoretical implications for patterns of testosterone secretion, mating systems, and breeding strategies. *The American Naturalist,* 136(6): 829–846.

Winston M. L. (1991) *The Biology of the Honey Bee.* Cambridge, MA: Harvard University Press.

Wittemyer G., and Getz W. M. (2007) Hierarchical dominance structure and social organization in African elephants, *Loxodonta africana. Animal Behaviour,* 73: 671–681.

Wong J. W. Y., Meunier J., and Kölliker M. (2013) The evolution of parental care in insects: The roles of ecology, life history and the social environment. *Ecological Entomology,* 38(2): 123–137.

Wray M. K., Mattila H. R., and Seeley S. D. (2011) Collective personalities in honeybee colonies are linked to colony fitness. *Animal Behavior,* 81(3): 559–568.

Wright C. M., Skinker V. E., Izzo A. S., Tibbetts E. A., and Pruitt J. N. (2017) Queen personality type predicts nest-guarding behaviour, colony size and the subsequent collective aggressiveness of the colony. *Animal Behavior,* 124: 7–13.

Yang A. S., Martin C. H., and Nijhout H. F. (2004) Geographic variation of caste structure among ant populations. *Current Biology,* 14(6): 514–519.

Zayed A., and Robinson G. E. (2012) Understanding the relationship between brain gene expression and social behavior: Lessons from the honey bee. *Annual Review of Genetics,* 46: 591–615.

Index

acacia tree, ant mutualism, 192
African gray parrot. *See Psittacus erithacus*
agent-based modeling, 42, 47, 50, 57
aggression, 1, 26, 91–97, 122, 134, 151, 212
  contexts associated with eusociality, 215
  coupled with foraging activity, 225
  defensive, 216
  and developmental constraint, 220
  female–female, 175
  and group size, 172, 220
  intragroup, 218–223
  lethal, 151
  maternal, 91, 216
  outgroup, 94
  and reproduction, 221
  role in social evolution, 225
Allee effect, 171
allogrooming. *See* grooming
altruism, 10, 46, 66, 68, 137, 187
  competitive, 57
ambrosia beetle, 195
anarchy, 7
ant
  differences in developmental trajectories, 221
  eusociality, 67
  farming, 195
  leaf-cutting, 69
  morphological subspecialization, 223
  mutualism with aphids, 194
  mutualism with lycaenid butterflies, 192
  mutualism with plants, 188, 192, 194
  selection against cheating, 195
  unicolonial, 218, *See also* insects, super colonies
anthropomorphism, 91
anxiety, 147, 152
aphids, 68, 190, 194
Apis mellifera, 214, 216, 219–226
architectonics, 112
arginine vasopressin. *See* vasopressin
arms race. *See* evolutionary arms race, *See* security
  dilemma
attachment, 134
attention, 120
autism spectrum disorder, 137, 148

baboon, 191. *See Papio cynocephalus*
bacteriocytes, 68
balance of power, 7
balanced polymorphism. *See* evolution, balanced
  polymorphism
Begin, Menachem, 15
biological market theory, 57, 185, 189–194, 199
  environmental impacts, 203
black-backed jackal. *See Canis mesomelas*
Bombus terrestris, 214, 220–221
bonobo. *See Pan paniscus*
Botryllus schlosseri, 72
breeding conflict, 174
Buchnera, 68
Buphagus erythrorhynchus, 197
bumblebee. *See Bombus terrestris*
bystander interventions. *See* policing

Caenorhabditis elegans, 74
California mouse. *See Peromyscus californicus*
Callicebus cupreus, 94
Callithrix jacchus, 94, 115, 144
Canis mesomelas, 171
capuchin monkey. *See Sapajus [Cebus] apella*
Carter, Jimmy, 15
cellular aggregation, 71
cellular fusion, 72
Ceratina calcarata, 220
challenge hypothesis, 222
cheater, 49, 66, 69, 74, 177, 195
cheating, 68–72, 186, 193
  selected against, 195
  by shirking, 194, 204
  and society, 72
Cheilinus undulatus, 204
chimerism, 70–72
  and kin recognition, 73
chimpanzee. *See Pan troglodytes*
chromosomal inversion, 77
cichlid fish, 193
civil war. *See* war, intrastate
cleaner fish. *See Labroides dimidiatus*
cleaning gobies. *See Elacatinus*
coalitions, 193

coercion, 73, 218
cognition, 1, 26
  affective, 119
  comparative, 201
  fish, 201–202
  semantic, 119
co-inheritance. *See* evolution, co-inheritance
Cold War, 14
collective action problem, 171, *See also* social
    dilemmas
collective behavior, 26, 217
colonial power, 13
Columba livia, 202
commensalism, 185, 190
commitment device, 17
communication, 66, 94, 104, 134
  alarm pheromone, 217
  avoiding misunderstanding, 11
  chemical, 217
  coordination signal, 26
  costly signal, 17
  honest signaling, 217, 219
  honeybee dance, 214
  and neuropeptides, 134
  referential, 204
  scent marking, 96
  social cues (gaze and pointing), 148
comparative approach, 3, 136, 161
comparative neuroanatomy, 108–113, 150
comparative neuroscience, 106
compensatory trait loss, 192
conceptual knowledge, 117
conceptual processing, 108
configural face processing, 115
conflict, 4, 134, 212
  contexts associated with eusociality, 215
  cryptic, 74
  definition, 186
  and group dynamics, 178
  and group size, 172
  human, 3
  importance of, 179
  intergroup, 1, 170, 175–179, 217
  intergroup vs. intragroup, 4
  and international relations, 9
  interspecies, 216
  intragroup, 66, 169, 174, 218–221
  and intragroup cooperation, 172
  intra-organismal, 71
  among kin, 168
  lethal, 176
  management, 12
  post-conflict costs, 176
  social, 74
  and society, 4
  types of, 169
conflict of interest, 45, 185, 203

conflict resolution. *See* reconciliation
conflict speciation. *See* evolution, conflict speciation
constraint, 218
cooperation
  assortment, 68
  by-product benefit, 186
  and cognition, 4, 199
  and common knowledge, 58
  competition over partners, 57
  cooperation–conflict balance, 1, 3, 66, 80, 92,
      153, 163, 167, 179, 218, 226
  costly, 45
  definition, 186
  effect of context, 54
  effect of group size, 168, 171, 191
  evolution of, 2
  evolved vs. ontogenetic roles, 188
  genetic basis, 80
  and global conflict, 11
  and hierarchy, 9
  human, 3, 26, 45, 104, 122, 190, 203
  importance of mutualisms, 185
  ingroup favoritism, 57
  through institutionalization, 12
  international, 7
  interspecific. *See* mutualism
  interspecific vs. intraspecific, 67, 188–190,
      194
  intragroup, 172, 218
  intra-organismal, 71
  intraspecific, 187, 203
  large scale, 26
  measuring costs and benefits, 179
  moderation by conflict, 180
  mutual, 46
  and negativity bias, 57
  problem of, 66
  prosocial preferences, 28
  and punishment, 26–28
  and relatedness, 67
  and reputation, 48, 53–59
  and reputation management, 202
  sanctions vs. reputation, 55
  and selfish behavior, 68–72
  shared economic interests, 7
  and society, 1
  and trust, 46
cooperation dilemma. *See* social dilemma
cooperative breeding, 4, 163, 171–174, 193
  conflict, 167, 179
cooperative brood care, 213, *See* also cooperative
    breeding
cooperative hunting, 204
coordination, 7, 66
coral bleaching, 202
cortisol, 198
cost/benefit analysis, 11, 27

costly behavior, 45, 51
costly signaling, 57
counter-adaptation, 74
counter-adaptations. *See* evolution, counter-
adaptations
critical group size effect, 171–174
cross-cultural effects, 3, 9, 59
cytoplasmic male sterility, 70, 73, 78

damselfish, 192
Darwin, Charles, 9, 66
Dawkins, Richard, 76
de Klerk, F. W., 15
de Waal, Frans, 15
deception, 118, 122, *See also* tactical deception
decision-making, 120
    moral and ethical judgments, 114
    preferences, 27
    strategic, 122
defection, 66, 186, 197, *See* cheating
defector. *See* cheater
democracy, 8, 11
democratic peace, 8
development, 146
    brood manipulation, 220
    developmental constraint, 220
    face recognition, 116
    human, 52
    human reputation management, 54
Dicrurus adsimilis, 196
Dictyostelium discoideum, 71, 74
dispersal, 171
division of labor, 213, 224
dolphin. *See Tursiops truncatus*
dominance, 139, 151, 170, 193, 226
    and dispersal, 172
    displays, 96
    female dominance, 151
    hierarchy, 96, 193, 214, 218
dopamine, 97
downstream reciprocity. *See* reciprocity, indirect
drive gene. *See* gene, drive gene
drongo. *See Dicrurus adsimilis*
Drosophila, 70, 79, 224
Drosophila pseudoobscura, 75

economic games, 45
    by-product benefits game, 191
    coordination games, 26
    dictator game, 46
    donation game, 49, 52
    prisoner's dilemma game, 7, 55, 96, 191, 193,
        195, 200
    public goods game, 29, 53, 55
    snowdrift game, 191
    strategies used in, 45
    trust game, 51–52, 55

ultimatum game, 47
veto game, 46, 193, 199
ectoparasites, 197
El Niño, 202
Elacatinus, 197
emergent phenomena, 26
emotion, 1
    fear, 121
    role in decision-making, 19
emotional processing, 106, 114, 120
empathy, 1, 89, 114, 137
endosymbionts, 68, 73
Enlightenment, 8
environment, effects on behavior, 9, 202, 213, 217,
    222
epigenetics, 96
Europe, 11
European Union, 8
eusociality, 1, 67, 69, 212–221, *See also* insects, *See*
    also reproductive division of labor
eviction. *See* dispersal
evolution, 2
    balanced polymorphism, 78
    co-inheritance, 68
    conflict speciation, 75
    convergence, 161
    counter-adaptation, 74
    domain generality, 19
    domain specificity, 19, 114
    gene-culture coevolution, 26, 42
    human, 2, 20, 24, 108–113, 122, 161
    major transitions, 67, 185, 194, 212
    mathematical modeling, 27
    misuses, 10
    natural selection, 3, 66
    proximate mechanisms, 1, 28
    purifying selection, 78
    segregation distorter, 70, 78
    segregation distorter, cryptic, 75
    selective pressure, 161
    selective processes for cooperation, 2
    selective sweep, 78
    sexual selection, 3
    social evolution theory, 43, 66
    species divergence, 78
    ultimate mechanisms, 1
    uniparental inheritance, 73
    vertical transmission, 68
        with bottlenecks, 73
evolutionarily stable strategy, 30, 47, 168
evolutionary arms race, 74, 79

face pareidolia, 115
face perception, 108, 113–116, *See also* face
    pareidolia, *See also* prosopagnosia
fairness, 47, 72, *See also* justice
    distributive, 53

false belief tasks. *See* theory of mind, false belief
  tasks
farming, 195
fig wasp, 192
fire ant. *See Solenopsis invicta*
forgiveness, 16–24
free rider, 45, 49, 54, 171, 177
friendship, 191

game theory, 17, 185, *See* also economic
  games
gene culture coevolution. *See* evolution, gene-
  culture coevolution
generalized rule learning, 201
generosity, 52
genes, 67
  associated with aggression, 224
  cryptic selfish gene, 74
  drive gene, 75
  Gp-9, 76
  greenbeard gene, 76
  pha-1, 74
  polymorphisms, 135
  selfish gene, 69–70, 78
  and social behavior, 135
  supergene, 77
genomics, 4, 76, 135, 224
gibbon, 113
globalization, 14
goal disengagement, 41
gonadotropin, 221
gorilla. *See Gorilla gorilla*, *See Gorilla beringeri
  beringei*
Gorilla beringei beringei, 177
Gorilla gorilla, 113
gossip. *See* language, gossip
grooming, 145, 193
group living, benefits of, 193
group size
  and cooperation, 178
  and habitat, 178
group size paradox, 171
grouper. *See Pessuliferus*
guilt, 27, 42
Gymnothorax javanicus, 204

habituation, 168
Haldane, J. B. S., 66
Hamilton, William, 67
haplodiploidy, 69, *See* also insects, haplodiploidy
helpers at the nest, 168, 193, 219, *See* also
  cooperative breeding
  concessions model, 177
  direct vs. indirect benefits of, 172
helping behavior, 186, 222
  and critical group size, 171
Heterocephalus glaber, 67

hitchhiking traits, 42
Homo sapiens, 1, 7–24, 94, 107–122
honest signaling. *See* communication, honest
  signaling
honey ant. *See Myrmecocystus mimicus*
honeybee. *See Apis mellifera*
honeyguide. *See Indicator indicator*
human nature
  biological underpinnings, 24
  scientific perspective, 8–11
human rights, 7
  failures, 12
human uniqueness, 106
humanities, 1
humans. *See Homo sapiens*
hymenoptera, 216

image scoring. *See* reputation
imitation, 118
immigration and group size, 172
inclusive fitness, 67, 91, 187, 219, *See* also kin
  selection
Indicator indicator, 204
indirect reciprocity. *See* reciprocity, indirect
individual differences in behavior, 134
individual recognition, 106, 201
infanticide, 176
insects, 4, 212
  brood care, 222
  castes, 213
  division of labor. *See* division of labor
  eusociality, 218, *See* also eusociality
  genetics of castes, 76
  haplodiploidy, 67
  monogyny vs. polygyny, 76, 218
  morphological specialization, 223
  nest defense, 217, 223
  nest robbing, 217
  ovarian activation, 221, 224
  ovarian suppression, 219
  queen, 67, 76, 213
  super colonies, 76, 218
  worker castes, 223
interdependence, 190, 203, 212
  economic, 11
interdisciplinary. *See* multidisciplinary
international relations
  political constructivism, 8, 10
  political liberalism, 7
  political realism, 7
  scientific approach, 9
interspecies mutualism. *See* mutualism

joint attention. *See* theory of mind
justice, 20, *See* also fairness
  retributive, 22
juvenile hormone, 221–223

Kant, Immanuel, 8
kidnapping, 177
kin recognition, 73, 188
kin selection, 45, 168, 187, 203

Labroides dimidiatus, 52, 185, 196–203
language, 48, 104, 110, 118–120, *See* also
    communication
  ape language, 118
  chunking, 202
  comprehension, 119
  gossip, 48–49, 53–56
  reading, 119
  semantic processing, 108
laterality, 112
  and face processing, 115
legumes, 67
loner strategy, 69
long-tailed macaque. *See Macaca fascicularis*
lycaenid butterfly, 192, 196
Lycaon pictus, 171

Macaca fascicularis, 113
Macaca mulatta, 106, 109–116, 147, 152
Macaca nemestrina, 73
Macaca spp.
  amygdala, 121
  theory of mind, 118
major transitions in evolution. *See* evolution, major
    transitions
Mandela, Nelson, 15
marmoset. *See Callithrix jacchus*
mate guarding, 91
maternal care, 213, *See* also parental care
meadow vole. *See Microtus pennsylvanicus*
meerkat. *See Suricata suricatta*
meiosis, 69
meiotic drive, 69
Melipona, 69
memory, 92, 120, 201
  autobiographical, 112
Mesocricetus auratus, 96
Microtus montanus, 95
Microtus ochrogaster, 93, 95, 161
mirror self-recognition. *See* theory of mind, mirror
    self-recognition
mitochondria, 67, 194–195
  selfish variant, 78
mobbing, 169
model species, 121, 161, 227
  preclinical model species, 90
modularity. *See* evolution, domain specificity
molecular biology, 1
monogamy, 95, 137, 161
  queen monogamy, 67
monogyny. *See* insects, monogyny vs. polygyny
montane vole. *See Microtus montanus*

moray eels. *See Gymnothorax javanicus*
mother–infant bond, 95
motivation, intrinsic vs. extrinsic, 27
mountain gorillas. *See Gorilla beringei beringei*
mouse. *See Mus musculus*, *See Peromyscus
    californicus*
multidisciplinary, 1
multisensory processing, 114
Mus musculus, 224
music, 119
mutualism, 4, 67, 162, 185–201, 203
  ant-flowering plant mutualism, 194
  without brains, 185
  and cognition, 203
  evidence for, 187
  evolutionary origin of, 194
  between human and wild animal, 204
  marine cleaning mutualism, 189, 194, 196–201
  mixed-species associations, 189
  partner choice, 204
  partner fidelity, 68
  partner interdependence, 190
  plant pollinator mutualism, 192
  trading services, 188
mycorrhizal fungi, 189, 192
Myrmecocystus mimicus, 217
Myxococcus xanthus, 74

naked mole rat. *See Heterocephalus glaber*
Napoleon wrasse. *See Cheilinus undulatus*
Nash equilibria, 26
Native Americans, 15
NATO, 12
natural selection. *See* evolution, natural selection
negativity bias, 56–57
nematode. *See Caenorhabditis elegans*
nest defense, 216
neuroanatomy
  amygdala, 92, 120–122
  and behavioral output, 92
  cerebral cortex, 104
  default mode network, 117
  human temporal lobe distinctiveness, 107
  interconnectivity, 92
  mesolimbic dopamine system, 95, 97
  neuropil distribution, 112
  primate temporal lobe, 108
  social behavior neural network, 92
  social salience network, 93
  temporal lobe, 4, 107–113, 119
  temporal lobe and social concepts, 117
  temporal lobe architectonic asymmetry,
      112
  terminology and definitions, 105
neuropeptides. *See* oxytocin and vasopressin
neuroscience, 1
  affective, 20

norms, 3, 7, 26–40
  evolution of, 42
  internalization of, 28, 40
  internalized, 27
  international, 11
  violation of, 54

octopus. *See Octopus cyanea*
Octopus cyanea, 204
oophagy, 174, 179
  intragroup, 177
optimality theory, 195
orangutan. *See Pongo pygmaeus*
organismal fusion, 73
other-regarding behavior. *See* prosocial behavior
oxidative phosphorylation, 225
oxpecker, 197. *See Buphagus erythrorhynchus*
oxytocin, 4, 93
  effects on behaviour and cognition, 134
  and food sharing, 95
  and New World primates, 136
  phylogenetic variation, 136
  and prior experience, 96
  and social bonds, 95
  and social reward, 97
  and the social behavior neural network, 94

pair bonding, 91, 94–95, 137, 161
Pan paniscus
  amygdala, 121
  AVPR1A gene polymorphism, 144
  symbolic representation, 118
Pan species comparison, 121, 150
Pan troglodytes, 15, 52, 95, 107–122, 137–153, 202
  amygdala, 121
  AVPR1A gene polymorphism, 139–144
  AVPR1A receptor polymorphism, 137
  joint attention, 148–150
  oxytocin receptor single nucleotide
    polymorphism, 142
  personality, 137–139
  personality pathology, 146
  symbolic representation, 118
paper wasp. *See Polistes dominulus, See Polistes*
Papio cynocephalus, 191
parasitism, 185, 190, 196, 203
parental care, 92, 189, 222, *See* also maternal care
parthenogenesis, 69
partner choice, 192, 199
partner control, 186, 195, 199
partner switching, costs of, 203
Peromyscus californicus, 94, 161
personality, 54
  AVPR1A genotype, 151
  AVPR1A polymorphism, 134–139
  boldness, 217
  nonhuman primate, 137–139

pathology, 145
  and rearing history, 146
perspective taking. *See* theory of mind
Pessuliferus, 204
phenotypic plasticity, 217
Phoeniculus purpureus, 177
pied babbler. *See Turdoides bicolor*
pigeon. *See Columba livia*
pig-tailed macaque. *See Macaca nemestrina*
play, 97, 145
pleiotropy, 225
poison–antidote system, 70, 75, 196
policing, 66, 73, 218
Polistes, 214, 220–221, 224
Polistes bellicosus, 216
Polistes dominulus, 219
Polistes exclamans, 216
Polistes fuscatus, 220
Polistes metricus, 217
political science, 2
polyethism, age or temporal, 213, 223
polygyny. *See See* insects, monogyny vs.
    polygyny
Pongo pygmaeus, 113, 202
prairie voles. *See Microtus ochrogaster*
predation, 185
predator defense, 216
prosocial behavior, 1, 41, 51, 89–92, 134,
    151, 162
  definitions, 89
  and neuropeptides, 90
  prosocial preferences, 28
prosopagnosia, 114, *See* also face perception
proximate mechanisms. *See* evolution, proximate
    mechanisms
Pseudomonas, 74
Pseudomonas aeruginosa, 77
Pseudomonas fluorescens, 72
pseudoreciprocity, 191
Psittacus erithacus, 202
psychology, 2
  evolutionary, 17, 20
  social, 89
psychopathy. *See* personality, pathology
public good, 27, 77
punishment, 4, 20, 26–40, 47–48, 54–56, 73, 177,
    200
  costly, 27
  external, 35, 54
  external vs. internal, 28
  internal, 32, 37
purifying selection. *See* evolution, purifying
    selection
pyoverdine, 77

rat. *See Rattus norvegicus*
rational choice, 17

Rattus norvegicus, 96–97, 202
reciprocal altruism. *See* reciprocity, reciprocal
    altruism
reciprocity, 9, 191
    bounded generalized, 57
    direct, 47
    human day-to-day interactions, 52
    indirect, 47, 52–53
    reciprocal altruism, 68
reconciliation, 15–24, 198
    internally motivated, 23
    process of, 20
    role of costly signals, 18
    truth and reconciliation effort, 23
relatedness, effects of, 28, 42, 72, 174–175, 177,
    188
reliable signaling. *See* communication, honest
    signaling
reproductive division of labor, 67, 71, *See* also
    eusociality
reproductive skew, 174
reproductive success
    and group size, 172
    impact of territorial conflict, 176
reputation, 3, 48–59, 200
    assessment errors, 50
    assessment strategies, 49
    biased, 56
    formation, 48
    and group size, 57
    image scoring, 49
    importance of the outgroup, 58
    influence of others' expectations, 27
    management, 53
    shunning, 50
    and social networks, 51
    standing strategy, 50
    stern judging, 50
resource competition, 96
revenge, 20
rhesus macaque. *See Macaca mulatta*
rhizobia, 67, 73, 185, 189
ritualized aggression, 217–219
ritualized displays, 218
Ropalidia marginata, 221

Sacks, Oliver, 114
Sadat, Anwar, 15
Saimiri spp., 96
Sapajus [Cebus] apella, 202
scent marking. *See* communication, scent
    marking
security dilemma, 13
segregation distorter. *See* evolution, segregation
    distorter
selective sweep. *See* evolution, selective
    sweep

self-control, 201
self-interest, 10
    and society, 1
selfish gene. *See* genes, selfish gene
sensory processing, 108
sequential hermaphrodite, 200
Sewell, Samuel, 15
sex change. *See* sequential hermaphrodite
sex difference, 97, 139–144, 200
    anxiety, 147
    joint attention, 150
    personality pathology, 146
    social behavior, 153
sexual reproduction, 189, 193
sexual selection. *See* evolution, sexual
    selection
Silene acaulis, 79
Silene vulgaris, 78
single nucleotide polymorphisms, 137
small-scale societies, 47
social behavior
    and *AVPR1A* genotype, 151
    context dependence, 134
    genetic basis of, 76
    manipulation, 145, 162
social bonding, 1, 94
social cognition, 104, 114, 117, 120, 134
    and *AVPR1A* genotype, 151
    cognitively simple strategies, 162
social concepts, 119
social context, impact on behavior, 92
social contract, 9
social dilemma, 26, 45, 55
social evolution theory. *See* evolution, social
    evolution theory
social exclusion, 57
social interactions, reward value, 97
social knowledge, 106, 114
social media, 58
social networks, 58
social neuroscience, 114
social norms. *See* norms
social preferences, 40
social sciences, 1
social structure, 212
social tool, 201
socialization, 27
Solenopsis invicta, 76
species divergence. *See* evolution, species
    divergence
speech, 110
    comprehension, 113
spite, 69
squirrel monkey. *See Saimiri spp.*
stress, 198
supply and demand, 192
Suricata suricatta, 187

symbiosis, 190, 195
Syria, 12
Syrian hamster. *See Mesocricetus auratus*

tactical deception, 200
    and reputation management, 202
termites, 67, 195, 218
territorial defense, 169, 177
territoriality, 91, 151
testosterone, 222
t-haplotype, 69
The Selfish Gene, 76
theory of mind, 106, 116–118
    false belief tasks, 118
    joint attention, 148–150
    mirror self-recognition, 118, 146, 201
thermoregulation, 214
titi monkeys. *See Callicebus cupreus*
tool manufacture, 151
tournaments. *See* ritualized displays
toxin–antidote system. *See* poison–antidote system
tragedy of the commons, 66, *See also* public good
transcriptomics, 76, 226
translational research, 153
transposable elements, 70
Trivers, Robert, 68
tunicate. *See Boytryllus schlosseri*
Turdoides bicolor, 168–171
Tursiops truncatus, 204

uniparental inheritance. *See* evolution, uniparental
    inheritance
United Nations, 7

values, existential, 13
vasopressin, 4, 93, 222
    effects on behavior and cognition, 134
    interaction with anxiety, 96
    and pair bonds, 95
    and prior experience, 96
    receptor polymorphism, 4, 95, 136–137
    and sex difference, 97
    and social bonds, 95
    and the social behavior neural network, 94
vertical transmission. *See* evolution, vertical
    transmission
visual processing, 107

war
    interstate, 11, *See also* conflict, intergroup
    intrastate (civil), 12–15, *See also* conflict,
        intragroup
    intrastate, causes, 14
    intrastate, cultural and ethnic conflicts,
        13
    intrastate, effect of decolonization, 14
    intrastate, internationalization of, 12
    intrastate, problems resulting from, 12
    intrastate, role of identity, 14
    shadow of, 7
    world wars, 11
wasp. *See Polistes*
wild dog. *See Lycaon pictus*
Wolbachia, 196
wood hoopoe. *See Phoeniculus purpureus*

Zootermopsis nevadensis, 218